Lecture Notes in Artificial Intelligence   3228

Edited by J. G. Carbonell and J. Siekmann

Subseries of Lecture Notes in Computer Science

Michael G. Hinchey
James L. Rash
Walter F. Truszkowski
Christopher A. Rouff (Eds.)

Formal Approaches to Agent-Based Systems

Third International Workshop, FAABS 2004
Greenbelt, MD, USA, April 26-27, 2004
Revised Selected Papers

 Springer

Series Editors

Jaime G. Carbonell, Carnegie Mellon University, Pittsburgh, PA, USA
Jörg Siekmann, University of Saarland, Saarbrücken, Germany

Volume Editors

Michael G. Hinchey
James L. Rash
Walter F. Truszkowski
NASA Goddard Space Flight Center
Greenbelt, MD 20771, USA
E-mail: {michael.g.hinchey, james.l.rash, walter.f.truszkowski}@nasa.gov

Christopher A. Rouff
SAIC, Advanced Concepts Business Unit
1710 SAIC Drive, McLean, VA 22102, USA
E-mail: christopher.a.rouff@saic.com

Library of Congress Control Number: 2004117653

CR Subject Classification (1998): I.2.11, I.2, D.2, F.3, I.6, C.3, J.2

ISSN 0302-9743
ISBN 3-540-24422-0 Springer Berlin Heidelberg New York

Springer is a part of Springer Science+Business Media

springeronline.com

© Springer-Verlag Berlin Heidelberg 2005
Printed in Germany

Typesetting: Camera-ready by author, data conversion by Scientific Publishing Services, Chennai, India
Printed on acid-free paper   SPIN: 11378839   06/3142   5 4 3 2 1 0

Preface

The 3rd Workshop on Formal Approaches to Agent-Based Systems (FAABS-III) was held at the Greenbelt Marriott Hotel (near NASA Goddard Space Flight Center) in April 2004 in conjunction with the IEEE Computer Society.

The first FAABS workshop was help in April 2000 and the second in October 2002. Interest in agent-based systems continues to grow and this is seen in the wide range of conferences and journals that are addressing the research in this area as well as the prototype and developmental systems that are coming into use.

Our third workshop, FAABS-III, was held in April, 2004. This volume contains the revised papers and posters presented at that workshop.

The Organizing Committee was fortunate in having significant support in the planning and organization of these events, and were privileged to have world-renowned keynote speakers Prof. J Moore (FAABS-I), Prof. Sir Roger Penrose (FAABS-II), and Prof. John McCarthy (FAABS-III), who spoke on the topic of self-aware computing systems, auguring perhaps a greater interest in autonomic computing as part of future FAABS events.

We are grateful to all who attended the workshop, presented papers or posters, and participated in panel sessions and both formal and informal discussions to make the workshop a great success. Our thanks go to the NASA Goddard Space Flight Center, Codes 588 and 581 (Software Engineering Laboratory) for their financial support and to the IEEE Computer Society (Technical Committee on Complexity in Computing) for their sponsorship and organizational assistance.

Springer once again undertook to publish the proceedings, for which we are grateful. We hope that the reader will find it a useful compilation of the state of the art in formal methods, agent-based technologies, and their intersection.

October 2004   Greenbelt, MD

Organization

Organizing Committee

Mike Hinchey, *NASA Goddard Space Flight Center*
Jim Rash, *NASA Goddard Space Flight Center*
Walt Truszkowski, *NASA Goddard Space Flight Center*
Chris Rouff, *SAIC*

Table of Contents

Poster Presentations

Ecology Based Decentralized Agent Management System

Maxim D. Peysakhov, Vincent A. Cicirello, and William C. Regli

Department of Computer Science, Drexel University,
Philadelphia PA 19104

Abstract. The problem of maintaining a desired number of mobile agents on a network is not trivial, especially if we want a completely decentralized solution. Decentralized control makes a system more robust and less susceptible to partial failures. The problem is exacerbated on wireless ad hoc networks where host mobility can result in significant changes in the network size and topology. In this paper we propose an ecology-inspired approach to the management of the number of agents. The approach associates agents with living organisms and tasks with food. Agents procreate or die based on the abundance of uncompleted tasks (food). We performed a series of experiments investigating properties of such systems and analyzed their stability under various conditions. We concluded that the ecology based metaphor can be successfully applied to the management of agent populations on wireless ad hoc networks.

1 Introduction

In a typical agent based system, a number of mobile agents cooperate to achieve a desired goal. The efficiency of the agent system in reaching the goal, and the completeness of the result depends on the number of agents in the system. Too few agents will not achieve the full potential of parallelism and will lead to decreased system efficiency. Too many agents can overburden the system with unnecessary overhead, and may also result in significant delays. The task of finding the optimal number of agents required to achieve the desired effect is difficult and problem-specific. In this paper, we propose an ecosystem-inspired approach to this problem. Similar to a real ecosystem, our solution exhibits properties of emergent stability, decentralized control, and resilience to possible disturbances. In our work, we propose to solve the technical problem of agent management using an ecological metaphor.

In Section 2 we describe the current state of research in the fields of simulated ecosystems and multi-agent control and stability. Section 3 introduces the problem of managing the number of agents populating a physical network and also explains a proposed solution. Lastly, Section 4 demonstrates the initial experimental results and conclusions.

M.G. Hinchey et al. (Eds.): FAABS 2004, LNAI 3228, pp. 1–11, 2005.

2   Related Work

2.1   Simulated Ecology

The majority of ecology-inspired systems are used to answer some question about real world ecosystems and its properties. For example, the RAM system has been used to study mosquito control [23]. There are two major approaches to simulating an ecosystem [6]. One is a species-based view of the system, where large classes of individuals interact in the simulation (i.e., modeling the dynamics of interaction of species rather than the interaction of individuals). Evolutionary game theory (e.g., [1] [18] [17]) and dynamical systems (e.g., [9] [15] [14]) are two approaches that often take the species-based view. The second approach is to simulate individuals and their interactions, a bottom up approach to construction of the ecological simulator.

We are most interested in *individual-based* simulations, since they are usually built with software agents. An example of an individual-based approach to ecosystems is a simulated habitat populated with synthetic organisms (agents) [19]. Often such systems are used to study the evolution (and co-evolution) of different species and testing their interactions and emergent behavior. Genetic Algorithms [8] and Genetic Programming [10] engines can be used in conjunction with synthetic ecosystems to allow species to evolve over time. Some of the most well known examples of synthetic ecosystems of this type are Evolve 1, 2 and 3 [4] [5] [21], "Artorg world" [3] and LAGER [19].

With this approach, global trends in the behavior of the system may *emerge* as a result of the low-level interactions of individual agents. The emergent behavior observed in an ecosystem may not be obvious given the individual behaviors of agents.

2.2   Agent System Stability

Service Replication. An increasing number of researchers are investigating the problems of reliability, robustness, and stability of multi-agent systems *(MAS)*. Most approaches toward improving system robustness revolve around the replication of agents and/or services on the MAS network. This direction has been taken by [7], [12], [16] and several others. Existing approaches focus on the methodology of agent/service replication.

Probabilistic Models. Another approach is the application of probabilistic models to the prediction of agent system stability and robustness. This research assumes some uncertainty in agent behavior or the agent's environment, and proposes mechanisms for estimating, evaluating and hopefully improving stability of agent systems. One of the first researchers to analyze probabilistic survivability in an MAS is Kraus in [11]. In that paper Kraus proposed a probabilistic model of MAS survivability based on two assumptions: (1) global state of the network is known at all times; and (2) the probabilities of host or connection failure are known. An alternative approach was proposed in [20, 2], where agents reason about the state of the network and security (insecurity) of their actions.

3   Problem Formulation

3.1   Motivation

In a typical dynamic ad hoc network there is limited, variable bandwidth between hosts, and the memory and CPU on each host is constrained. Given this dynamic and resource constrained environment, it is impractical to prescribe any pre-computed solution.

The solution we propose for such networks is to create a system that can control the number of agents dynamically, adapting to the ever-changing environment. In order to work in the context of an agent based system, a control system should be *distributed* and *decentralized*. By distributed, we mean that the system should be able to use the underlying network to parallelize problem solving on multiple hosts. By decentralized, we mean that the system should avoid reliance on a single node, and should allow each agent to act independently. The emergent behavior resulting from the individual localized control decisions ideally will yield an optimal, or near-optimal, solution at the global level.

3.2   Approach

Large ecosystems usually have several attractive qualities (such as dynamic decentralized control, self regulation, no single point of failure, robustness, and stability) that we require for our system. We propose a solution to the problem of determining the number of agents appropriate for a task at hand that is inspired by large ecosystems:

1. Each task in our system is associated with food.
2. Agents which successfully complete a task collect the associated food points.
3. Agents consume food points over time to sustain their existence.
4. Agents that exhaust their supply of food die.
5. An abundance of food can cause a new agent to spawn.

By this analogy, tasks can be thought of as plant life growing at some rate. Agents are associated with herbivore animals that perform tasks, therefore eating all the food provided by successfully completing a task. Upon completion of a task, an agent is forced to migrate to look for more food (tasks to complete). As time passes, agents consume food according to a predefined consumption function, analogous to a metabolic rate of an animal. Agents unable to find enough food (tasks) to sustain their existence over time will exhaust their food resources and will be terminated. Large amounts of food collected by a single agent or accumulated in a single location can force a new agent to spawn at this location. Agents procreate by division similar to a cell mitosis. However, this approach makes it impossible for the system to recover from a state with no agents. Therefore, we also allow tasks the ability to spawn a servicing agent whenever a certain threshold of accumulated food supply is reached. This control metaphor allows the system to dynamically adjust to the environment, while avoiding centralized control.

3.3   Formal Model

The set H denotes the set of producers h where $h \in H$, with the production rate defined by a function $F_h(t)$ for each individual producer h. The set A defines the set of consumers a $(a \in A)$, and each consumer has a predefined consumption function $f_a(t)$. The dynamic system of H producers and A consumers is considered to be in an equilibrium state over some period of time from t_1 to t_2, if and only if the amount of food produced during that period of time is equal to the amount of food consumed during that same period of time. This relationship can be expressed as:

$$\sum_{h \in H} \int_{t_1}^{t_2} F_h(t)\, dt = \sum_{a \in A} \int_{t_1}^{t_2} f_a(t)\, dt$$

At the simplest level, these principles can be modeled by a dynamic system of homogeneous producers and homogeneous consumers with constant production and consumption rates, c and d respectively. The equations below define the equilibrium state for this simple example:

$$\sum_{h \in H} \int_{t_1}^{t_2} F_h(t)\, dt = \sum_{a \in A} \int_{t_1}^{t_2} f_a(t)\, dt$$

$$\sum_{h \in H} \int_{t_1}^{t_2} c\, dt = \sum_{a \in A} \int_{t_1}^{t_2} d\, dt$$

$$|H| \times c \times (t_2 - t_1) = |A| \times d \times (t_2 - t_1)$$

$$|A| = |H| \frac{c}{d}$$

This is essentially a species-based analysis of our individual-based ecological control system.

4   Experimental Results

4.1   System Setup

In order to confirm our conclusions we implemented a series of experiments using a discrete event simulation. The control flow of an ecology based agent is shown in Figure 1(a). According to this control flow diagram, an agent first decreases it's internal food bank by $f_a(t)$ for each second that elapsed since the last decrement. Then, the agent completes the task and collects all food points associated with that task. Based on its current food resources, the agent may decide to die or to reproduce. Lastly, the agent migrates to another random host looking for food. We experimented with different ways for an agent to decide when to reproduce. We chose a *fuzzy threshold* method. Given the threshold value r, the probability of an agent reproducing is 0 if the amount of food is less then $r - \frac{r}{2}$. The probability of an agent reproducing is 1 if the

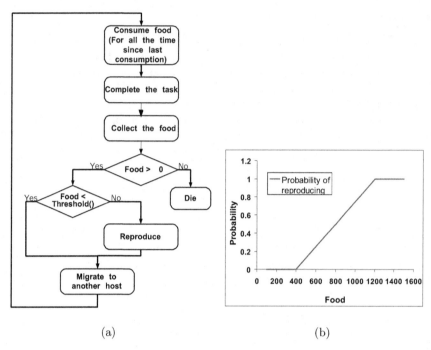

(a)   (b)

Fig. 1. Agents life cycle (a) and probability of reproduction based on the food level (b)

food level exceeds $r + \frac{r}{2}$. And the probability of reproducing grows linearly between these two points. A plot of the probability of reproducing is shown in Figure 1(b). The threshold needs to be fuzzy to avoid undesirable oscillations in the system.

If a small number of agents is desired on the network, it is possible for the system to go into an *extinction mode* — state with no agents on the network. In order to recover from this situation, we enable hosts to spawn new agents. The same fuzzy threshold rules apply to hosts as to agents. All of the experiments were performed on the completely connected network of statically placed hosts. All hosts grow food at the rate 1 unit per iteration. All experiments start with a single agent with initial food bank of 500 units. The reproduction threshold r was set to 800 resulting in 800 ± 400 range for hosts and agents. Additional experiments were performed using the real agent system EMAA [13] over a wired local area network.

4.2   System Behavior over Time

In this section, we investigate the changes in the number of agents over time. The consumption rate is set to 5 food units per iteration for all agents. Each experiment consists of 15 trials. A single trial consists of initializing the system and running it for 90,000 iterations. The number of agents is recorded every 10 iterations. Data is averaged across all trials to obtain the plots.

Constant Number of Hosts. Experiments were repeated for graphs of 35, 23, 15 and 8 hosts on the Figure 2 (top to bottom). Horizontal bold lines represent the targeted number of agents 7.0, 4.6, 3.0 and 1.6 respectively and the actual number of agents on the system is plotted by the thinner lines. It is easy to see that in all of the experiments, the actual number of agents oscillates close to the target value, however oscillations are somewhat higher for the network of 8 hosts. Figure 3 demonstrates the actual distribution of the numbers of agents during the experiments for 8 (a), 15 (b), 23 (c) and 35 (d) hosts with a normal distribution curve fitted to the data. Distribution is close to normal for all experiments but the one with 8 hosts. Such system behavior can be explained by the fact that the system with the small number of agents is prone to extinction of the population. Whenever the system recovers, it usually overshoots the targeted number of agents and oscillates for a while. These oscillations are repeated every time the system goes into extinction mode. More detailed analysis of this phenomenon is given in Section 4.3.

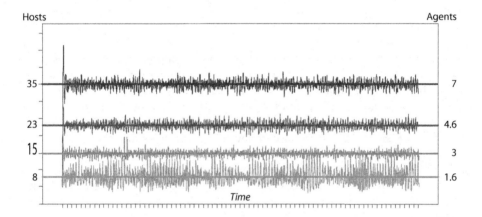

Fig. 2. System behavior over time

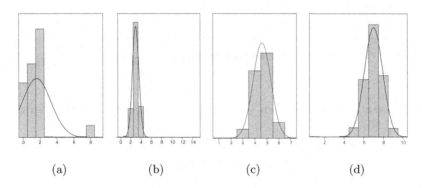

Fig. 3. Distribution of the number of agents for the network sizes of 8 (a), 15 (b), 23 (c) and 35 (d) hosts

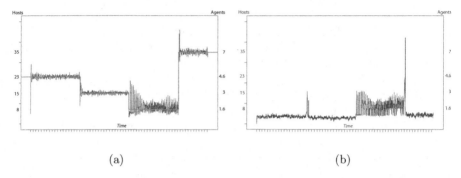

(a)   (b)

Fig. 4. System behavior over time (a) and Standard deviation (b)

Changing Number of Hosts. During this test we observed the system's ability to react to rapid unplanned changes in the number of hosts. Experiment was setup identically to the one described in Section 4.2 except that the number of hosts was changed every 30,000 iterations without reinitializing the system. The number of hosts was changed from 23 to 15 to 8 and back to 35 hosts. We feel that such drastic changes in the number of hosts approximate the process of islanding and merging in wireless mobile networks of lightweight devices carried on foot by police or military units. Whenever the hosts were shut down all of the agents on these hosts and agents traveling to these hosts were also shut down. Whenever brought back on line, hosts initially had no food or agents on them. That type of change introduces a high level of disturbance into the system. The number of agents over time is plotted in Figure 4(a). The bold red line represents the target number of agents at any given moment. The black thinner line shows the actual number of agents. One can see that the actual number of agents follows closely the target number in all segments of the plot except for the one that corresponds to 8 hosts.

The standard deviation of the number of agents is plotted in Figure 4(b). Standard deviation peaks when we change the number of hosts on the network due to the highly disruptive nature for the agent community of shutting down (or starting up) several hosts. Also standard deviation is higher at the segment corresponding to 8 hosts. We believe that such high standard deviation is caused by temporary extinction of agents and the oscillations that occur during recovery from it.

4.3 Dependency Between the Number of Agents and the Number of Hosts

In this Section, a single trial consisted of 100,000 iterations of the simulator. The number of agents is recorded every iteration and averaged across the trial to obtain a single data point. Trials were repeated for networks of sizes 3 to 35 hosts (odd numbers of hosts only). Experiments are plotted in Figure 5(a) with consumption rates set to be 3 times, 5 times and 7 times the production rate from top to bottom. Although all 3 graphs appear to be linear, they are composed

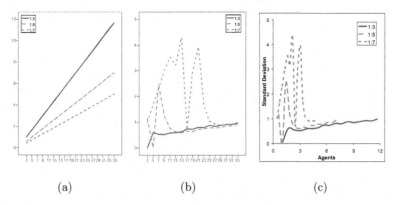

(a)   (b)   (c)

Fig. 5. Dependency between number of agents and number of hosts (a), between number of hosts and standard deviation (b), number of agents and standard deviation (c)

of 16 independently obtained data points. The experiment confirms that the system does what it is designed to do, namely maintain the given average ratio of hosts to agents, despite dynamic changes in the number of hosts. Figures 5(b) and 5(c) show the standard deviation of the number of agents in terms of the number of hosts and the number of agents respectively for all 3 experiments. Styles and colors of the plots correspond to the ones in figure 5(a). Although each is unique, the overall shape of the plots is similar. After the initial hump associated with the extinction mode and recovery from it, the plots level off in the area of 3 – 4 agents and then increase slightly. The linear increase can be explained by the linear increase in the number of agents. The only disturbance to that scheme is the point with target value of exactly 1 agent. For such systems it is possible to sustain a single agent for the duration of the whole experiment without ever going into the extinction mode resulting in *no* variance in the data.

4.4   Dependency Between the Number of Agents and the Link Quality

This set of experiments was set up exactly as the one described in Section 4.3 except that the changing parameter was link quality. A link of 100% quality implies that no artificial delay is introduced and migration only takes one iteration. Link of 0% quality means that maximum possible delay is introduced and migration takes 16 iterations (in simulator time). Figure 6(a) shows the target number of agents, actual number of agents and standard deviation of the number of agents changing based on the link quality. Although the actual number of agents slightly increases with decrease in link quality, it remains within 10% of the target value. Standard deviation however increases significantly as the speed of communication decreases. Some improvement of standard deviation at extremely low speeds can be explained by *consistently poor* performance of the system. Figure 6 (b) and (c) show the actual distribution of the number of agents for 10% and 90% respectively. The distribution for the higher link speeds

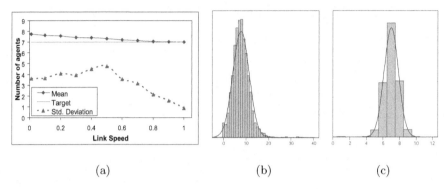

(a)   (b)   (c)

Fig. 6. Dependency between number of agents and standard deviation from liq speed (a), and distributions for 10% (b) and 90 % (c) link speed

is more compact and closer to normal. Such behavior of the system can be explained by the fact that at the lower values, agents cannot move from one host to another fast enough to collect enough food to sustain their existence. This causes extinction of agents and forces the system to re-stabilize after it recovers from the state with no agents.

5 Future Work and Conclusions

5.1 Future Work

In the future we are planning several extensions to this work.

1. We are planning more extensive set of live experiments utilizing the Secure Wireless Agent Testbed (SWAT) [22].
2. We would also like to create a more detailed mathematical model of such systems to be able to predict and control the emergent behavior of an agent system. This model should be used for parameter fine tuning, something that was done manually during current experiments.
3. We are also planning to introduce an on-line system for tuning such parameters as consumption and production rates, thresholds and fuzzy intervals, etc. Some of the techniques we are planning to try include machine learning, swarm based techniques and genetic algorithms.
4. It would be interesting to expand the model from plant — herbivore system to plant — herbivore — carnivore. That extension will allow us to create more complicated food chains resulting in more elaborate control over the populations of different types of agents.

All of these techniques promise to improve on the current research and provide a more stable decentralized ways to control the number of agents on a wireless ad hoc network.

5.2   Conclusions

This paper developed an ecology-based model for managing the number of agents on ad hoc wireless networks. We have discovered that an ecosystem based model can provide decentralized distributed robust control of agents in dynamic and uncertain network environments. Our approach involves a novel exploitation of properties of ad hoc networks, enabling mobile agents to automatically adapt to changes that affect their communication and migration. The capability to dynamically adjust to the state of their network provides new possibilities for stable MAS.

References

1. J. M. Alexander. Evolutionary game theory. In E. N. Zalta, editor, *The Stanford Encyclopedia of Philosophy*. Stanford University, Summer 2003. http://plato.stanford.edu/archives/sum2003/entries/game-evolutionary/.
2. Donovan Artz, Max Peysakhov, and William C. Regli. Network meta-reasoning for information assurance in mobile agent systems. In *Eighteenth International Joint Conference on Artificial Intelligence*, pages 1455–57, Aug 2003.
3. A. Assad and N. Packard. Emergent colonization in an artificial ecology. In F. Varela and P. Bourgine, editors, *Toward A Practice of Autonomous Systems: Proceedings of the First European Conference on Artificial Life.*, pages 143–152. MIT Press, 1992.
4. M. Conrad and H.H. Pattee. Evolution experiments with an artificial ecosystem. *J. Theoret. Biol.*, 28:393–409, 1970.
5. M. Conrad and M. Strizich. EVOLVE II: A computer model of an evolving ecosystem. *BioSystems*, 17:245–258, 1985.
6. Gary William Flake. *The Computational Beauty of Nature: Computer Explorations of Fractals, Chaos, Complex Systems, and Adaptation.* MIT Press, July 1998.
7. Felix C. Gartner. Fundamentals of fault-tolerant distributed computing in asynchronous environments. *ACM Computing Surveys*, 31(1):1–26, 1999.
8. D. Goldberg. *Genetic Algorithms in Search, Optimization and Machine Learning.* Addison-Wesley Pub Co, December 1989.
9. S. Goldenstein, E. Large, and D. Metaxas. Non-linear dynamical system approach to behavior modeling. *The Visual Computer*, 15:349–364, 1999.
10. J. R. Koza, F. H. Bennett III, F. H. Bennett, D. Andre, and M. A. Keane. *Genetic Programming III: Automatic Programming and Automatic Circuit Synthesis.* Morgan Kaufmann Publishers, 1999.
11. S. Kraus, V.S. Subrahmanian, and N. Cihan Tacs. Probabilistically survivable mass. In *Proceedings of the International Joint Conference on Artificial Intelligence (IJCAI-2003)*, pages 789–795, 2003.
12. Sanjeev Kumar, Philip R. Cohen, and Hector J. Levesque. The adaptive agent architecture: Achieving fault-tolerance using persistent broker teams. In *Proceedings of the Fourth International Conference on Multi-Agent Systems (ICMAS 2000)*, pages 159–166, July 2000.
13. R.P. Lentini, G. P. Rao, J. N. Thies, and J. Kay. Emaa: An extendable mobile agent architecture. In *AAAI Workshop on Software Tools for Developing Agents*, July 1998.

14. K. Lerman and A. Galstyan. A general methodology for mathematical analysis of multi-agent systems. Technical Report ISI-TR-529, USC Information Sciences, 2002.

15. K. Lerman and A. Galstyan. Macroscopic analysis of adaptive task allocation in robots. Submitted to IROS-03, 2003.

16. O. Marin, P. Sens, J. Briot, and Z. Guessoum. Towards adaptive fault tolerance for distributed multi-agent systems. In *Proceedings of the first international joint conference on Autonomous agents and multiagent systems: part 2*, pages 737 – 744, 2002.

17. J. Maynard-Smith. *Evolution and the Theory of Games*. Cambridge University Press, Cambridge, 1982.

18. J. Maynard-Smith and G. Price. The logic of animal conflict. *Nature*, 146:15–18, 1973.

19. R. L. Olson and A. A. Sequeira. An emergent computational approach to the study of ecosystem dynamics. *Ecological Modeling*, 79:95–120, 1995.

20. M. Peysakhov, D. Artz, E. Sultanik, and W. C. Regli. Network awareness for mobile agents on ad hoc networks. In *Proceedings of the Third International Joint Conference on Autonomous Agents and Multi Agent Systems (AAMAS-2004)*, July 2004.

21. M. Rizki and M. Conrad. EVOLVE III: A discrete events model of an evolutionary ecosystem. *BioSystems*, 18:121–133, 1985.

22. Evan Sultanik, Donovan Artz, Gustave Anderson, Moshe Kam, William Regli, Max Peysakhov, Jonathan Sevy, Nadya Belov, Nicholas Morizio, and Andrew Mroczkowski. Secure mobile agents on ad hoc wireless networks. In *The Fifteenth Innovative Applications of Artificial Intelligence Conference*, pages 129–36. American Association for Artificial Intelligence, Aug 2003.

23. Taylor, E. Charles, Turner, Scott, and Seth R. Goldman. Ram: Artificial life for the exploration of complex biological systems. In C.G. Langton, editor, *Artificial Life: SFI Studies in the Sciences of Complexity.*, pages 275–295. Addison-Wesley, Redwood City, CA, 1989.

From Abstract to Concrete Norms
in Agent Institutions

Davide Grossi and Frank Dignum

Utrecht University,
The Netherlands
{davide, dignum}@cs.uu.nl

Abstract. Norms specifying constraints over institutions are stated in such a form that allows them to regulate a wide range of situations over time without need for modification. To guarantee this stability, the formulation of norms need to abstract from a variety of concrete aspects, which are instead relevant for the actual operationalization of institutions. If agent institutions are to be built, which comply with a set of abstract requirements, how can those requirements be translated in more concrete constraints the impact of which can be described directly in the institution? In this work we make use of logical methods in order to provide a formal characterization of the *translation rules* that operate the connection between abstract and concrete norms. On the basis of this characterization, a comprehensive formalization of the notion of institution is also provided.

1 Introduction

Electronic institutions, such as auctions and market places are electronic counterparts of institutions that are established in our societies. They are established to regulate interactions between parties that are performing some transaction (see [6] for more details on the roles of institutions). Interactions are regulated by incorporating a number of norms in the institution which indicate the type of behavior each of the parties in the transaction should adhere to within that institution. The main concern of this work is to investigate what formal relation could be specified which accounts for how (abstract) norms can be incorporated in the (concrete) procedures constituting the institution, in such a way that agents operating within the institution either operate in accordance with those norms, or may be punished as they violate them.

That this relation is more complicated than just adding some constraints on the actions in the institution can be seen from the following example. The norm *"it is forbidden to discriminate on the basis of age"* can be formalized in deontic logic as "F(discriminate(x,y,age))" (stating that it is forbidden to discriminate between x and y on the basis of age). The translation of this formula would get down to something like that the action "discriminate(x,y,age)" should not occur. However, it is very unlikely that the agents operating within the

M.G. Hinchey et al. (Eds.): FAABS 2004, LNAI 3228, pp. 12–29, 2005.

institution will explicitly have such an action available. The action actually states something far more abstract. We claim that the level on which the norms are specified is more abstract and/or general than the level on which the processes and structure of the institution are specified. From an institutional standpoint norms need, in order to be incorporated in the institution itself, to be therefore "translated" to a level in which their impact on the institution can be described directly. A formal account of these "*translation rules*" constitutes the central aim of this work.

The work is organized in accordance with the following outline. In Section 2 some preliminaries about the notions of norms, normative systems and institutions are set forth; in Section 3 the issue addressed is made concrete by means of two examples, and our line of analysis of the problem is stated; in Section 4 a formal framework is proposed, which allows for formal definitions of the notions of abstract and concrete norms, and of translation rules; in Section 5 these definitions are used in order to provide a formal account of the notion of institution itself able to cope with the issue of abstractness of norms; in Section 6 this formal notion is shown to be embeddable in various formal argumentation systems, thus enabling the possibility of articulate institutional reasoning patterns; finally, in Section 7, some conclusions are drawn.

2   Some Preliminaries

The first concept to introduce is the concept of **norm**. As we will see later in Section 2.2, institutions are defined in terms of norms, which are therefore the basic building block, so to say, of our work. With the term norm we intend whatever in general indicates something ideal and which, consequently, presupposes a distinction between what is ideally the case and what is actually the case. In natural language norms are usually, but not always, expressed by locutions such as: "it is obligatory", "it is forbidden", "it is permitted", etc..

In this paper we will assume norms to be conditional, because that is the form in which they mostly appear in statutes and regulations governing institutions. In conditional norms we recognize the condition of application of the norm, and its normative effect, i.e. the normative consequence the norm subordinates to its condition: "under condition A, it is obligatory (respectively, permitted or prohibited) that B"

Another important concept we will come to take into consideration, though not in detail, is the concept of **procedure**. Here a procedure is seen as an algorithm-like specification describing how a certain activity is carried out. The difference between a norm and a procedure is of extreme relevance for our purposes (see Section 2.2): a norm states that something ought to be the case under certain conditions, while a procedure describes only a way of bringing something about; semantically, norms incorporate a concept of ideality, whereas for procedures it is instead central a notion of transition.

2.1   Normative Systems

In [14] normative systems are defined as follows:

> "a normative system is any set of interacting agents whose behavior can [...] be regarded as norm directed".

According to this view, a normative system is thus a norm directed agency. In this sense, a set of norms meant to direct an agency constitutes a form of (normative) specification of that agency; in other words, a set of norms addressed to a given agency determines that agency as a normative system. As such, normative systems are therefore amenable of formal description in terms of logical theories containing normative expressions[1].

There is wide agreement upon the fact that all normative systems of high complexity, like for example legal systems, cannot be regarded simply as sets of norms ([14, 13]). Besides norms, they consist also of definitional components yielding a kind of contextual definition: "A means (counts as) B in context i". An example: "signing form 32 counts as consenting to an organ donation, in the context of Spanish transplant regulation [26][2]". Normative components of this type are known in legal and social theory as *constitutive norms*, while purely normative components, i.e. what we called norms, are known as *regulative norms* (see for example [12, 19, 25]). Both these components will be logically represented (Section 4) by means of rules: regulative norms via rules having a deontic consequent *normative rules*; constitutive norms via *translation rules*. Concepts introduced are recapitulated in Table 1.

Table 1. Normative systems' components

COMPONENTS	regulative norms	constitutive norms
REPRESENTATION	normative rules	translation rules

2.2   Institutions

The term institution is quite ambiguous. Following [17] we distinguish two senses of the term, which are of significance for our purposes.

- First, an institution can be seen a the set of agents with specific roles, private and common objectives, the activities of which are procedurally determined. We speak in this case about institutions seen as **organizations**. As an example, the agents operating Utrecht Hospital, and the set of procedures according to which their activity is planned, constitute an organization.
- Second, an institution can be seen as the set of norms (constitutive and regulative) an organization can instantiate implementing them. We use in

---

[1] This is precisely how normative systems are conceived in [1], where they are analyzed as sets of sentences deductively connecting normative conditions to normative effects.

[2] These examples have been chosen on the basis of work carried out on the regulations from which they are excerpted.

this case the term **institutional form**. In this sense the set of regulations holding at Utrecht Hospital defines an institutional form. Also the set of regulations concerning hospitals in The Netherlands defines an institutional form, namely a general institutional form, say, "hospital". The organization of Utrecht Hospital instantiates both these institutional forms.

This distinction between organizations and institutional forms lies in the aforementioned distinction between norms and procedures. While analyzing institutions as organizations emphasizes the procedural aspects involved in operating institutions, an analysis of them in terms of institutional forms stresses instead the normative nature of institutions specifications. This last perspective on institutions is the one underpinning the analysis of abstract and concrete norms that will be carried out in the next sections. Viewing institutions as institutional forms, that is to say, as sets of constitutive and regulative norms, allows for an application of a normative system perspective ([13, 14]) to their analysis and will lead, in Section 5, to a formal definition of institutions as sets of rules[3].

It is instructive to spend still some more words on the distinction proposed. The relation between these two conceptions of institutions constitutes a very interesting issue, which is also of definite relevance in relation with the general problems addressed here. What is at stake is the understanding of how an organization implements an institutional form, or in other words, how can a set of procedures implement a set of norms, what is the formal link between norms and procedures. Answering these questions would lead to a deeper understanding of the variety of aspects characterizing institutionalized agencies. This problem forms nevertheless a separate issue, which will not be explicitly dealt with in the present paper[4].

3   Abstractness of Norms

3.1   Abstract Norms and Concrete Norms

The issuing of norms, as it appears in various statutes or regulations specifying constraints over institutions, has the characteristic of stating norms in such a form that allows them to regulate a wide range of situations and to be stable for a long period of time. The vaguer or abstract norms are, the easier it becomes to keep them stable. The downside of this stability is that normative formulations seem to be less well defined. In law it is even an explicit task of the judges to interpret the law for specific situations and determine whether someone violated it or not.

It is our thesis that abstract and concrete notions are described within different ontologies. Concrete norms are described in terms of the concepts that are

---

[3] The formal analysis of *organizations*, i.e. procedural description of agencies, is therefore left aside in this work. In what follows we will use the terms institution and institutional form interchangeably.

[4] See [7] for some first thoughts on this topic.

used to specify (possible) procedural descriptions of the concrete institutions. Abstract levels are instead described using a more general ontology.

In order to precisely illustrate the problem we are concerned with, we discuss two examples. The first one is taken from the Dutch regulation about personal data treatment within police registers ([8]). In the mentioned regulation the following norm is stated: "*the inclusion of personal data in a severe criminality register occurs only when it concerns: a) suspect of crimes; b) etc.*" (Article 13a). This norm states that, under certain conditions, personal data may be included in a specific kind of police register. Suppose now that an electronic institution for that register has to be built which fully complies with the norms regulating the use of that register ([5]). The following question comes naturally about: "*what can be concretely included in the register*", that is "*what is classified to be* personal data *in the context of* [8]"? That this is more than just a definitional issue can be seen from the fact that more data may be included as they regard suspects and less as they regard persons which are indirectly connected with a crime: the notion of personal data varies. These "variations" are specified in the model regulations on police registers ([16]).

The second example is instead taken from the Spanish regulation on organ transplantation ([26]): "*a living donor must consent before a transplantation may take place*" (Article 9). An analogous question can be raised: "*what is understood as* consent *in the context of* [26]"? This example shows that abstraction takes place over data (first example) as well as over actions. The *consent* action can be implemented by *signing form 32* within the context of the transplant regulation in Spain. However, this way of implementing consent is only "valid" within that context.

On what basis are we entitled to consider the above translations as complying with the abstract ones? Signing a form seems a reasonable implementation of giving consent, whereas we would probably not accept *wearing a red hat* as a way of implementing consent. What does the connection between abstract and concrete normative formulations consist of, from a formal point of view? This is the central question we are here addressing.

3.2 Connecting Abstract and Concrete Norms

The model regulation on severe criminality registers ([16]) is explicitly conceived to lead to an application of the law in the context of the usage of severe criminality registers. The following norm is stated: "*[In a severe criminality register] the following kinds of data can at most be included: financial and corporate data; data concerning nationality; etc.*" (Article 6). Basically, this article provides the list of data that are allowed to be included in the register, and it therefore consists of a concrete version of Article 13a cited in Section 3.1. Such a "translation", as we called it, is possible because an interpretation of the notion of *personal data* occurring in Article 13a, is somehow presupposed: "*personal data are financial and corporate data; data concerning nationality; etc.*". This rule, defining the notion of personal data within the context of the usage of severe criminality registers, states that if something is a datum concerning the nationality of, for instance, a suspect, then this datum is a personal datum and it can therefore be legally included in the register. We claim these rules to constitute the connection between abstract and concrete norms.

In this example, being a personal datum is an abstract fact exactly because something can be a personal datum in many ways, depending on the context: in the context of the regulation of severe criminality registers, data as specified in Article 6 count as personal data, but within a different context, for example in the regulation about so called provisional police registers, something else can count as a personal datum. Abstract constraints are stable and hold for many situations because they are made concrete in several, possibly different, ways. The contextual nature of these translation rules led us to the logical framework we are going to expose in Section 4.

To understand this contextual nature of institutions it seems useful to see them as regulating facts that hold on specific levels of abstractness: concrete levels are the levels on which facts hold that can be directly handled by the procedures an institution is organized through (something is a datum concerning nationality); abstract levels are the levels on which more abstract facts hold (something is a personal datum), and to which many more concrete levels can be seen to converge via translation rules. We therefore understand institutions as sets of norms and translation rules which regulate facts holding on levels of abstractness[5]. Such a perspective also shows how more particular institutions, such as the ones operating severe criminality registers, are nested in more general ones, such as the one regulating the use of police registers in general. This nesting takes place through the abstractness layering. Picture 1 below provides a graphical account of the intuitions just exposed.

Analogous considerations may be carried out in relation with the second example mentioned in Section 3.1.

4   Formal Framework

4.1   A Logic for Levels of Abstractness

Before presenting a proposal to formally capture the notion of level (context) we have in mind, it is necessary to identify, in further detail, the features of this concept that we would like to be able to express in our formalism.

1. In our view, levels constitute a structure ordered according to the relation "i is strictly less abstract than j". This relation is, reasonably, irreflexive, asymmetric and transitive. Moreover, it seems intuitive to assume it to be partial. There might be levels i and j both strictly less abstract than a given level k, but such that they remain unrelated with respect to each other[6].
2. Levels are such that what holds in a level holds irrespectively of the level from which that fact is considered: if at level i the donor expresses his/her

---

[5] See section 2.

[6] Notice that these are precisely the properties also of the *conventional generation* relation analyzed in [10].

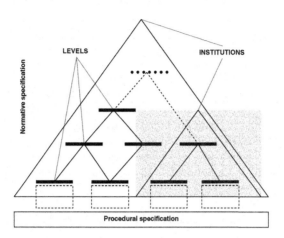

Fig. 1. Institutions and levels

consent, then at level j it holds that at level i the donor expresses his/her consent and vice versa.

3. No inconsistency holds at any level, levels are coherent.
4. Finally, there exists a trivial "outermost level", representing the absence of context, that is, the level of logical truths.

To capture these features we use a multi modal logic $KD45_n^{i-j}$ ([15]) which corresponds to a propositional logic of n contexts (PLC) with: consistency property (corresponding to feature 3), flatness property (feature 2), outermost context (feature 4) and total truth assignments (see $[18, 4, 3])^7$.

Language. The alphabet of language $\mathcal{L}^L$ for levels of abstractness expands the language for propositional logic and contains the following sets of symbols: the set of logical connectives $\{\neg, \wedge, \vee, \rightarrow\}$; the set of propositional constants $\mathbb{P}$; and the set of modal operators $\{\square_i\}_{i \in L}$ where L is the set of indexes denoting levels of abstractness, and $||L|| = n$, that is to say, there are as many modal operators as levels of abstractness. The set of well formed formulas $\mathbb{F}$ is then defined as follows:

---

^{7} We deemed a multi modal formalism to be better readable than a propositional context logic one. This is the reason why we chose for using a modal logic formulation instead of a contextual logic one. The correspondence result we claimed is guaranteed by results proved in [3]. A word must be spent also about the use of propositional context logic with total truth assignments. In fact, partial truth assignments are one of the most relevant features of context logics as introduced in [4, 3]. However, it has been proved in [18] that every propositional context logic system with partial truth assignments is equivalent to one with total truth assignments. For this reason this aspect has been here disregarded.

$$\mathbb{F} := \mathbb{P} \cup (\neg \mathbb{F}) \cup (\mathbb{F} \wedge \mathbb{F}) \cup (\mathbb{F} \vee \mathbb{F}) \cup (\mathbb{F} \rightarrow \mathbb{F}) \cup (\Box_i \mathbb{F}).$$

By means of this language it is possible to express statements about what holds on a level (in a context) via modal formulas.

Semantics. As a semantics for this system we can use very simple models $M = (W, L, <, c, v)$ such that for every level of abstractness (or context) $i \in L$ function c associates a non-empty subset of W ($c : L \longrightarrow Pow^+(W)$), v is the usual valuation function assigning truth values to propositions in worlds. Ordering $< \subseteq L \times L$ is an irreflexive, asymmetric and transitive ordering on L, the intuitive reading of which is: $i < j$ means that i is less abstract than j (feature 1). Using these models we can define the semantics of the levels of abstractness as follows:

$$M, w \vDash \Box_i A \text{ \textbf{iff} } \forall w' \in c(i) : M, w' \vDash A$$

We omit here the obvious clauses for satisfaction of propositional formulas. Notice that the truth value of $\Box_i A$ does not depend on the world where it is evaluated. This reflects the intuition that whether A is true at level i does not depend on the place from which you evaluate it. It only depends on the truth of A in that specific level (in this precisely consists the aforementioned flatness property corresponding to feature 3). With respect to the other requirements, we have that: feature 2) is guaranteed by the fact that c delivers non-empty subsets of W, and feature 4) is guaranteed by the fact that there can be worlds not belonging to any $c(i)$[8]. Noticeably, this semantics implements in a straightforward way the thesis developed in context modeling according to which contexts can be soundly represented as sets of possible worlds ([27]).

A final aspect worth stressing is that the ordering of the levels does not play any role in the semantics. One could imagine that the ordering on L imposes an ordering on the sets W_i. E.g. $i < j \Rightarrow W_i \subseteq W_j$. This would imply the following validity: $\Box_j A \rightarrow \Box_i A$ iff $i < j$ i.e. a kind of inheritance from more abstract levels to more concrete levels. We have chosen not to include this property because it would impose many restrictions on the relation between levels, which are not really necessary. We will come back to this point later on in Section 5 where we will indicate some ideas about more subtle relations between levels of abstractness.

Axiomatization. $KD45_n^{i-j}$ is obtainable via the following axioms and rules schemas:

---

[8] It is instructive to notice that this semantics is equivalent with a more standard relational semantics for $KD45_n^{i-j}$ given in terms of Kripke models with a family of accessibility relations $\{R_i\}_{i \in L}$ which are serial, transitive, and i-j euclidean ($wR_iw', wR_jw'' \Rightarrow w'R_iw''$). The proof can be obtained once the family $\{R_i\}_{i \in L}$ is defined to be such that wR_iw' iff $w' \in c(i)$. The whole proof is worked out in [15].

(P) all tautologies of propositional calculus

(K) $\Box_i(A \to B) \to (\Box_i A \to \Box_i B)$

(D) $\neg\Box_i\bot$

(4^{i-j}) $\Box_i A \to \Box_j \Box_i A$

(5^{i-j}) $\neg\Box_i A \to \Box_j \neg\Box_i A$

(MP) $A,\ A \to B\ /\ B$

(N) $A\ /\ \Box_i A$

The system at issue is then a multi modal homogeneous $KD45$ with the two interaction axioms 4^{i-j} and 5^{i-j}[9]. This axiomatization is sound and complete with respect to the semantics presented (see [15]).

4.2 A Logic for Translation Rules

Informally, A *counts as* B iff A at a level i determines the truth of B at a level j, where $i < j$ (see Section 3.2).

Theoretically, our proposal consists in understanding translation rules as *bridge rules* in the sense of theory of contexts (see for example [18]). Translation rules connect truth among different levels of abstractness, and more precisely from more concrete to more abstract levels. In addition, we consider translation rules to be defeasible. The reason for this choice is that different translation rules could have contradictory consequents, and therefore the antecedent of a translation rule cannot be strenghtened: "signing form 32 counts as consenting for organ donation" but "signing form 32 while being under threat does not count as consenting for organ donation".

To model this notion of translation rule we make use of normal prioritized default logic ([2]) defining a normal prioritized default theory $\mathcal{T}_T$ on the system $KD45_n^{i-j}$ for language $\mathcal{L}^L$:

$$\mathcal{T} = (F, D_T, \prec_T)$$

where F is a (possibly empty) set of assumptions, D_T is a set of defaults, $\prec_T$ is a priority ordering on defaults of D_T. By means of this logical machinery the following definition of translation rule can be stated:

Definition 1. (Translation Rules)
A translation rule is a default rule of this form:

$$\Box_i A \rightsquigarrow \Box_j B \quad \text{with}\ i < j.$$

Here "$\Box_i A \rightsquigarrow \Box_j B$" is a shorthand for $\Box_i A : \Box_j B / \Box_j B$, i.e. a normal default, the meaning of which is that the truth of B can be derived on level j from the truth

---

[9] Instead of 4^{i-j}, it would be sufficient to assume a simple 4 axiom: $\Box_i A \to \Box_i \Box_i A$ (see [15]).

of A at level i if the truth of B on level j does not result in an inconsistency. This account has several advantages: it has a clear theoretical grounding on context theory; it has a neat semantics; it enables easy non monotonic derivations; it can rely on a broadly investigated logic. Thus, the fact that "signing form 32" is a way of "consenting for organ donation" in a certain hospital can now be formally represented as:

$$\Box_i \; signing\_form\_32 \rightsquigarrow \Box_j \; consent$$

where i is a more concrete level of abstraction within the institution of "hospital" than j.

In order to deal successfully with defeasibility we also introduced in definition 1 explicit prioratization ordering $\prec_T$ on the set of defaults:

$$d_1 : \; \Box_i A \rightsquigarrow \Box_j B$$
$$d_2 : \; \Box_i(A \wedge C) \rightsquigarrow \Box_j \neg B$$

One prioratization criterion is that more specific defaults have the precedence according to a strict partial ordering. So, this means $d_2 \prec_T d_1$.

Note that this prioritization orders only conflicting defaults such that either the prerequisites of the first imply the prerequisites of the second or vice versa. It does not supply a tool for deciding among conflicting defaults the prerequisites of which are logically unrelated. It may be useful, for example, to include a prioritization based on concreteness of the antecedent. This can be used in the following case:

$$d_1 : \; \Box_i A \rightsquigarrow \Box_j B$$
$$d_2 : \; \Box_k A \rightsquigarrow \Box_j \neg B$$
$$k < i$$

obtasining that $d_2 \prec d_1$.

We deem important to stress that specificity and concreteness are only two of the many ways of deciding about conflicting defaults. In normative reasoning especially, conflicts are often decided on the basis of authority hierarchies subsisting on norms, or on the basis of the time of their enactment ([21]). Moreover, conflicts between priority ordering themselves can arise. The specificity and concreteness criteria should therefore only be seen as an exemplification of this range of possible criteria.

4.3   A Logic for Normative Rules

Having defined levels of abstractness and their relations in the previous sections, we now turn to defining the norms themselves that operate on levels. To do this, we have to: first, enable a representation of deontic notions within the framework defined in Section 4.1; then, introduce suitable rules to model the conditional aspect of norms, which has been stressed in Section 2.

Let us focus on the first point. To handle deontic notions (obligation, permission, prohibition), the standard deontic logic system KD (see [28]) suffices our needs here. We can therefore define a fusion[10] $KD \otimes KD45_n^{i-j}$ on a common language $\mathcal{L}^{LO}$ containing the language for expressing the abstractness layering $\mathcal{L}^L$, and the language of standard deontic logic $\mathcal{L}^O$.

Language. The language is a propositional logic language the alphabet of which is expanded with an O-operator and a set of indexed $\square_i$-operators. The set of well found formulas $\mathbb{F}$ is defined as follows:

$$\mathbb{F} := \mathbb{P} \cup (\neg \mathbb{F}) \cup (\mathbb{F} \wedge \mathbb{F}) \cup (\mathbb{F} \vee \mathbb{F}) \cup (\mathbb{F} \to \mathbb{F}) \cup (\square_i \mathbb{F}) \cup (\square_i O(\mathbb{F}))$$

Note that we allow deontic modalities to operate only within $\square_k$-formulas and we do not allow deontic operators to have $\square_k$ formulas in their scope if they are not under the scope of another $\square_k$-operator. This expressive limitation is dictated by the fact that we do not want deontic operators to occur if not in the scope of a $\square_k$-operator. This to capture the idea according to which normative consequences of certain conditions are supposed to be always holding at certain levels of abstractness: normative consequences are always localized.

Semantics. Semantics for $\mathcal{L}^{LO}$ is given on structures $M = (W, L, <, c, R, v)$ such that $(W, L, <, c, v)$ is a model for $\mathcal{L}^L$ (see Section 4.1), and (W, R, v) is a model for $\mathcal{L}^O$ with R being a serial accessibility relation on W. We omit here the obvious clauses for satisfaction of propositional formulas. The semantics of $\square_k$-operators remains the same described in Section 4.1. As to the semantics for the O-operator we use the usual clause obtaining the following expanded clause for formulas in $\square_i O(\mathbb{F})$:

$$M, w \vDash \square_i O(A) \text{ iff } \forall w' \in c(i), \forall w'' \in W : R(w', w'') \Rightarrow M, w'' \vDash A$$

Permission (P-operator) and prohibition (F-operator) can be defined in terms of obligation: $P(A) \equiv \neg O(\neg A)$ and $F(A) \equiv O(\neg A)$.

Axiomatization. Logic $KD \otimes KD45_n^{i-j}$ can be easily axiomatized by the union of the set of axioms for $KD45_n^{i-j}$ and the set of axioms for KD. Axiomatization $KD45_n^{i-j}$ (Section 4.1) should thus be extended as follows:

(P) all tautologies of propositional calculus

$(K_\square)$ $\square_i(A \to B) \to (\square_i A \to \square_i B)$

$(D_\square)$ $\neg \square_i \bot$

$(4_\square^{i-j})$ $\square_i A \to \square_j \square_i A$

$(5_\square^{i-j})$ $\neg \square_i A \to \square_j \neg \square_i A$

(MP) $A, A \to B / B$

$(N_\square)$ $A / \square_i A$

(K_O) $O(A \to B) \to (OA \to OB)$

(D_O) $\neg O \bot$

(N_O) A / OA

---

[10] For a detailed exposition of the concept of fusion we refer to [9]. Intuitively, a fusion of two logics is the simple join of them.

Notice that no interaction axioms between $\Box_i$ and O operators are stated. As proved in [9], fusions of systems preserve soundness and completeness, therefore system $KD \otimes KD45_n^{i-j}$ is sound and complete with respect to the semantics presented.

To enable a representation of the aspect of conditionality of norms, and then of *normative rules*, we make again use of normal prioritized default logic defining a normal prioritized default theory $\mathcal{T}_N$ on the system $KD \otimes KD45_n^{i-j}$ for language $\mathcal{L}^{LO}$:

$$\mathcal{T}_N = (F, D_N, \prec_N)$$

where F is a (possibly empty) set of assumptions, D_N is a set of defaults, $\prec_N$ is a priority ordering on defaults of D_N. By means of this logical machinery the following definition of normative rules can be stated:

Definition 2. (Normative Rules)
A normative rule is a default rule of the form:

$$\Box_i A \rightsquigarrow \Box_j OB \quad \text{with} \quad i < j.$$

Here "$\Box_i A \rightsquigarrow \Box_j OB$" is a shorthand for $\Box_i A : \Box_j OB / \Box_j OB$, i.e. a normal default, the meaning of which is that the truth of OB can be derived on level j from the truth of A at level i if the truth of OB on level j is not leading to an inconsistency.

Conditional permission and prohibition are easily defined by replacing the O-operator by the P and F operators respectively. All remarks underlined in Section 4.2 about prioritizing defaults formalizing translation rules hold also for defaults formalizing normative rules. Given the above definition we can represent the norm that consent is required in order to perform a transplantation, as follows:

$$\Box_i\ consent \rightsquigarrow \Box_i P\ transplant$$

At this point, it is worth remarking that translation rules and normative rules share the same type of defeasibility. This representational choice captures an important analogy which we deem to subsist between the two types of rules composing institutions:

- Translation rules *connect* truth on a level to truth on a more abstract level, and this connection takes place in a defeasible way.
- Normative rules *connect* truth on a level to ideality on another, possibly the same, level, and also this connection takes place defeasibly.

That connection is what they share and what we represented here by means of normal defaults [11].

Within this framework, definitions of abstract and concrete normative rules, representing respectively abstract and concrete norms, can be also stated:

---

[11] In this respect, our approach is close to the proposal in [11], though we carried it out by means of different formal tools.

Definition 3. (Concrete Normative Rules)
A concrete normative rule is a default $\Box_i A \rightsquigarrow \Box_j O$ (B) *s.t. there is no default* $\Box_h C \rightsquigarrow \Box_k D$ *with* $h < k$ *s.t.* $A \equiv D$ *and* $i = k$ *or* $B \equiv D$ *and* $j = k$.

Definition 4. (Abstract Normative Rules)
An abstract normative rule is a normative rule which is not concrete.

In the next section we put this articulate framework at work, providing the reader with an example.

4.4   An Example

The example we are going to model is chosen again from [8, 16].

Example 1. (Personal data in severe criminality registers)
Part of the abstract norm *"the inclusion of personal data in a severe criminality register occurs only when it concerns: a) suspect of crimes; b) etc."* can be modeled as follows:

$$\Box_a(personal(datum) \wedge suspect(datum)) \rightsquigarrow \Box_c P \ include(datum)$$

Part of the concrete norm *"personal data are financial and corporate data; data concerning nationality; etc."* might be represented as follows:

$$\Box_c(nationality(datum) \wedge suspect(datum)) \rightsquigarrow \Box_c P \ include(datum)$$

The translation rule *"personal data are financial and corporate data; data concerning nationality; etc."* is representable as follows:

$$\Box_c nationality(datum) \rightsquigarrow \Box_a \ personal(datum)$$

where $c < a$.

The first norm is more abstract because it operates between level a and level c. The second one is instead more concrete. The connection among the two of them is expressed by the translation rule connecting c to a with respect to the states of affairs $nationality(datum)$ and $personal(datum)$[12]. It may be worth noticing a reasoning pattern straightforwardly available on the basis of this representation: assuming $\Box_c(nationality(datum) \wedge suspect(datum))$, by means of default $\Box_c nationality(datum) \rightsquigarrow \Box_a personal(datum)$ and validities for $\Box$, we can infer $\Box_a(personal(datum) \wedge suspect(datum))$; we can then infer the normative consequence $\Box_c P \ include(datum)$ by means of default $\Box_a(personal(datum) \wedge suspect(datum)) \rightsquigarrow \Box_c P \ include(datum)$[13].

---

[12] Notice that we presupposed the state of affairs $include(datum)$ to be a concrete one.
[13] Notice that this argument is nothing but a normal defaults proof.

5   Institutions Defined Formally

On the basis of the formal analysis just presented we are now in a position to provide a formal definition of the concept of institution in terms of default theories. However, before getting to this, a related issue should be considered, that is: how to rigorously relate institutions and levels of abstractness. In other words, at what level of abstractness does the institution end? If one includes only the levels explicitly specified for the institution, then the norms possibly coming from more abstract levels would not come to belong to the institutional theory. I.e. if $i < j$ and j is a level that does not belong to the institution then the norms operating on level j also are not "inherited" by the institution. On the other hand, incorporating all levels of abstractness connected to the levels explicitly defined within the institution would include the complete layering in which the institution is merged.

We therefore choose to propose two definitions, one corresponding to an "explicit" view on institutional theories and one corresponding to the "implicit" one.

Let us consider the default theory $\mathcal{T} = (F, D_N \cup D_T, \prec_N \cup \prec_T)$, i.e., a default theory for both translation and normative rules, and let L be the set of abstractness levels and $<$ their ordering. Let then L_I be the set of levels of abstractness on which institution I works. Let then $<_{L_I}$ be the sub-ordering of $<$ on L_I. The following definitions can be stated.

Definition 5. (Explicit Institutional Theories)
An explicit institutional theory I^{expl} is defined as a triple $(N_I, T_I, \prec_I)$ where:

$$N_I \equiv \bigcup_{i \in L_I} N_i$$

with $N_i \equiv \{\Box_i A \rightsquigarrow \Box_j O\ B \mid \Box_i A \rightsquigarrow \Box_j O\ B \in D_N\ \&\ j \in L_I\}$. And where:

$$T_I \equiv \bigcup_{i \in L_I} T_i$$

with $T_i \equiv \{\Box_i A \rightsquigarrow \Box_j B \mid \Box_i A \rightsquigarrow \Box_j B \in D_T\ \&\ j \in L_I\}$. The third element of the triple consists in the prioritization ordering $\prec_I \subseteq \prec_N \cup \prec_T$ on defaults in N_I and T_I.

Intuitively, an institution is described as the set of all normative and translation rules defined between the levels explicitly belonging to that institution.

Definition 6. (Implicit Institutional Theories)
An implicit institutional theory I^{impl} is defined as a triple $(N*_I, C*_I, \prec *_I)$ where:

$$N*_I \equiv N_I \cup \bigcup_{k \in L} N_k$$

with $N_k \equiv \{\Box_k A \rightsquigarrow \Box_l O\ B \mid \Box_k A \rightsquigarrow \Box_l O\ B\ \in D_N\ \&\ \exists j \in L_I, j < k\}$. *And where:*

$$T *_I \equiv T_I \cup \bigcup_{k \in L} T_k$$

with $N_k \equiv \{\Box_k A \rightsquigarrow \Box_l B \mid \Box_k A \rightsquigarrow \Box_l B\ \in T_N\ \&\ \exists j \in L_I, j < k\}$. *The third element of the triple consists in the prioritization ordering* $\prec *_I\ \subseteq \prec_N \cup \prec_T$ *on defaults in* $N *_I$ *and* $T *_I$.

Intuitively, an implicit theory of an institution I is nothing but a sort of closure of the explicit theory I^{expl} of I along the abstractness ordering $<$, leading the explicit theory to incorporate every normative and translation rules defined between more abstract levels than the levels explicitly belonging to I. From definitions 5 and 6 obviously follows that: $N_I \subseteq N *_I$ and $T_I \subseteq T *_I$. Let us consider now a simple example excerpted again from [26].

Example 2. (Rules inheritance within institutions)
In order to extract an organ from a living donor each hospital in Spain ought to ascertain the legal age of the donor. The state of affairs *legal_age* is not a concrete one; let the level of abstractness it holds on to be s_3. The institution "hospital in Spain" I_S inherits a rule from Spanish general law according to which *legal_age* supervenes on *being_eighteen_years_old*. Neither this last state of affairs can be properly seen as concrete; let its level be s_2. Then the institution "Valencia hospital" I_V contains another rule according to which *being_eighteen_years_old* supervenes on *ID_testifies_legal_age*. This can be deemed as concrete; let its level be s_1. We then have three ordered levels and two institutions constituted by rules operating on those levels. One institution is general, namely I_S, and it works between levels s_1, s_2 and s_3, the other one, namely I_V, is more particular and it operates between s_1 and s_2.
Theory I_S^{expl} would be a triple $(N_S, T_S, \prec_S)$ such that:

$$\Box_{s_1} extract \rightsquigarrow \Box_{s_2} O\ (being\_eighteen\_years\_old) \in N_S,$$
$$\Box_{s_2} being\_eighteen\_years\_old \rightsquigarrow \Box_{s_3} legal\_age \in T_S$$

Theory I_V^{expl} would instead be a triple $(N_V, T_V, \prec_V)$ such that, basically:

$$\Box_{s_1} (ID\_testifies\_legal\_age) \rightsquigarrow \Box_{s_2}(being\_eighteen\_years\_old) \in T_V.$$

To understand the sense of this rule in the context of I_V it is necessary to consider the explicit account I_V^{impl} of this institution: $(N *_V, C *_V, \prec *_V)$. We then obtain what follows:

$$\Box_{s_1} extract \rightsquigarrow \Box_{s_2} O\ (being\_eighteen\_years\_old) \in N *_V,$$
$$\Box_{s_2} being\_eighteen\_years\_old \rightsquigarrow \Box_{s_3} legal\_age \in T *_V$$

This means that I_V^{impl} and I_S^{expl} share something: in this case $N *_V \cap N_S \neq \emptyset$ and $T *_V \cap T_S \neq \emptyset$. This exactly shows how I_V inherits rules from I_S, and more noticeably how I_V concretizes norms belonging to I_S by means of translation rules.

6   Reasoning with Institutional Theories

In this section we show how our formal approach to institutions, that led to Definitions 5 and 6, can be straightforwardly merged in formal argumentation frameworks specifically developed to account for legal reasoning, such as [24, 20, 22]. This will display some guidelines on how to enable articulate reasoning patterns within our approach.

Logical systems for argumentation formalize "a particular group of patterns of inferences, namely those where arguments for and against a certain claim are produced and evaluated, to test the tenability of the claim" ([23]). In [24] an argumentation framework is presented, which is based on normal default logic and which accounts for reasoning with both what we called, in Section 2, regulative and constitutive norms of normative systems. Within this setting, the central concept on which the argumentation system is based is the concept of *deontic context*, that is, a set of facts on which the set of default rules can be applied inferring the relevant normative consequences to that set of facts. In that work, anyway, no attention is given to the issue of abstractness and concreteness of norms, and consequently the logic on which default theories are built upon is just a standard deontic logic system KD. Defaults are therefore rules of this type: $A \rightsquigarrow B$ and $A \rightsquigarrow O\ B$. If we assume the multi-modal system exposed in Section 4.3 as the logic on which to apply normal defaults, and recalling Definitions 5 and 6, this useful notion can be adapted to our approach and modified as follows.

Definition 7. (Institutional Contexts)
An explicit institutional context $\mathcal{I}^{expl} = (F, I^{expl})$ consists of a set F of propositional sentences on a language $\mathcal{L}^{LO}$, and an explicit institutional theory I^{expl}. An implicit institutional context $\mathcal{I}^{impl} = (F, I^{impl})$ consists of a set F of propositional sentences on a language $\mathcal{L}^{LO}$, and an implicit institutional theory I^{impl}.

By means of these notions of institutional contexts, scenarios in which an institution I is made operative on the set of facts F can be formalized: through the rules of which institution I consists normative consequences at different levels of abstractness can be defeasibly established from F. The whole formal argumentation machinery exposed in [24] can then be put at work on *institutional contexts* instead of on deontic contexts, thus providing definitions of the notions of: *argument*, *conflict* and *defeat relations* between arguments, and *justified*, *defensible* and *overruled* arguments[14].

Analogous observations can be carried out in relation with the argumentation framework for legal reasoning presented in [20, 22], which is also based on normal default logic and therefore, in principle, perfectly suitable to handle our notion of institutional theory.

---

[14] For an exhaustive account of the role of these concepts in argumentation logics we refer to [23].

7   Conclusions and Future Work

In this work we discussed the problem of incorporating abstract norms into institutions that regulate the interactions between agents. We have shown by means of several examples that the level of abstraction of the norms is different from that of the procedures operating the institution. For this reason it does not suffice to just formalize the norms and procedures and then validate or verify the procedures against the norms. We therefore proposed to use explicit translation rules (formalized by normal defaults), corresponding to the so-called constitutive rules in legal and social theory, to formally characterize this translation. In order to capture the idea of a translation from the abstract level to the concrete level we chose to represent those levels explicitly, modeling them as contexts. Translation rules played then a kind of bridging role between levels/contexts.

Two research lines are particularly worth investigating in order to further develop the results presented here. First, as underlined in Section 2.2, an adequate understanding of the relation of implementation of a set of norms via a set of procedures deserves an accurate analysis in order to fully understand how norms are translated to an operational dimension, and therefore how institutions are instantiated by specific organizations. Secondly, although the logical formalism proposed gives the tools to describe the relations between norms on different abstraction levels, it does not in itself account for the restrictions which apply to this relation. As already noticed in Section 3.1, "wearing a red hat" is probably not acceptable as an implementation of "consenting for organ donation", or analogously the "daily temperature" can not count as a "personal datum". We intend to use formal ontological descriptions to account for this kind of restrictions constraining translation rules.

References

1. C. E. Alchourrón and E. Bulygin. *Normative Systems*. Springer Verlag, Wien, 1986.
2. G. Antoniou. *Nonmonotonic Reasoning*. MIT Press, Cambridge, 1997.
3. S. Buvač, S. V. Buvač, and I. A. Mason. The semantics of propositional contexts. *Proceedings of the eight ISMIS. LNAI-869*, pages 468–477, 1994.
4. S. V. Buvač and I. A. Mason. Propositional logic of context. *Proceedings AAAI'93*, pages 412–419, 1993.
5. F. de Jonge, W. Wiegerinck, E. Akay, M. Nijman, J. Neijt, and A. van Beek. Proanita: A multi-agent solution for legitimate information retrieval. In *Proceedings BNAIC'03, Nijmegen*, pages 453–454, 2003.
6. F. Dignum. Agents, markets, institutions, and protocols. In *Agent Mediated Electronic Commerce, The European AgentLink Perspective.*, pages 98–114. Springer-Verlag, 2001.
7. F. Dignum. Abstract norms and electronic institutions. In *Proceedings of the International Workshop on Regulated Agent-Based Social Systems: Theories and Applications (RASTA '02), Bologna*, pages 93–104, 2002.
8. Wet politieregisters (dutch police records act). Staadsblad, 1990.

9. D.M. Gabbay, A. Kurucz, F. Wolter, and M. Zakharyaschev. *Many-dimensional modal logics. Theory and applications.* Elsevier, 2003.
10. A. I. Goldman. *A Theory of Human Action.* Princeton University Press, Princeton, 1976.
11. G. Governatori, J. Gelati, A. Rotolo, and G. Sartor. Actions, institutions, powers. preliminary notes. In *International Workshop on Regulated Agent-Based Social Systems: Theories and Applications(RASTA'02)*, pages 131–147, 2002.
12. H. L. A. Hart. *The Concept of Law.* Clarendon Press, Oxford, 1961.
13. A. J. I. Jones and M. Sergot. Deontic logic in the representation of law: towards a methodology. *Artificial Intelligence and Law 1*, 1992.
14. A. J. I. Jones and M. Sergot. On the characterization of law and computer systems. *Deontic Logic in Computer Science*, pages 275–307, 1993.
15. A. Lomuscio and M. Sergot. Deontic intepreted systems. *Studia Logica 75*, pages 63–92, 2003.
16. Modelreglement register zware criminaliteit. Sttatscourant, 2002.
17. D. C. North. *Institutions, Institutional Change and Economic Performance.* Cambridge University Press, Cambridge, 1990.
18. L. Serafini P. Bouquet. wo formalizations of context: a comparison. *Modeling and Using Contexts. LNAI 2116*, pages 468–477, 2001.
19. A. Peczenik. *On Law and Reason.* Kluwer, Dordrecht, 1989.
20. H. Prakken. A logical framework for modeling legal argument. In *Proceedings of the Fourth International Conference on Artificial Intelligence and Law.* ACM Press, 1993.
21. H. Prakken. Two approaches to the formalization of defeasible deontic reasoning. *Studia Logica 57*, 1996.
22. H. Prakken. *Logical Tools for Modelling Legal Arguments.* Kluwer, 1997.
23. H. Prakken and G. Vreeswijk. Logics for defeasible argumentation. *Handbook of Philosophical Logic vol. IV*, pages 218–319, 2002.
24. L. Royakkers and F. Dignum. Defeasible reasoning with legal rules. *Defeasible Deontic Logic*, pages 263–286, 1997.
25. J. Searle. *The Construction of Social Reality.* Free Press, 1995.
26. Ley 30/1979, de 27 de octubre, sobre extracción y transplante de órganos. Boletín Oficial del Estado 266, 29th april 1986.
27. R. Stalnaker. On the representation of context. In *Journal of Logic, Language, and Information 7*, pages 3–19. Kluwer, 1993.
28. G. H. Von Wright. Deontic logic. *Mind 60*, pages 1–15, 1951.

Meeting the Deadline: Why, When and How

Frank Dignum, Jan Broersen, Virginia Dignum, and John-Jules Meyer

Institute of Information and Computing Sciences
Utrecht University
{dignum, broersen, virginia, jj}@cs.uu.nl

Abstract. This paper defines a possible semantics for deadline obligations. Also, we make explicit several choices to be made in defining the semantics of deadline obligations. We characterize deontic deadline operators in CTL, minimally extended with propositional violation constants. The advantage of this reduction approach is that formal reasoning can be performed in CTL.

1 Introduction

A normative system is defined as any set of interacting agents whose behavior can usefully be regarded as norm-directed [9]. Most organizations, and more specifically institutions, fall under this definition. Interactions in these normative systems are regulated by normative templates that describe desired behavior in terms of deontic concepts (obligations, prohibitions and permissions), deadlines, violations and sanctions. Agreements between agents, and between an agent and the society, can then be specified by means of contracts. Contracts provide flexible but verifiable means to integrate society requirements and agent autonomy, and are an adequate means for the explicit specification of interactions [14]. From the society perspective, it is important that these contracts adhere to the specifications described in the model of the organization. If we want to automate such verifications, we have to formalize the languages used for contracts and for the specification of organizations.

In [13] we presented the logic LCR, which is based on deontic temporal logic. LCR is an expressive language for describing interaction in multi-agent systems, including obligations with deadlines. *Deadlines* are important norms in most interactions between agents. Intuitively, a deadline states that an agent should perform an action before a certain point in time. The obligation to perform the action starts at the moment the deadline becomes active. E.g. when a contract is signed or approved. If the action is not performed in time a violation of the deadline occurs. It can be specified independently what measure has to be taken in this case.

In previous work, we have advocated the use of declarative deadline specifications, as it facilitates the check for compliance to a deadline and enables reasoning about norms before the planning process determines the next sequence of actions [5]. In this paper we investigate the deadline concept in more detail.

M.G. Hinchey et al. (Eds.): FAABS 2004, LNAI 3228, pp. 30–40, 2005.

The paper is organized as follows. Section 2 defines the variant of CTL we use. In section 3, we discuss the basic intuitions of deadlines. Section 4 presents a first intuitive formalization for deadlines. In section 5, we look at a more complex model for deadlines trying to catch some more practical aspects. Finally, in section 6 we present issues for future work and our conclusions.

2   Preliminaries: CTL

The reader can find the definitions for the branching time logic CTL in the literature (e.g. [3, 7, 4]). But, since we need a specific variant of the until operator, we define CTL here explicitly.

Well-formed formulas of the temporal language $\mathcal{L}_{\mathrm{CTL}}$ are defined by:

$$\varphi, \psi, \ldots := p \mid \neg\varphi \mid \varphi \wedge \psi \mid E\alpha \mid A\alpha$$
$$\alpha, \beta, \ldots := \varphi U^e \psi \mid X\varphi$$

where φ, ψ represent arbitrary well-formed formulas, and where the p are elements from an infinite set of propositional symbols $\mathcal{P}$. Formulas $\alpha, \beta, \ldots$ are called 'path formulas'. We use the superscript 'e' for the until operator to denote that this is the version of 'the until' where φ is not required to hold for the point where ψ, i.e., the point where ϕ is excluded. However, the present state is not excluded, which means that our until operator is reflexive. This gives us the following informal meanings of the until operator:

$E(\varphi U^e \psi)$: there is a future for which eventually, at some point m, the condition ψ holds, while φ holds from now until the moment before m

We define all other CTL-operators as abbreviations. Although we do not use all of the LTL operators X, F, and G in this paper, we give their abbreviations (in combination with the path quantifiers E and A) in terms of the defined operators for the sake of completeness. We also assume the standard propositional abbreviations.

$$\begin{aligned} EF\varphi &\equiv_{def} E(\top U^e \varphi) & AG\varphi &\equiv_{def} \neg EF\neg\varphi \\ AF\varphi &\equiv_{def} A(\top U^e \varphi) & EG\varphi &\equiv_{def} \neg AF\neg\varphi \\ A(\varphi U\psi) &\equiv_{def} A(\varphi U^e(\varphi \wedge \psi)) & E(\varphi U\psi) &\equiv_{def} E(\varphi U^e(\varphi \wedge \psi)) \end{aligned}$$

The informal meanings of the formulas with a universal path quantifier are as follows (the informal meanings for the versions with an existential path quantifier follow trivially):

$A(\varphi U\psi)$: for all futures, eventually, at some point the condition ψ will hold, while φ holds from now until then

$AX\varphi$: at any next moment φ will hold

$AF\varphi$: for all futures, eventually φ will hold

$AG\varphi$: for all possible futures φ holds globally

A CTL model $M = (S, \mathcal{R}, \pi)$, consists of a non-empty set S of states, an accessibility relation $\mathcal{R}$, and an interpretation function π for propositional atoms. A full path σ in M is a sequence $\sigma = s_0, s_1, s_2, \ldots$ such that for every $i \geq 0$, s_i is an element of S and $s_i \mathcal{R} s_{i+1}$, and if σ is finite with s_n its final situation, then there is no situation s_{n+1} in S such that $s_n \mathcal{R} s_{n+1}$. We say that the full path σ starts at s if and only if $s_0 = s$. We denote the state s_i of a full path $\sigma = s_0, s_1, s_2, \ldots$ in M by σ_i. Validity $M, s \models \varphi$, of a CTL-formula φ in a world s of a model $M = (S, \mathcal{R}, \pi)$ is defined as:

$$
\begin{aligned}
&M, s \models p && \Leftrightarrow s \in \pi(p) \\
&M, s \models \neg\varphi && \Leftrightarrow \text{not } M, s \models \varphi \\
&M, s \models \varphi \wedge \psi && \Leftrightarrow M, s \models \varphi \text{ and } M, s \models \psi \\
&M, s \models E\alpha && \Leftrightarrow \exists \sigma \text{ in } M \text{ such that } \sigma_0 = s \text{ and } M, \sigma, s \models \alpha \\
&M, s \models A\alpha && \Leftrightarrow \forall \sigma \text{ in } M \text{ such that } \sigma_0 = s \text{ it holds that } M, \sigma, s \models \alpha \\
&M, \sigma, s \models X\varphi && \Leftrightarrow M, \sigma_1 \models \varphi \\
&M, \sigma, s \models \varphi U^e \psi && \Leftrightarrow \exists n > 0 \text{ such that} \\
&&& \quad (1)\ M, \sigma_n \models \psi \text{ and} \\
&&& \quad (2)\ \forall i \text{ with } 0 \leq i < n \text{ it holds that } M, \sigma_i \models \varphi
\end{aligned}
$$

Validity on a CTL model M is defined as validity in all states of the model. If φ is valid on a CTL model M, we say that M is a model for φ. General validity of a formula φ is defined as validity on all CTL models. The logic CTL is the set of all general validities of $\mathcal{L}_{\text{CTL}}$ over the class of CTL models.

3   Basic Choices for the Formalization of Deadlines

In this section we study some choices to make when developing a formal model for deadlines. The deontic aspect of deadlines is formalized by introducing a set $\mathcal{A}$ of agent identifiers and a propositional constant $Viol(a)$ for each $a \in \mathcal{A}$ in $\mathcal{L}_{\text{CTL}}$. The general idea is that the violation condition holds (i.e., the propositional constant $Viol(a)$ is true) at those moments where agent a violates a deontic deadline. This enables us to reason about violations explicitly, and about what to do if they occur, which is a distinctive feature of deontic reasoning. We model deadline conditions as propositions. This seems a reasonable choice given that we do not want to model a deadline in a logic of explicit time (real time). Our view is more abstract, and a deadline is simply a condition true at some point in time. We use the symbols δ and γ to denote deadline propositions.

Although the basic idea of a deadline is very simple it appears that the details are intricate. We suggest that one of the reasons is that in order to model deadlines, we need to model a *causal* relation between non-fulfilment of an obligation and, so called, 'violation conditions'. Causal relations are notoriously hard to formalize. Figure 1 pictures the situation.

The figure shows several possible futures from a point where a deadline is in force. In some futures the required action does not take place and a violation results after the deadline is reached. For other futures, the action does take place before the deadline is reached, and no violations appear after the action.

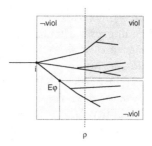

Fig. 1. The semantics of deadlines

We denote a deadline for agent a saying that it is obliged to achieve the condition ρ before δ holds, by the formula $O_a(\rho \leq \delta)$. We will give a formal definition of the semantics of this formula after, in the next sections, we have discussed some basic choices to make.

3.1   Do Obligations Persist After the Deadline?

A first distinction we make is between deadline obligations that are discharged by a failure to meet the deadline, and deadline obligations where the obligation is not discharged at the deadline. For a deadline of the first type it makes no sense to perform the action after the deadline passes. E.g., submitting a paper after the deadline of a conference has no effect. An example of the second type is the situation where one has to pay a fine for some traffic offense by the end of the month. Also when one does not pay, the obligation to pay persists (see also the work of Brown on 'standing obligations' [2]). Yet another category are the 'repetitive obligations', where the same deadline obligation is repeated over a period of time. For example monthly mortgage payments.

3.2   What if the Deadline Is Never or Immediately Met?

We first consider the case where δ equals $\bot$. Clearly, $\bot$ is a condition that will be never met. A natural question is, whether it is actually possible to have a deadline obligation for a deadline that never occurs. One could choose to say that this is impossible, which leads to the optional property (1) $\models \neg O_a(\rho \leq \bot)$. This is the case for our deadline definition is section 5, because, in the definition given there, we assume that a deadline obligation can only be in force if the deadline condition actually occurs at some point in the future. Another possibility is to say that for any condition ρ such an obligation is actually always valid, but void, i.e, without any 'force'. This corresponds to the property (2) $\models O_a(\rho \leq \bot)$. Such obligations can be considered void, because they cannot be violated; since the deadline never occurs, there will never be a point in time where non-compliance is evaluated. It might be argued that a similar situation occurs in standard deontic logic [15], where we have $\models O\top$, which corresponds with the void obligation for a tautology (also something that can never be violated). Our formalization in section 4 satisfies this property.

Obviously, the third possibility is that neither property (1), nor property (2) is satisfied. For instance, one could argue that an obligation for a deadline that never occurs, i.e., $O_a(\rho \leq \bot)$, is not void, but should be interpreted as follows: the impossibility of the deadline condition means that the deadline is ill-defined, but this does not imply that the agent is free to postpone his duty forever: he has to comply at some future point anyway (where that point can be arbitrarily far in the future). The corresponding formula is: (3) $\models O_a(\rho \leq \bot) \rightarrow AF\rho$.

Now consider the case where δ equals $\top$. This means that the deadline condition is met trivially, in the current state. One possible view is that in this case, we can still comply to the obligation by ensuring that also ρ is met in the current state. The corresponding property is: (4) $\models O_a(\rho \leq \top) \rightarrow Viol(a) \vee \rho$.

Alternatively, we might argue that it is impossible to comply to a deadline for which the deadline condition is true *now*. For an agent, it takes some time to decide whether or not to comply, and to bring about the condition ρ the obligation is concerned with. Then, if the deadline condition is true now, there is no time left for this process, and the agent will inevitably violate the obligation. In our definitions of section 4 and 5, we take this aspect into account. The corresponding property is (5) $\models O_a(\rho \leq \top) \rightarrow Viol(a)$, which is satisfied by the deontic deadline definition in sections 4 and 5. Note that under this view, the violation is not avoided if accidentally the condition ρ is true in the present state. This is because under this view, conditions are linked to agents that bring them about, which is a decision they make in the previous state, as we explain later on.

Finally one short comment about the thought that we have to account for the situation that a deadline condition might have been true in the past. Clearly we do not have to consider this situation, because it is impossible to have an obligation to do something *before* something that occurred in the past.

3.3   What if the Accomplishment Is Accidentally, Never or Trivially Achieved?

First we address the question whether it counts as compliance to a deadline obligation when the condition that is obliged occurs 'accidentally'. It is possible that the state ρ occurs without any effort or intention of the agent for whom the obligation holds. E.g. if a person is obliged to write the introduction of a paper, fails to do so, but a co-author writes the introduction (because he is tired of waiting for that person). Did the person fulfill his obligation or not? If obligations are personal, should it not be the case that also the achievements ρ are personal? After all, we do not want that if another agent, or 'nature', brings about the achievement, the agent with the obligation has complied. We encounter a basic choice to make here. If we do not want our obligations to be personal, we do not have to personalize the achievements. But, if we do want our obligations to be personal, we somehow have to link achievements to agents. There is a vast amount of literature about personalizing the achievement of conditions [10, 1, 8, 6]. Usually, such theories are called 'logics of action and/or agency'. Inspired by the work of Pörn [10], we use the *stit* operator $E_a\rho$, to denote that agent a achieves condition ρ. A difference with the *stit* operator of Pörn is that in our temporal setting, performing a 'seeing to it' action takes one time-step. That is, our stit-operator obeys $\models E_a\rho \rightarrow X\rho$, and not $\models E_a\rho \rightarrow \rho$, which holds for most other agency operators.

Our next question concerns the case where the achievement can never be reached. For instance, one might think of a personal obligation for a condition not under control of an agent. An example is the condition $\bot$. Again, a first option is to say that obligations of the form $O_a(\bot \leq \delta)$ are impossible or inconsistent. After all, it seems reasonable to take the position that one can never be obliged to achieve the impossible. This leads to the optional property (6) $\models \neg O_a(\bot \leq \delta)$, which is similar to standard deontic logic's D-axiom $\neg O\bot$ [15]. However, we might also take the position that one can have an obligation to achieve the impossible. But, since $O_a(\bot \leq \delta)$ expresses that we have to achieve the impossible before the deadline condition δ occurs, we have to conclude that this leads to the view that there will certainly be a violation whenever δ occurs for the first time. This leads to the optional property: (7) $\models O_a(\bot \leq \delta) \rightarrow \neg E(\neg\delta U^e(\delta \wedge \neg Viol(a)))$.

Finally we consider the case where the accomplishment is $\top$. How to deal with this situation depends on whether we consider the obligation to be personal or not. As discussed, for the personal case, we have to use an agency operator. In most logics of agency, $\top$ cannot be achieved by any agent ($\models \neg E_a\top$). This motivates the optional property (8) $\models \neg O_a(\top \leq \delta)$. However, if obligations are not personal, this is not necessarily intuitive. At this point we might not want to digress from standard deontic logic, where the obligation for a tautology is always valid. Thus we have the optional property (9) $\models O_a(\top \leq \delta)$.

4   A Simple Formalization

After having discussed some choices for modelling deadlines in the previous section we will present a first logical formalization.

As mentioned, $E_a\rho$ indicates that the agent a sees to it that ρ becomes true. If $E_a\rho$ is true at some point in time, then ρ is true at the next point in time. We use the symbols ρ and σ for propositions that embody some kind of accomplishment being established before a deadline condition occurs.

Let M be a CTL model, s a state, and $\sigma = \sigma_0, \sigma_1, \sigma_2, \ldots$ a full path in M. A straightforward modal semantics for the operator $O_a(\rho \leq \delta)$ is then defined as follows:

$$M, s \models O_a(\rho \leq \delta) \Leftrightarrow \forall\sigma \text{ with } \sigma_0 = s, \forall j :$$
$$\text{if } M, \sigma_j \models \delta$$
$$\text{and } \forall i \text{ with } 0 \leq i < j : M, \sigma_i \models \neg E_a\rho,$$
$$\text{then } M, \sigma_j \models Viol(a)$$

This says: if at some future point the deadline occurs, and until then the result has not yet been achieved, then we have a violation at that point. This semantic definition is equivalent to the following definition as a reduction to CTL:

$$O_a(\rho \leq \delta) \equiv_{def} \neg E(\neg E_a\rho \, U^e(\delta \wedge \neg Viol(a)))$$

This formula just expresses the negation of the situation that should be excluded when a deontic deadline is in force. In natural language this *negative* situation is: 'δ becomes true at a certain point, the achievement has not been met until then, and there is *no* violation at δ'. This shows that it is fairly easy to show the equivalence of the semantic definition and the definition in terms of CTL (details left to the reader). The above defined deadline operator persists after reaching the deadline, and satisfies properties 2, 5, and 7 discussed in the previous section.

However, despite the nice properties and the simple and elegant representation of the concepts, the definition does not cover the intuitions of figure 1 completely. This becomes apparent when we look at a situation in which an agent a achieves ρ before a certain condition δ becomes true. Whenever this appears to be true it follows that a has the obligation to achieve ρ. I.e., the fact that an agent will achieve something implies that he is obliged to achieve it.

We suggest that the source of this problem might be that we have failed to formalize the 'causal link' that intuitively relates failures to comply to the obligation and occurrences of the violation condition. In the truth condition above, we have only dealt with one direction of the implicative relation between non-compliance and violation: we have captured that when there is non-compliance, there is also a violation. But we have failed to capture a reverse implicative direction saying that only if there is non-compliance there can be violations.

In the next section we will propose an extended definition that tries to establish this causal link between non-achievements and violations.

5   The Causal Approach

In [13] we have already attempted to capture some aspects of the causal link between non-achievement and violations. However that formalization did not force the condition that there can never be a violation of the obligation before the deadline condition holds. It also allows situations where ρ is achieved while there is still a violation after the deadline condition. Somehow we have to 'close' the possible worlds in a way that either we have the achievement and no violation after that or a violation and no achievement before the deadline. In this way we approach most closely that the achievement of ρ *causes* the $\neg Viol(a)$.

The definition given below differs from the one in section 4 on three important points. First of all, for a deadline obligation to be valid, it now requires that the deadline condition actually occurs at some point in the future. A second crucial difference is that we strengthen the 'if' construction in the truth condition to an 'if-and-only-if' condition, by which we attempt to capture the causal relation between non-compliance and violation. This 'if-and-only-if' condition takes the form of a disjunction (the 'or' in the truth condition below) saying that either $E_a\rho$ holds (in time), meaning that there is compliance, or $E_a\rho$ does not hold before δ, in which case there is non-compliance. Note that the disjunction is exclusive, because either ρ is achieved or not, but not both. Finally, we require violations to persist ones they have occurred, and we require non-violations to persist when the achievement is accomplished in time, or if no deadline or achievement condition has yet occurred.

$$M, s \vDash O_a(\rho \leq \delta) \text{ iff } \forall \sigma \text{ with } \sigma_0 = s : \exists j > 0 :$$
$$M, \sigma_j \vDash \delta \text{ and } \forall 0 \leq k < j : M, \sigma_k \vDash \neg Viol(a) \wedge \neg \delta \text{ and}$$
$$(\exists 0 \leq k < j : M, \sigma_k \vDash E_a \rho \wedge AG \neg Viol(a) \text{ or}$$
$$(\forall 0 \leq k < j : M, \sigma_k \vDash \neg E_a \rho \text{ and } M, \sigma_j \vDash AG\, Viol(a)))$$

We can express this semantic definition in terms of a CTL formula as well:

$$O_a(\rho \leq \delta) \equiv_{def} A($$
$$(\neg Viol(a) \wedge \neg \delta) U^e \delta \wedge$$
$$(\neg \delta U^e (\neg \delta \wedge E_a \rho \wedge AG \neg Viol(a)) \vee$$
$$(((\neg E_a \rho \wedge \neg \delta) U^e (\delta \wedge AG\, Viol(a)))))))$$

The lines of the formula correspond to the lines of the truth condition. The second line expresses that δ becomes true at a specific point in the future, that we consider the first time this happens, and that there cannot be a violation of the obligation until then. The third line expresses one side of the exclusive disjunction, saying that $E_a \rho$ occurs before the first δ, and that there cannot be a violation afterwards. The fourth line expresses the other side of the disjunction, saying that $E_a \rho$ has not occurred before the first δ, and that starting from the point where δ, violations persist forever. The latter condition expresses that the information that the obligation is violated, is preserved.

In the above definition, the obligation is always discharged by the occurrence of a deadline condition. So, for this variant, the obligation does not persist until after the deadline. Furthermore, the definition obeys the properties 1, 5 and 7 of section 3.

6   Practical Aspects of Deadlines

In this section we briefly discuss a few aspects that start playing a role when looking at more concrete aspects of deadlines.

The first aspect is the violation constant. In this paper the $Viol$ constant has only one parameter, the agent a. However, we would actually like to tie the violation to a specific obligation incurred at a specific moment in time. This is necessary to distinguish two obligations for the same agent that might only differ in the timing. E.g. the obligation to pay the rent before the end of the month occurs every month. But each month it is a different obligation. This can be achieved through the addition of a unique identifier for each obligation. This definition provides a very operational means to deal with violations, as it gives explicit information about what has caused the violation and can therefore enable to reason about what are the consequences and sanctions related to the violation.

However, at the same time this unique identifier would eliminate any logical relations between obligations that are connected. E.g. someone might have an obligation to pay a conference fee while (due to budget restrictions that became clear only later) it is from now on prohibited to pay for any conference. The two norms relate to the same person and have opposite effects on the action of paying. However, if each would be modelled with a violation constant with a different identifier they could not be related and the intuitive contradiction between the two would not exist.

As a solution to this problem we could introduce violations that have the same parameters as the obligations to which they are linked. In this way it becomes possible to specify logical relations between violations of which the actor, the deadlines and the situation to be achieved are related. However, this has as consequence that the violations are now also modal operators!

A second point that comes up right away is which logical relations should hold between the violations? Do we have

$$(V_a(\rho < \delta) \wedge (\rho' \to \rho)) \longrightarrow V_a(\rho' < \delta)$$

and/or

$$(V_a(\rho < \delta) \wedge (\delta' \to \delta)) \longrightarrow V_a(\rho < \delta')$$

Of course these properties are directly coupled to the properties that we would like to have for the obligation operator. A complete investigation into this issue warrants a separate paper and therefore will not be pursued here. However we would like to point to [11] for some related work in this area.

Closely related to the above item is the point that we made violations (and non-violations) persistent over time. Once a deadline is violated, this violation will never disappear again. This seems a bit contradictory to common practice where sanctions are defined as obligations, conditional on the occurrence of a violation, in order to make it possible for violations to be redeemed. So, we make a difference between a violation that has not been "made up for" yet and one for which a sanction has been exercised already. This aspect could be modelled by not having the violation persistent, but have an axiom that triggers a sanction (obligation) whenever a violation occurs.

A second item that is important in practice is that obligations are often conditional and/or repeated. The above example on paying the rent is a very typical case of a repeated obligation. The whole obligation to pay rent, however, can be made conditional on the fact that the house is properly maintained by the owner. Related to this aspect is that more temporal conditions can be specified for the achievement. E.g. the salary should be paid between the 25th and the end of each month.

Although we represent the deadline condition as a proposition in this paper, often it contains a relative temporal expression such as "the book should be paid within one week after delivery". In order to express this type of conditions one should have a more powerful language in which explicit reference to time can be made.

A last item to mention here is the use of discrete time in our model. This is particularly important to decide on the exact moment when a violation arises. In a model with continuous time the achievement of a fact (an action) has to have a duration (whereas the achievement in our model is always in one time step). So the definition of $E_a\rho$ has to be changed. On the other hand we can in this model with continuous time determine a violation before the deadline if it is impossible to achieve the required state before the deadline condition anymore.

7   Conclusions

In this paper we have shown that the use of a violation constant is in principle enough powerful to account for the deontic aspect of the deadlines. Of course a temporal logic is needed to account for the temporal aspects. Finally we used the *stit* operator E_a to relate the achievement of a state to an agent. This is important, because we consider the deadlines to be directed towards an agent and thus this agent has the responsibility to fulfill it. We do not use dynamic logic to model explicit actions in order to keep the model as abstract as possible. However, an obvious connection between the operator presented and dynamic logic can be made through the use of Segerberg's *bringing it about* operator [12].

We have also shown that a correct definition of deadlines in the formalism requires a modelling of the intuitive causal relation between the occurrence of the action before the deadline and the violation state. This causal relation makes the formal definition of a deadline quite complicated, although the simple intuitive picture of the semantics (given in section 2) is still valid.

References

1. N. Belnap and M. Perloff. Seeing to it that: A canonical form for agentives. *Theoria*, 54:175–199, 1988.
2. M.A. Brown. Doing as we ought: towards a logic of simply dischargeable obligations. In M.A. Brown and J. Carmo, editors, *Deontic logic, agency, and normative systems. Proceedings DEON '96*, Workshops in Computing, pages 47–65. Springer, 1996.
3. E.M. Clarke, E.A. Emerson, and A.P. Sistla. Automatic verification of finite-state concurrent systems using temporal logic specifications. *ACM Transactions on Programming Languages and Systems*, 8(2), 1986.
4. E.M. Clarke, O. Grumberg, and D. Long. Verification tools for finite-state concurrent systems. In *A decade of concurrency*, volume 803 of *Lecture Notes in Computer Science*, pages 124–175. Springer, 1993.
5. F. Dignum and R. Kuiper. Obligations and dense time for specifying deadlines. In *Proceedings of thirty-First HICSS, Hawaii*, 1998.
6. D. Elgesem. The modal logic of agency. *Nordic Journal of Philosophical Logic*, 2(2):1–46, 1997.
7. E.A. Emerson. Temporal and modal logic. In J. van Leeuwen, editor, *Handbook of Theoretical Computer Science, volume B: Formal Models and Semantics*, chapter 14, pages 996–1072. Elsevier Science, 1990.

8. R. Hilpinen. On action and agency. In E. Ejerhed and S. Lindström, editors, *Logic, Action and Cognition: Essays in Philosophical Logic*, pages 3–27. Kluwer Academic Publishers, 1997.

9. A.J.I. Jones and M. Sergot. On the characterization of law and computer systems: The normative sytems perspective. In J.-J.Ch. Meyer and R.J. Wieringa, editors, *Deontic Logic in Computer Science: Normative System Specification*, pages 275–307. John Wiley and Sons, 1993.

10. I. Pörn. *Action Theory and Social Science: Some Formal Models*. D. Reidel Publishing Company, 1977.

11. J.-J.Ch. Meyer F. Dignum R. Wieringa, H. Weigand. The inheritance of dynamic and deontic integrity constraints. *Anals of Mathematics and Artificial Intelligence*, 3.

12. K. Segerberg. Bringing it about. *Journal of Philosophical Logic*, 18(4):327–347, 1989.

13. F. Dignum H. Weigand V. Dignum, J.J. Meyer. Formal specification of interaction in agent societies. In *2nd Goddard Workshop on Formal Approaches to Agent-Based Systems (FAABS), Maryland*, 2002.

14. E. Verharen, H. Weigand, and F. Dignum. Dynamic business models as a basis for interoperable transaction design. *Information Systems*, 22(2).

15. G.H. von Wright. Deontic logic. *Mind*, 60:1–15, 1951.

Multi-agent Systems Reliability, Fuzziness, and Deterrence

Michel Rudnianski[1] and Hélène Bestougeff[2]

[1] University of Reims, ARESAD, 39 bis Avenue Paul Doumer, 75116 Paris –France
michel.rudnianski@wanadoo.fr
[2] CODATA – Boulevard Montmorency 75016 Paris - France
hbest@magic.fr

Abstract. The problem of MAS reliability is approached through representing the functioning of a MAS as a system of logical implications, and then interpreting this system as a game of deterrence. The game solutions provide indicators for the agent's reliability, and enable in case of an agent's failure, to select a search direction for determining the origin of the failure. The MAS reliability is increased by duplicating some agents. The impact of the duplicate's positioning in the MAS is analysed on a particular case.

1 Introduction

Multi-agent systems are comprised of several entities, organized to work together in order to collectively solve problems. A system's failure may occur because of one component's failure, which propagates throughout the agent network. Conversely, the occurrence of a particular agent's failure may have different reasons.

First, the source of an agent failure may be internal, i.e. occur because one of its elements doesn't work properly. It is then possible to, either remove the agent from the system and physically replace it by an equivalent one, or simply by-pass it, the taskflow running through some alternate device or agent.

This would be the case for instance if two similar agents work in parallel, their workloads being shared. Assume that agent A breaks down. If the work load has been properly dimensioned on the basis of both the total inflow and, say, the probability of an agent failure, then agent B can add to its own load, agent's A load, at the possible cost of some delay.

But the source of the agent failure may also be external, i.e. one of the neighbouring agents at least, doesn't function properly. If the neighbouring agent failure comes itself from the failure of an agent which is neighbour to this neighbouring agent, finding the ultimate source of the system's failure, implies developing some recursive process, and hence building some causality chain.

An interesting case is the one when failure derives from a specific state of the agent, incompatible with given states of neighbouring agents.

This is of course just a particular case of the more general inference problem, that has been addressed using different techniques like finite state machines, cognitive maps, qualitative probabilistic networks, or structural analysis [1,2,6].

M.G. Hinchey et al. (Eds.): FAABS 2004, LNCS 3228, pp. 41–56, 2005.

In line with our previous developments on multi-agent systems, we propose here to model inference by using the game of deterrence approach in the fuzzy case.

On the one hand, this approach will match the finite state machine approach, while on the other it will provide an inference system that will give an assessment of the role of a particular agent in the well- functioning, or on the opposite ill-functioning of another agent.

We shall begin by recalling some basic games of deterrence definitions before associating these games, through graphs of deterrence, with systems of logical implications.

We shall then see how introducing fuzziness in the strategies playabilities leads to the development of an inference network, such that each node is associated with an inference value, based on the playability index associated with that node.

The method will be illustrated with two examples. The first one will consider the elementary case, where no internal failure occurs. The second, generalizing the first one, will introduce the possibility of internal failures.

Eventually, we shall propose on a particular example of Multi-Agent System, a straightforward method to find the origin of an agent's failure.

2 Deterrent Agents and Games of Deterrence

Deterrent agents can only distinguish between two states of the world: acceptable (noted 1) and unacceptable ones (noted 0). All that they want is to be in an acceptable state.

If they have a strategy that guarantees them an acceptable situation, whatever the other agents do, they should by all means play it because it is *safe*. But this is not always the case. Sometimes selection of a strategy could put them in an unacceptable situation if other agents would select some particular strategies. In that case the strategy is clearly *dangerous*. But that does not mean that it is not *playable*. Suppose for instance that Erwin and Roger are agents such that, when Erwin and Roger select the strategic pair (e,r), Erwin gets a 0, and Roger's strategy r is not playable. Then, e is playable albeit dangerous. We shall say that e is *positively playable*.

In the case where Erwin has no positively playable strategy since he has to take a decision, any strategy will do, albeit poorly. We shall then say that such a strategy is *playable by default*. A strategy neither positively playable nor playable by default will be termed *not playable*.

Moreover, we shall say that Erwin's strategy e is *deterrent* vis-à-vis Roger's strategy r if the three following conditions apply :

1) e is playable
2) implementation of strategic pair (e,r) leads to an unacceptable situation for Roger
3) Roger has an alternative strategy r' which is positively playable

It has been shown [3] that a strategy r of Roger is not playable if and only if there is a strategy e of Erwin deterrent vis-à-vis r.

Let us illustrate these concepts with the following elementary example.

	r_1	r_2
e_1	(1,1)	(1,1)
e_2	(1,1)	(1,0)

We see that both strategies of Erwin and strategy r_1 of Roger are safe, while r_2 is dangerous. Furthermore e_2 is deterrent vis-à-vis r_2, which is thus not playable.

Now it has been shown [ibid] that every matrix game of deterrence can be associated with a bipartite graph such that, if (e,r) is a strategic pair, then there is an arc of origin e (resp.r) and extremity r (resp.e) if and only if the outcome of Roger (resp. of Erwin) is 0.

3 Associating a System of Logical Implications with a Game of Deterrence

Causality problems analysis usually resorts to an oriented graph representation such that, given an arc linking two concepts, its origin is a causal factor of its extremity.

We propose here to revisit the problem by bridging causality with games of deterrence, through the common graph approach.

3.1 Logical Representation of a Game of Deterrence

Given a two player game of deterrence, let us consider the following set of logical formulae :

1) A finite set of propositions $\zeta(s)$ indicating that a strategy *s* is a playable strategy
2) A finite set of propositions $J(s)$ indicating that a strategy *s* is a positively playable strategy

Only a particular set S of formulae will be used for this representation:

(i) propositions and negations of propositions as defined above;
(ii) logical implication built on these elementary formulae.

3.2 Non Fuzzy Graph of Deterrence and Representation of a Logical System

To build the graph associated with a given set S of formulas, consider the following implication $P \Rightarrow \neg Q$, denoted by $N(P,Q)$

With each subset S_1 of formulas, we can associate a bipartite graph defined as follows :

1. the graph vertices are propositions or negations of propositions
2. arcs are pairs of formulas (P,Q) such that $N(P,Q)$ is true
3. If $P \rightarrow Q$ denotes the arc of origin P and extremity Q, then $P \rightarrow Q$ iff $P \Rightarrow \neg Q$.

Then, as $P \Rightarrow Q$ writes $N(P, \neg Q)$, $P \Rightarrow Q$ can be represented by the path $P \to \neg Q \to Q$ iff $P \Rightarrow Q$ is true (since obviously $\neg Q \Rightarrow \neg Q$ is true).

As $\neg Q$ and Q are both vertices of the graph, and hence represent two strategies, we must add a consistency condition which discards the game of deterrence solutions for which both strategies are playable. This condition will provide a safe strategy for player II, implying that $\neg Q$ is not playable, and that Q is.

This can be done by adding to the set S_1, the proposition $\neg (\neg Q . Q)$ called *first order consistency condition for Q*, which defines a vertex with neither predecessor nor successor in the graph.

We can then associate with the set $S_1' = S_1 \cup \{\neg (\neg Q , Q)\}$ a matrix game of deterrence G with two abstract players I and II such that :

1. $\{P,Q\}$ is the strategic set of player I
2. $\{\neg (\neg Q , Q), \neg Q\}$ is the strategic set of player II
3. The graph here above is the graph of deterrence of G
4. with every strategic pair (X,Y) we associate the binary outcome pair $(a(x,y),b(x,y))$ such that $a(x,y) = 1$ (resp. $b(x,y) = 1$) unless there is an arc of origin Y (resp X) and extremity X (resp Y).

In other words, implication $P \Rightarrow Q$ is equivalent to the above game of deterrence. This conclusion can be extended straightforwardly to any set of bivalent implications.

3.3 Fuzzy Graph Representation of a Logical System

Similarly, for every set ψ_L of propositions of a bivalent logical system L, there exists a unique fuzzy matrix game of deterrence G such that ψ_L is the logical representation of G.

Indeed, the construction procedure introduced above is still valid, since the only difference between fuzzy and non fuzzy matrix games of deterrence lies in the domains of playability and positive playability indices, the matrices being the same.

The correspondence between ψ_L and G is derived from the matrix, with the exception of the consistency condition, which does not provide here a safe strategy for player II, but a circuit or more generally a graph with no paths (i.e. no roots).

Such a consistency condition, which will be called the *second order consistency condition for Q*, must ensure absence of contradiction on the one hand, and the possibility of non binary values for positive playability indices on the other.

Consider for instance : $\{N(\neg (\neg Q . Q), (\neg Q . Q)) ; N((\neg Q . Q)), \neg (\neg Q . Q))\}$, which, in terms of graph representation, is equivalent to the circuit :

$$\neg (\neg Q . Q) \leftrightarrow (\neg Q . Q)$$

This second order condition introduces not one, but two extra strategies that are adjacent vertices of a graph of deterrence. Therefore, we need to allocate each one of them to a different player.

For the sake of simplicity, let us denote $(\neg Q . Q)$ by a, and $\neg (\neg Q . Q)$ by a'.

It is clear that on the circuit $J(a) = (1-J(a'))v$ and $J(a') = (1-J(a))v$, where $v = 1-j_{II}$ and j_{II} is the index of playability by default of player II[1].

---

[1] i.e. $j_{II} = 1$ if positive playability indices of all strategies of player II equal 0.

Solving this elementary system of two equations leads straightforwardly to :

$$J(a) = J(a') = v / (1+v)$$

Consequently, a and a', having the same positive playability, may be associated with either player, provided of course that both are not allocated to the same player.

It stems from the demonstration here above that this result does not depend on which particular strategy has been selected to build the second order consistency condition. In turn, this means that a variety of such conditions can be chosen, depending on the particular case under consideration.

It also means that with each vertex X, which is not a root of the graph, we shall associate a second order consistency condition for X, defined by a circuit comprised of two strategies, the positive playability value of which is $v/(1+v)$.

3.4 Example

Let us consider for instance the logical system defined by : $P \Rightarrow Q \Rightarrow R$

The set of propositions can be translated into the following graph:

$$P \rightarrow \neg Q \rightarrow Q \rightarrow \neg R \rightarrow R \;\; ; \;\; \neg(\neg Q . Q) \leftrightarrow (\neg Q . Q) \;\; ; \;\; \neg(\neg R . R) \leftrightarrow (\neg R . R)$$

It can be shown that there is a unique non binary solution for which :

$$J(P) = 1 \; ; J(\neg Q) = 0 \; ; \; J(Q) = .81 \; ; \; J(\neg R) = .16 \; ; \; J(R) = .68;$$
$$J(\neg (\neg Q . Q)) = J(\neg Q . Q) = J(\neg (\neg R . R)) = J(\neg R . R) = .45$$

To analyze the exact meaning of these positive playability indices associated with logical propositions, we need to come back to the meaning of positive playability indices associated with strategies.

More precisely, let us consider the case of a path. It has been shown that [4] :

– positive playability indices of strategies of odd rank decrease with the rank
– positive playability indices of strategies of even rank increase with the rank
– when the length of the path tends toward the infinite, the value of the positive playability index tends toward .5

The interpretation is quite straightforward. Given that the root is the only safe strategy for player I, the "likelihood" of player I selecting the root is very big. Therefore, the "likelihood" of player II selecting strategy 2 can be considered negligible. In turn, this means that the "likelihood" of player 1 selecting strategy 3, while being smaller than for the root (after all, selecting the root presents absolutely no danger, and it is the only strategy which displays this property), can be quite large. In turn, the "likelihood" of player II selecting strategy 4, although small (because the likelihood of player I selecting strategy is large), is bigger than the "likelihood" of player II selecting strategy 2, and so on.

How does this translate in the case where strategies are logical propositions ?

The almost trivial idea that immediately comes to mind is that the above three-proposition system is nothing more than a causality chain. Consequently, if we limit the reasoning to the path once again (discarding the consistency conditions), the value of the playability index for the vertices of odd rank somehow describes the "inference value" of the root with respect to the other vertices.

Considering the vertices of even rank, which represent the negations of the propositions associated with the vertices of odd rank that follow, we can say the same.

In more explicit terms :

- the effect of P on the occurrence of Q can be measured by J(Q), which in the above case equals .81
- the effect of P on the non-occurrence of ¬Q, can be measured by 1- J(¬Q), which in the above case equals 1.

The difference between these two values comes from the fact that ¬Q and Q are vertices associated with different ranks on the graph. Although it might look weird at first glance, it seems that for Q to occur, it is not enough that P does, but one must *moreover state* that ¬Q doesn't.

4 Application to MAS Reliability : Example 1

As already stated, a failure at an agent level can diffuse throughout the network, causing the failure of other agents, and possibly of the entire system.

Moreover, it can happen that partial but simultaneous failure of different agents may generate identical phenomena.

To avoid that, one usually resorts to a two step method :

1) analyze the effect of the particular agent failure on the system
2) redesign the system to improve its global reliability

We propose to revisit that method, with the help of the above results.

4.1 Example 1: The Three Agent System with no Internal Failure

First, considering an agent network means considering an inference network.

To give an elementary illustration, let us consider for instance an information line, with three agents, p, q and r, such that p transforms some input, then passes the transformed input to agent q, which does the same with agent r. It is clear that agent r, in order to be able to fulfil its task, needs to get the result of agent q's work, which in turn requires the data transformed by agent p.

Let P, Q and R be three logical bivalent propositions defined as follows:

P : agent p functions properly
Q : agent q functions properly
R : agent r functions properly

Let us assume that : $P \Rightarrow Q \Rightarrow R$, which means that q and r never know any internal failure.

It stems from the previous paragraph that this double implication is equivalent to the graph

$$P \rightarrow \neg Q \rightarrow Q \rightarrow \neg R \rightarrow R \; ; \; \neg(\neg Q \, . \, Q) \leftrightarrow (\neg Q \, . \, Q) \; ; \; \neg(\neg R \, . \, R) \leftrightarrow (\neg R \, . \, R)$$

associated to a fuzzy matrix game of deterrence, such that there is a unique non binary solution defined by :

$$J(P) = 1 \; ; J(\neg Q) = 0 \; ; \; J(Q) = .81 \; ; \; J(\neg R) = .16 \; ; \; J(R) = .68;$$
$$J(\neg (\neg Q . Q)) = J(\neg Q . Q) = J(\neg (\neg R . R)) = J(\neg R . R) = .45$$

As already stated, the value of the positive playability index associated with a vertex located on the path can be interpreted as an "inference value" of the root with respect to the vertex under consideration.

More specifically, if we consider the vertices of odd rank, the positive playability index indicates how the well-functioning of agent p influences the well-functioning of agents q and r

In other words, J(X) indicates the "likelihood" of agent x functioning properly, given :

- – the system's structure
- – that the agent associated with the root functions properly.

These two elements define the information scheme on which the likelihood is based.

Hence with a different information scheme, the likelihood may take a different value.

In particular, $J(X) = 1$ is associated with a specific information scheme, for which X may be considered as a root of the sub-graph derived from the original graph, by deleting all predecessors of X.

4.2   Exploiting Information About Agents' Playability

Assume that $J(X) = 1$.

It stems from the playability equations that $J(\neg X) = 0$, which implies that if Y is a direct predecessor of ¬X on the original graph, $J(Y) = 1$, which implies in turn that $J(\neg Y) = 0$, and so on.

So, if we assume for instance that $J(R) = 1$, then $J(Q) = J(P) = 1$.

At first sight, this conclusion seems to contradict the results obtained when assessing the values of positive playability indices, for it was found that no other vertex than the root could be associated with a positive playability index equal to 1. But in reality this contradiction is only apparent, since one should remember that any playability system always has an integer solution (i.e. there is always a distribution of integer values of the positive playability indices satisfying the playability system).

Now, given the initial double implication which discards the possibility of agents' internal failures, the above conclusion becomes trivial: It just states that for the last agent to function properly, all agents that are predecessors of this last agent (here p and q) must function properly.

Conversely, if agent x doesn't function properly, proposition X is not true, and, because propositions considered here are bivalent, proposition ¬X is true.

In other words, proposition ¬X can be associated with $J(X) = 0$

But on the other hand, by the construction method developed here above, ¬X is a vertex of the graph of deterrence, and it is the only adjacent predecessor of X on this graph, which means that either $j(\neg X) = 1$ (v = 0) or $J(\neg X) = 1$.

In the first case, all strategies of player II, and especially all strategies of type ¬X, are playable by default, with the consequence that no strategy of player I, except for the root, is playable. Hence the system can never work. Therefore this case can be discarded, for it has no interest (this is precisely the reason why we have introduced a second order consistency condition enabling fuzzy playabilities).

So let us examine the case where $J(\neg X) = 1$.

By using the same backward induction as above, one comes to the conclusion that agent p does not work, which simply states the double implication : $\neg R \Rightarrow \neg Q \Rightarrow \neg P$, strictly equivalent to the original double implication.

On the whole, in the very elementary example considered here, the only possible source of failure of agent r, or of agent q, is the failure of agent p, and the source of failure of an agent can be derived from the graph structure.

We shall now apply these conclusions to more complex cases.

5 Introducing the Possibility of Internal Failures: Example 2

5.1 Graph Structure and Solution of the Playability System

Everything else being equal, let us relax the assumption of no internal failure.

This means that now the cause of failure of agent x can be:

- either an internal failure
- or a failure of one of agent x predecessors

Of course, the above double implication between P, Q and R is no longer valid.

To represent the MAS by a system of logical implications, we need to introduce propositions describing states of internal failure for each agent, (with a possible exception for the root).

So let us introduce the proposition I_x : "there is an internal failure of agent x".

Agent x will function properly if and only if y functions properly *and* there is no internal failure of x, which writes : $X \Rightarrow Y \wedge \neg I_x$, or : $\neg Y \vee I_x \Rightarrow \neg X$

It can be noticed that in the graph of deterrence representation of a bivalent logical system, the logical operator $\vee$ is equivalent to two arrows pointing at the same vertex.

So, the above implication can be represented by :

$$\neg Y \;\rightarrow\; X \rightarrow \neg X$$
$$\uparrow$$
$$I_x$$

To avoid I_x being a root of the graph, which would mean that agent x would always have an internal failure, we need to introduce the possibility of no internal failure, and write that $\neg I_x$ and I_x cannot occur simultaneously, which also gives the second order consistency condition associated with agent x.

The structure of the graph around X becomes :

$$\ldots \; Y \;\rightarrow\; \neg Y \;\rightarrow\; X \;\rightarrow\; \neg X \ldots$$
$$\uparrow \qquad\qquad\quad \uparrow$$
$$I_y \leftrightarrow \neg I_y \qquad\quad I_x \leftrightarrow \neg I_x$$

The graph of deterrence representing the MAS is then the following :

$$P \; \rightarrow \; \neg P \; \rightarrow \; Q \; \rightarrow \; \neg Q \; \rightarrow \; R \; \rightarrow \; \neg R$$
$$I_q \leftrightarrow \neg I_q \qquad I_r \leftrightarrow \neg I_r$$

The reason why there is no consistency condition associated with P is two fold :

- just as in the previous example, we try to assess how the proper functioning of p affects the proper functioning of the MAS other agents
- were it not the case, the consistency condition would be partially redundant with proposition P, because agent p has no predecessor in the MAS

The players strategic sets are :

- for player I : $P, Q, \neg I_q, \neg I_r$, and R
- for player II : $\neg P, I_q, \neg Q, I_r$, and $\neg R$

The playability system writes:

$J(P) = 1 ; J(\neg P) = 0 \; ; \; J(Q) = [1-J(\neg P)][1-J(I_q)]v \; ; \; J(\neg Q) = [1-J(Q)]v$
$J(R) = [1- J(\neg Q)][1-J(I_r)]v; \; J(\neg R) = [1-J(R)]v$
$J(I_q) = \; [1-J(\neg I_q)]v \; ; \; J(\neg I_q) = [1-J(I_q)]v; \; J(I_r) = \; [1- J(\neg I_r)]v; \; J(\neg I_r) = [1-J(I_q)]v \; ;.$
$1-v = (1- J(\neg P)) \; [1- J(\neg Q)][1- J(\neg R)][1- J(I_q)][1- J(I_r)]$

We know that $J(I_q) = J(\neg I_q) = J(\neg I_r) = J(\neg I_r) = v / [1+v]$

This means that the likelihood of an internal failure is the same for agent q and agent r (or more generally, for any subsequent agent in the MAS). Of course this result is associated with a particular state of information, for which the only thing known about the MAS is its structure : no difference being made between the agents, it is only natural that the likelihood of their internal failure is the same.

Then : $J(Q) = \; J(\neg Q) = v/[1+v] \; ; \; J(R) = v/[1+v]^2 \; ; \; J(\neg R) = v[1 - [v / (1+v)^2]]$,
And $1 -v = [1/(1+v)]^3[1 - v[1 - [v / (1+v)^2]]] = [1+v-v^3] / [1+v]^5$

One can show that this equation has a single non binary solution : v= .973.
The playability system's solution is then:

$J(P) = 1 ; J(\neg P) = 0 \; ;$
$J(I_q) = J(\neg I_q) = J(Q) = J(I_r) = \; J(\neg I_r) = J(\neg Q) = .493$
$J(R) = .25 \; ; \; J(\neg R) = .73.$

5.2 Interpretation and Generalization

The likelihood of internal failure is the same for the two agents Q and R, about .5, which simply states that in absence of further information, internal failure and well-functioning are equally likely.

The properties found in the case with no failure are still valid :

- positive playability of vertices with odd rank decrease with the rank
- positive playability of vertices with even rank increase.

The main difference with the no internal failure case is the magnitude of the variation, which is much greater in the present case.

This greater magnitude stems precisely from the possibility of internal failure : the likelihood of an agent x functioning properly sharply decreases with the distance of X from the root, since the proper functioning of x now requires two conditions, while only one was required in the case of no internal failure.

This conclusion can be generalized. Let us consider a MAS comprised of N agents $x_1, x_2, ..x_n$, positioned in line.

The expression of the positive playability in terms of v :

- is the same for X_2 than for Q in the previous example
- is the same for X_3 than for R in the previous example

It follows that $J(X_3) \leq J(X_2)$

Since $J(\neg X_2) = [1-J(X_2)]v$, and $J(\neg X_3) = [1-J(X_3)]v$, we can then conclude that $J(\neg X_3) \geq J(\neg X_2)$

Furthermore, $J(X_3) = [v/(1+v)][1- J(\neg X_2)]$ and $J(X_4) = [v/(1+v)][1- J(\neg X_3)]$

It then stems from the inequality here above that $J(X_4) \leq J(X_3)$

More generally given two agents x_k and x_{k+1} on the MAS, such that:

- x_{k+1} is the immediate successor of x_k
- $J(X_k) \leq J(X_{k+1})$,

by using the same method, one can show that:

- $J(\neg X_{k+1}) \geq J(\neg X_k)$
- $J(X_{k+1}) \leq J(X_{k+2})$

Where x_{k+2} is the immediate successor of x_{k+1} on the MAS
Whence, the conclusion.

5.3 Finding the Origin of a Network Dysfunction

Consider for instance that agent r doesn't function properly, i.e. that $J(R) = 0$.

We know that there are two possible reasons for that : either there is an internal failure, or the predecessor q of r doesn't function properly itself.

Given the elementary case under consideration, such a conclusion is obvious. Nevertheless, it is of interest to note that one can come to such a conclusion by looking at the correspondence between the graph of deterrence and the system of logical implications with which it is associated.

Indeed, the graph structure around R is :

$$Q \rightarrow \neg Q \rightarrow R \rightarrow \neg R$$
$$\uparrow$$
$$I_r \leftrightarrow \neg I_r$$

which translates into the implication $Q \wedge \neg I_r \Rightarrow R$, equivalent to $\neg R \Rightarrow \neg Q \vee I_r$

Because here, Q and $\neg I_r$ play a similar role (i.e. both are simultaneously needed for R to be true, and $J(I_r) = J(\neg Q) = .493$), there is no particular direction toward which one should look *first* to see the origin of the ill-functioning of R.

We shall see in the sequel that it might not always be the case.

6 Increasing the System's Reliability

The second stage of the method consists of redesigning the graph in order to increase the playability indices values of nodes with odd rank.

The new graph must satisfy some requirements with respect to the agent network.

Thus, in the above example, there must be a way to go from agent p to agent q, and from agent q to agent r. At the same time, it would be meaningless to draw a direct "path" from agent p to agent r, since the latter needs the result of agent q's work to fulfil its own task. This means that in the example under consideration, it seems that no "redesign" is possible.

Of course this conclusion does not apply to more complex systems, in particular, systems in which some of the agents can work in parallel.

6.1 Graph Representation of Agents' Parallelism

Let us assume that agents p, q and r are substitutable, and that in order to increase the system's reliability, we add another agent x that can replace any one of the three. Assume furthermore that only one agent can be added for reasons say, of available place (this could be the case for a system embarked on a space flight).

The question is then, where should x be positioned ?

To answer, we first need to give an interpretation of the parallelism between agents x and y in terms of the graph of deterrence.

What is meant here by parallelism is that if agent x or agent y functions properly, then the system comprised of these two agents will function properly. So the proper-functioning of this system can be associated with proposition $X \lor Y$, and conversely, the ill-functioning of the system can be associated with $\neg X \land \neg Y$

In turn, this means that the graph of deterrence structure around X and Y is

$$X$$
$$\downarrow$$
$$Y \;\; \rightarrow \; \neg X \land \neg Y \rightarrow$$

6.2 Duplicating the Root

Let us first assume that x is parallel to p. The network is then the following :

$$X$$
$$\downarrow$$
$$P \;\; \rightarrow \; \neg X \land \neg P \rightarrow Q \; \rightarrow \; \neg Q \; \rightarrow \; R \rightarrow \neg R$$
$$\uparrow \qquad\qquad\qquad \uparrow$$
$$I_q \leftrightarrow \neg I_q \qquad\quad I_r \leftrightarrow \neg I_r$$

X has the same positive playability as P in the graph associated with the previous MAS, while $\neg X \land \neg P$ has the same positive playability as $\neg P$ in the graph associated with the previous MAS. So, $J(\neg X \land \neg P) = 0$.

It follows that $J(Q)$ can be expressed by the same function of v as in the previous case, and, because $J(\neg X) = 0$, v has the same value as previously. Consequently, it stems from the graph structure that the same goes with the positive playabilities of the remaining vertices.

On the whole, putting agent x in parallel with agent p does not affect the well-functioning of agents located further down the MAS. The reason for this stems directly from the assumption of agent x's well-functioning : P is the root of the graph of deterrence, which means that agent p is assumed to always function properly, in which case there is no interest to duplicate this agent.

6.3 Duplicating Agent Q

So let us consider now that x is parallel to Q. The corresponding graph is:

$$I_x \leftrightarrow \neg I_x$$
$$\downarrow$$
$$\rightarrow X \text{————}$$
$$| \qquad\qquad \downarrow$$
$$P \rightarrow \neg P \rightarrow Q \rightarrow \neg X \wedge \neg Q \rightarrow R \rightarrow \neg R$$
$$\uparrow \qquad\qquad\qquad \uparrow$$
$$I_q \leftrightarrow \neg I_q \qquad\qquad I_r \leftrightarrow \neg I_r$$

We can see from the graph structure that the positive playability of P, $\neg P$, Q, I_q, $\neg I_q$, I_r, and $\neg I_r$ are the same as in the case with no duplicate.

I_x, and $\neg I_x$ have the same positive playability as I_q, while X has the same positive playability as Q : $J(X) = v/(1+v)$

Furthermore, $J(\neg X \wedge \neg Q) = v/(1+v)^2$

For the sake of simplicity, let us momentarily denote $J(\neg X \wedge \neg Q)$ by w.

Then : $J(R) = [v/(1+v)][1-w]$, and $1- J(R) = [1+vw]/[1+v]$

Similarly, $J(\neg R) = v[1-J(R)] = [v/(1+v)][1+vw]$, and $1- J(\neg R) = [1-v^2 w]/[1+v]$

And $1-v = [1-J(\neg P)] [1-J(\neg X \wedge \neg Q)] [1-J(\neg R)] [1-J(I_q)] [1-J(I_x)] [1-J(I_r)]$
$= [1-w][(1-v^2 w)/(1+v)] / [1+v]^3 = [1-w][1-v^2 w] / [1+v]^4$

Replacing w with its value in terms of v, and solving the equation, leads to v = .971, which implies w = .25 and $J(R) = .37$, $J(\neg R) = .612$

So, putting agent x in parallel with Q increases the positive playability of R by nearly 50%.

6.4 Variation of the Positive Playability Along the Graph

One can easily show that the positive playability variation properties still hold for the vertices located along the "spinal cord" (i.e. the main path) of the graph.

Indeed, let us again consider a MAS originally comprised of n agents, and then add an agent x, positioned parallel to agent x_2.

It stems from above that :
$J(X_3) = [v/1+v][1- J(\neg X \wedge \neg X_2)]$, and $J(\neg X_3) \geq J(\neg X \wedge \neg X_2)$.
Furthermore, $J(X_4) = (v/1+v)[1-J(\neg X_3)]$.
It follows that $J(X_4) \leq J(X_3)$.
We then know from the above paragraph, that for all $k \geq 3$:

- $J(\neg X_{k+1}) \geq J(\neg X_k)$
- $J(X_{k+1}) \leq J(X_{k+2})$

where x_{k+2} is the immediate successor of x_{k+1} on the MAS.

6.5 Origin of a Network Dysfunction

Let us consider once more that agent r doesn't function properly. We see from the graph that this situation can be associated with the implication $\neg R \Rightarrow (\neg Q \wedge \neg X) \vee I_r$.

Since all agents are equivalent by assumption, the time needed to check the well-functioning of an agent is the same, no matter the agent. Let us take this time as the unit of time. It means that looking in the direction of R's internal failure takes one unit of time, while looking in the alternative direction (i.e. ill functioning of Q and of X) takes two units.

Let us consider, in line with the assumptions about the available information, that the probability of having any one of the three agents not function properly is the same, and furthermore, for the sake of simplicity, let us consider that only one agent doesn't function properly.

One can easily establish that whatever the direction selected first, the mean time required to find the origin of the failure is the same.

Consequently, with no information available (which means that the likelihood of each one of the three propositions is the same), there is no rationale to select one direction of research over another[2].

On the contrary, if we take into account the "structural" information derived from the graph of deterrence, since $J((\neg Q \wedge \neg X) = .25$, while $J(I_r) = .49$, the likelihood of occurrence of I_r is greater than the likelihood of $\neg Q \wedge \neg X$, and hence, it seems preferable to begin by looking at R's possible internal failure.

On the whole, we see that the graph of deterrence representation of the MAS provides extra information, which can be used to select a direction of research for determining the origin of the network's dysfunction.

7 Positioning the Duplicate

We have seen in the above paragraph that the exact positioning of the duplicate impacts R's playability. So can we optimize the positioning of the duplicate ?

To answer, we need to define what we mean by "optimize".

---

[2] The values of the mean time required to find the origin of the problem are directly derived from the assumption that there is only one agent that doesn't function properly, which means that only two checks need to be made in both cases.

At first sight, two different meanings can be envisaged:

- maximize the reliability of the last agent (just as in the above example)
- maximize the reliability of all agents.

If, in theory, the second goal is the most desirable one, it is not certain that it can be reached, in which case it will be necessary to consider the optimization problem as a multi-criteria one, or as an N-player non cooperative game of Nash.

Now, considering the first meaning, the fact that the last vertex of odd rank has the smallest positive playability (among vertices of odd rank), implies that maximizing the latter amounts to increasing a positive playability "threshold". This is certainly not equivalent to optimizing the positive playability of all vertices of odd rank, but this is already a step in the right direction. Therefore, for the sake of simplicity, we shall hereby restrict our attention to that first meaning.

In the MAS without duplicate, $J(X_2) = v/(1+v)$.

For the sake of simplicity, let us denote this expression by a. It follows that:

$J(\neg X_2) = v[1-a]$; $J(X_3) = a-av+a^2v$;
$J(\neg X_3) = v[1-a][1+av] = v[1-a][1-a^2v^2] / (1-av)$

Let us assume that $J[\neg X_k] = v[1-a][1-a^{k-1}v^{k-1}] / [1-av]$.
Then $J(X_{k+1}) = a[1-J(\neg X_k)] = a-av+a^2v-a^2v^2+...-a^{k-1}v^{k-1}+a^kv^{k-1}$
Whence $J(\neg X_{k+1}) = v[1-J(X_{k+1}] = v[1-a][1+av+a^2v^2+... a^{k-1}v^{k-1}]$
$= v[1-a][1-a^kv^k] / [1-av].$

Suppose now that a duplicate X is introduced in the MAS, and positioned in parallel with X_k. We know that $J(X) = J(X_k)$, which implies $J(\neg X_k \wedge \neg X) = v[1-J(X_k)]^2$, and $J(X_{k+1}) = a[1- J(\neg X_k \wedge \neg X)]$.

Again, for the sake of simplicity, let us denote $J(X_{k+1})$ by b.

$J(\neg X_{k+1}) = v[1-b]$; $J(X_{k+2}) = a-av+avb$;
$J(\neg X_{k+2}) = v[1-a+av-avb] = v[1-a+av-a^2v+a^2v-avb] = v[1-a][1+av] + av^2[a-b]$
$= v[1-a][1-a^2v^2] / (1-av) + av^2[a-b]$

Just as above, it can be shown with some elementary algebraic computation that more generally, $J(\neg X_{k+i}) = v[1-a][1-a^iv^i] / (1-av) + a^{i-1}v^i [a-b].$

Let us then consider that $k+i = n-1$

Then : $J(\neg X_{n-1}) = v[1-a][1-a^{n-k-1}v^{n-k-1}] / (1-av) + a^{n-k-2}v^{n-k-1}[a-b]$

It follows that the positive playability of the last agent writes:

$J(X_n) = a[1-[v[1-a][1-a^{n-k-1}v^{n-k-1}] / (1-av) + a^{n-k-2}v^{n-k-1}[a-b]]$

Let us take for instance a MAS comprised of five agents $x_1, x_2...x_5$.

The reasoning developed in the three agent case still applies, and shows that positioning the duplicate at the root level does not impact the positive playability associated with the proper functioning of the last agent.

Similarly, one can disregard positioning the duplicate at the level of the last agent, since such positioning would not impact the functioning of the last agent.

What remains is to consider positioning the duplicate in parallel with agent 2, agent 3 and agent 4, respectively.

To do that, all we need is to compute the positive playability in each one of the three corresponding cases.

The results are given in the array here below :

k	2	3	4
v	.99	.94	.997
$J(X_5)$	.18	.32	.49

We see that the positive playability of X_5 increases significantly with k, the highest value being obtained when the duplicate is positioned parallel to X_4, that is just before X_5.

Now if we consider that agent x_5 provides the general output of the MAS, it means that the duplicate should be positioned as near from the output as possible.

Of course, this result has been obtained in the case of a particular example. Its generalization still needs to be explored.

8 Conclusions

In this paper we have used a three layer approach to analyze multi-agent systems.

The first layer is the MAS itself. Each agent of the MAS has been associated with logical propositions. The set of these propositions has then been structured with the help of appropriate implications (layer 2). In turn, we have shown that with this system of implications, we can associate the graph of a game of deterrence (layer3), the vertices of which are the propositions of layer 2. This graph represents an inference scheme in which "inference values" are given by the positive playability indices associated with its vertices.

Three applications are in the developing process :

1) Determination of the graph associated with a given MAS
2) Possible selection of a first direction to explore in order to minimize the time necessary to find the origin of a network dysfunction
3) Optimal positioning of a duplicate agent, in order to maximize the network's reliability.

These applications and the corresponding properties have been developed in the framework of an elementary MAS. Extending these applications and properties could pave the way for designing a MAS while simultaneously taking into account reliability problems, and building efficient redundancy subsystems.

References

1. Aguilar J., " A Dynamic Fuzzy Cognitive Map Approach, Based on Random Neural Networks" , *In International Journal of Computational Cognition*, Vol 1, N°4, pp 91-107, Décember 2003.
2. Kosko B., "Fuzzy Cognitive Maps", *International Journal of Man-Machine Studies*, 24, pp 65-75, 1986

3. Rudnianski M. ,"Deterrence Typology and Nuclear Stability, a Game Theoretical Approach", *Defense Decision Making*, R. Avenhaus, H. Karkar, M.Rudnanski. Eds, Springer Verlag, Heidelberg, 1991, pp137-168.
4. Rudnianski M. ," Deterrence Fuzzyness and Causality", ISAS 96, pp 473-479
5. Rudnianski M., Bestougeff H., Modeling Task and Teams through Game Theoretical Agents, in *Proceedings of the Nasa First Goddard Workshop on Formal Approaches to Agent Based Systems*, Springer Verlag, Lecture Notes in Computer Science, 1871, pp 235-249 series, Berlin 2001
6. Wellmann M. ,"Inference in Cognitive Maps", in *Qualitative Reasoning and Decision Technologies*, N. Piera Carreté, M.G. Singh, Eds, CIMNE, pp 95-104, Barcelona, 1993

Formalism Challenges of the Cougaar Model Driven Architecture

Shawn A. Bohner[1], Boby George[1], Denis Gračanin[1],
and Michael G. Hinchey[2]

[1] Department of Computer Science,
Virginia Polytechnic Institute and State University,
Blacksburg, VA 24061, USA
{sbohner, boby, gracanin}@vt.edu
[2] NASA Goddard Space Flight Center, Greenbelt,
MD 20771, USA
Michael.G.Hinchey@nasa.gov

Abstract. The Cognitive Agent Architecture (Cougaar) is one of the most sophisticated distributed agent architectures developed today. As part of its research and evolution, Cougaar is being studied for application to large, logistics-based applications for the Department of Defense (DoD). Anticipating future complex applications of Cougaar, we are investigating the Model Driven Architecture (MDA) approach to understand how effective it would be for increasing productivity in Cougar-based development efforts. Recognizing the sophistication of the Cougaar development environment and the limitations of transformation technologies for agents, we have systematically developed an approach that combines component assembly in the large and transformation in the small. This paper describes some of the key elements that went into the Cougaar Model Driven Architecture approach and the characteristics that drove the approach.

1 Introduction

Software development can be thought of as the evolution of abstract requirements into a concrete software system. Starting with requirements that must be refined and elaborated, the system's evolution is achieved through a successive series of transformations. For non-trivial systems, this can be complex, time consuming, and prone to errors as software engineers work together to develop the requisite components, assemble them, and verify that they meet specifications. Model Driven Architecture (MDA), also known as Model Driven Development, represents an emerging approach for organizing this evolution and its resulting artifacts. Through a successive series of computational independent, platform independent and platform specific model transformations, MDA facilities generation of software systems.

With the relentless advancement of technology, complexity and integration issues often dominate modern computing. To respond to the sheer volume of software and consequential complexity, the software community has increasingly

M.G. Hinchey et al. (Eds.): FAABS 2004, LNAI 3228, pp. 57–71, 2005.

embraced architecture principles. Software architecture provides a framework to understand dependencies that exist between the various components, connections, and configurations reflected in the requirements. Some situations lend themselves to what is called an agent-based architecture.

As software grows in complexity and autonomy, manifold dependencies between critical elements of software increasingly drive many software architectures. Agent-based software systems address this complexity particularly where components may have autonomous properties (i.e., complex information and task-intensive situations) and require mechanisms to control these and other properties in a predictable way. The task orientation coupled with intelligent agents provides a strategic and holistic environment for designing large and complex computer-based systems.

This research concentrates on understanding and applying the MDA approach in an Agent-Based Architecture — specifically, Cougaar. The goal is to explore ways to use MDA to facilitate domain and software engineering staff developing Cougaar applications, to move up and program at the higher level, the domain level. We investigate how to compose Cougaar components into a General Cougaar Application Model (GCAM) and develop a General Domain Application Model (GDAM) for specifying and generating software applications. While the scope of this research focuses on the establishment of the GCAM and GDAM, it also provides example recipes for transforming the models into relevant software artifacts such as requirements, design, code, and test documents.

1.1   Agent Based Systems

While there are several definitions for software agents [1],we simply define an agent as a software entity that perceives its environment and responds through action(s) or tasks to fulfill a designed purpose. This broad definition covers a wide range of software agents, where agent types are characterized by properties, such as autonomous, interactive, adaptive, sociable, cooperative, competitive, proactive, intelligent, and mobile. By combining these properties in different ways, researchers have defined different agent types and, depending on the criteria, organized these agent types into taxonomies.

An "agent system is a platform on which agents are deployed"[2]. Software agent systems, also known as frameworks, need not be large systems, requiring enterprise-class machines to execute. Some agent systems are characterized by a large footprint and require considerable resources to execute. Others are lightweight and can execute in an embedded architecture.

A general agent platform architecture consists of three major components: a platform manager, an advertisement registry and a set of agents. Key characteristics of this general agent platform are that (a) there is some mechanism by which agents are managed (i.e., created, deleted, suspended, resumed, etc.), registered and also discovered by other agents; and (b) there is a communication mechanism. The platform manager is responsible for managing the agents, handling operations such as the creation, deletion, suspension and resumption of agents. The advertisement registry contains descriptions of the agents in the system and facilitates discovery of those agents. Implied in this architecture is that agents can communicate with each other, with the platform manager and with the advertisement registry.

Some interesting examples of software agent systems include Grasshopper [3], JACK [4], Cougaar [5], and JADE [6]. There are also several more agent systems that are compliant with Foundation for Intelligent Physical Agents (FIPA) specifications [7].

1.2   Cougaar

The Cognitive Agent Architecture (Cougaar) is an open source, distributed agent architecture [8] resulting from over eight years of research and development, and over $150 million investment by the Defense Advanced Research Projects Agency (DARPA) under the Advanced Logistics Program (ALP) and the Ultra*Log program [9]. Cougaar is a Java-based architecture for the construction of large-scale distributed agent-based applications characterized by hierarchical task decompositions. ALP demonstrated the feasibility of using advanced agent-based technology to carryout rapid, large scale, distributed logistics planning. Ultra*Log is developing information technologies to enhance the survivability of these distributed agent-based systems operating in extremely chaotic environments. Over the last four years, fault tolerance, scalability and security have become the focus of evolving this platform for more robust applications.

The Cougaar environment enables developers to build intelligent applications that can recognize and accept high level tasking, determine suitable processes and activities, and allocate appropriate resources to complete the tasking. From an information systems workflow perspective, Cougaar agents can accomplish various tasks based on the functional business processes with which they are configured.

Cougaar agents are organized into a society that collectively solve(s) problem(s). A society can encompass one or more communities of agents that share functional purpose or organizational commonality. A Cougaar node refers to a single Java Virtual Machine (JVM) running on a single server that contains one or more agents. A society may be deployed across several nodes. Agents on the same node may compete for resources including CPU, the memory pool, disk space, and network bandwidth.

Figure 1 illustrates the Cougaar agent structure consisting primarily of a blackboard and, a set of plugins and logic providers that are referentially uncoupled. The blackboard is a container of objects that adheres to publish/subscribe semantics. Plug-ins provide business logic. Logic providers translate both incoming and outgoing messages. When an agent receives a message, it publishes it to the blackboard where a logic provider observes its addition and transforms it into an object that plugins work on. Plugins publish/remove objects, or publish changes to existing objects. Plugins create subscriptions to get notified when objects of its interest are added, removed or changed in the blackboard.

Agents collaborate with other agents, however they do not send messages directly to each other. Instead, a task is created. Each task creates an "infor-

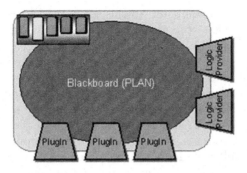

Fig. 1. Cougaar Agent Structure [8]

mation channel" flowing through the society, for requirements passing down and responses going back. In order to send an object or resource, to another agent, the developer must first associate the object or resource with the task. Cougaar uses the concept of asset to represent objects or resources used by task. Only instances of the Asset class can be associated with the task (i.e., all multi-agent objects must be defined as assets).

Once the task is created, then the task to be allocated must be located. This is typically accomplished by creating a subscription that examines the roles or property groups of organizations in the local blackboard. Once the proper organization is found, the task containing the object to be sent to the other agent is allocated to that organization by creating an allocation and publishing it to the blackboard. The Cougaar communication infrastructure then ensures that the task is sent to the specified organization and the specified agents blackboard.

A relationship between two agents can either be a superior/subordinate or customer/provider relationship. The superior/subordinate relationship supports long-standing transactions where a superior gives high-level tasks to the subordinate, which then performs the task and then reports aggregate and trend information back to the superior on a periodic basis. Cougaar supports dynamic re-planning and execution monitoring, based on these aggregate/trend information. A customer/provider relationship on the other hand is for task-order services between agents on a peer-to-peer basis and may result in large scale discrete data flows between the agents.

1.3 Model Driven Architecture

In some sense, MDA is a natural progression from previous advances in computer science. Using models in the development of a system has been practiced for decades, and even for centuries in other engineering disciplines (e.g., Building Architecture). Perhaps the most telling transition in mindset is how modeling in MDA takes a model (typically an abstraction of a reality) and creates an executable form through a series of predictable transformations. Since the computer uses a conceptual medium developed by a software engineer (i.e., a model or series of models), transforms now make abstractions of the real world accessible and even executable on a computer. In this sense, models are no longer simply an aid in understanding — the model can now become something much more concrete.

Like other engineering disciplines, software architecture helps us deal with the inherent complexities of building today's software systems. Systematically, separating concerns, formalizing the interfaces through standards and the like, provides better leverage for developing and evolving the software we employ. Software architecture — the structure or structures of the system, which encompass software components, their externally visible properties, and the relationships among them[10] — addresses the aforementioned growing complexity by providing a structure for thinking about and communicating key relationships between components, whether they are commercial-off-the-shelf software (COTS), middleware, or custom developed.

MDA endeavors to achieve high portability, interoperability, and reusability through architectural separation of concerns. In some respects, MDA is an advanced perspective on well-known essential systems development concepts practiced over the years (albeit frequently practiced poorly). MDA hinges on the long-established concept of separating the operational specification of a system from the details of how that system implements those capabilities on its respective platform(s). That is, separate the logical operational models (external view) from the physical design for the platform implementation.

Starting with an often abstract computation independent model (CIM) such as a business process workflow or functional description, the platform independent model (PIM) is derived through elaborations and mappings between the original concepts and the PIM renderings. Once the PIM is sufficiently refined and stable, further platform specific models (PSM) are derived through a series of elaborations and refinements into a form that can be transformed into a completed operational system.

The CIM layer is where vernacular specific to the problem domain is defined, where constraints are placed on the solution, and where specific requirements reside. Artifacts in the CIM layer focus largely on the system requirements and their environment to provide appropriate vocabulary and context (e.g., domain models, use case models, conceptual classes). The CIM layer contains no processing or implementation details. Instead, it conveys non-functional requirements such as budgetary constraints, deployment constraints, and performance constraints as well as functional constraints.

The PIM provides the architecture, the execution plan, but not the execution of the plan in a tangible form. Beyond high level services, the problem domain itself must be modeled from a processing perspective. The PIM is where the logical components of the system, their behaviors, and interactions are modeled. PIM artifacts focus on modeling what the system should do from an external or logical perspective. Structural and semantic information on the types of components and their interactions (e.g., design classes, interaction and state diagrams) are rendered in UML, the defacto modeling language for MDA.

Mapping from the PIM to the PSM, is a critical element of the MDA approach. The mappings from platform independent representations to those that implement the features or functions directly in the platform specific technologies are the delineation point where there is considerable leverage in MDA. This mapping allows an orderly transition from one platform to another. But the utility does not stop there. Like the PIM, there is the opportunity to have layers within the PSM to produce intermediate-transformations on the way to the executable system. These models can range from detailed behavior models to physical source code used in the construction of the system.

Direct PIM to PSM mappings are only possible in relatively simple situations today. Today's modeling languages are not sufficient to express all possible processing mechanisms. While UML 2.0 is attempting to address this limitation, it's too early to measure its impact. Therefore, in this research effort, we have attempted to glean the the benefits of the MDA approach while avoiding, to the extent possible, its inherent limitations.

The MDA approach specifies a system independently of the platform that supports it, specifies the platform(s), chooses platforms for the system, and transforms the system specification into those for particular platforms. While this approach is still evolving, we are encouraged by its progress and skeptical of some claims made by proponents. Therefore, we have adopted an approach that incorporates the more stable concepts supported by tool technology and delayed others that are still in question as far as implementation potential in the next year.

2   Cougaar Model Driven Architecture

The objective of this research project is to improve the productivity of Cougaar system developers by applying Object Management Groups MDA approach. The productivity enhancement is achieved by automatic generation of partial sets of software artifacts such as requirements, design, code and test cases. While technologically, this has not been accomplished before, the Cougaar Model Driven Architecture (CMDA) Project endeavors to inspire solutions toward fully automated generation of software artifacts.

The CMDA system simplifies Cougaar-based application development by providing two important abstraction layers namely Generic Domain Application Model (GDAM) and Generic Cougaar Application Model (GCAM). The GDAM represents the PIM and encompasses the representation of generic agent and domain specific components found in the domain workflow. The GCAM layer, upon which the GDAM layer is built, reflects the PSM or Cougaar architecture, its specifications and environment. The user specifies the workflow of the intended Cougaar system using workflow components and the system is then detailed using GDAM and GCAM models.

CMDA approach uses a combination assembly and transform approaches to assemble components specified in GDAM and GCAM models and then transform them into intended Cougaar-based systems. The GDAM and GCAM engines assemble the respective models and the transform engine parses through

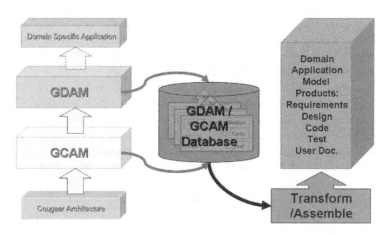

Fig. 2. Basic CMDA Approach

the assembled set of models to produce the actual software artifacts such as requirements, design, code and test cases.

Figure 2 depicts how all the pieces fit together conceptually. The CIM is realized through the GDAM/PIM, which is realized through the GCAM/PSM. While this is not a fully implemented MDA approach in every detail, it does conceptually reflect the key principles.

To a large extent, the CMDA systems capabilities are dependent on the effectiveness and efficiency of the transformation process. The transformer generates the system requirements by parsing mostly components present in the workflow layer, as the system's flow of execution and related constraints are described at that layer. While generating the requirements, the transformer also examines the components in the GDAM layer. Such examination is warranted due to the influence or tailoring some GDAM components have on the requirements that are being generated. Further it should be noted that the requirements, which are generated automatically, are partial in nature. The low-level design of the intended Cougaar system is to be elicited from the assembled GCAM components. The low-level design encompasses the GCAM model of the system, which includes (but is not limited to) UML class diagrams, sequence diagrams, state transition diagrams and deployment diagrams. The code and test cases are generated and/or assembled from the GCAM model, whose model representation will be in a suitable representation that provides the required completeness and correctness.

2.1   Formal Method Approach Selection

Cougaar is a highly complex system that implements concepts such as "time phased locality of reference" and "managed inconsistency." Hence, testing and finding errors using traditional testing methods such as testing for all possible states or artificially reducing the states by discerning selection, may be grossly inadequate. In such complex systems, formal methods are the chosen methods to assure correct operation [11]. Formal methods, whose underlying basis is mathematical notations and techniques, offer capabilities to fully specify the system using mathematical models. The completeness and correctness of the system is verified by validating the equivalent mathematical model of the system. However for most applications, due to time constraints, it is not advisable or even economically feasible to apply formal methods to fully specify the entire system. Frequently in real-world projects, formal methods are applied to a small subset of components that have the necessity for formal treatment [11].

The transformation processes for the CMDA system encompass significant challenges. While researchers have conducted transformations before, we are yet to come across any example that has attempted to perform transformations to this scale or depth. While other parts of the system such as mapping between GCAM and GDAM components are significantly difficult, the transformation is beset with some interesting challenges. The transformation challenges include:

1. Difficulties arising due to correctness and completeness errors in the input model,
2. Need for accurate depiction of the complex input model in the generated software artifacts (verifiability), and
3. Need to provide consistent output when repeated transformations (with same input) are performed.

These are particularly important for the portions of the CMDA system where equivalence and rewrite rules are applied. The degree to which these challenges are not met are proportional to the degree to which "human in the loop" will be necessitated. A major decision taken while deciding on the transformation approach was on adopting the assembly approach or synthesis approach. Given the complexities involved, it was decided to follow a combined assembly/transformation approach - thereby leveraging the simplicity of assembly approach and the efficiency of transformation approach. Further, the existence of many-to-many or at the least many-to-one mapping between components in two different levels makes a purely synthesis approach very difficult and highly error prone. In particular, the many-to-many mapping relationships between GDAM and GCAM components could result in a complex and unwieldy system, if synthesis approach or fully automated software artifact generation technique is used.

The following were identified as the key transformation requirements for the CMDA system.

1. Assembling the systems intended external behavior, specified using the workflow and GDAM semantics, into English requirement statements,
2. Assembling the system design represented using GDAM and GCAM components into system design in UML representation,
3. Generating code and test cases from the GCAM model by means of assembly approach,
4. Verification and validation of code generated.

3   Formal Methods

Formal methods, a combination of specification language and formal reasoning, can be classified into three categories: (1) Mainstream formal methods, (2) Theorem provers and (3) Customized formal methods. A brief description of the three categories is given in this section to give a flavor of the decision space available for the CMDA system.

Mainstream formal methods use rigorous mathematical models to specify the system. The foundations for mainstream models are usually based on set theory and first order predicate calculus. Examples of mainstream formal methods capabilities include Z, B, CSP, VDM, RAISE.

Theorem provers use rigorous mathematical proofs to describe software systems. Examples theorem provers include Nqthm, PVS, OBJ, and Isabelle. While theorem provers can be very effective, they may suffer poor usability, unintuitive development environments and graphical user interfaces. Further, development of systems using theorem provers can be difficult.

Custom formal methods are essentially extensions and adaptations of mainstream formal methods and theorem provers. Examples of these include VDM++, Temporal PetriNets, and Timed CSP. Formal methods are extended to support specific development paradigm such as object-oriented systems. Hybrid formal methods, a type of custom formal methods, are formed by combining two or more different types of formal methods.

3.1   Formal Methods in Transformation

The capabilities of the formal methods were understood by conducting an in-depth survey on some of the important formal methods that were used for specifying agent-based systems. Table 3.1 depicts the representative formal methods surveyed based on their Object-Oriented (OO) modeling support, usability, tool support and concurrency support. The rows of the table lists the different formal methods that were surveyed ranked in the increasing order of preference for the CMDA system. The columns of the table indicate the comparison criteria with decreasing order of importance (as far as CMDA system is concerned) as one move from left to right. The criteria were selected keeping in mind the transformation requirements, which necessitate representation notations that have adequate support for representing components and their constraints, scalability to represent large and complex systems and tool support for the assembly approach.

The support for representing objects is the most important selection criterion as Cougaar is an object-oriented system. The OO support criterion includes ability to represent objects and their constraints such as pre-conditions and post-conditions. The tool support is another important criterion for selection since CMDA is to be interfaced with eclipse IDE platform. The tool support should include GUI interfaces to perform consistency checks, type checking and code generation. The usability criterion gives an indication on the amount of difficulty in learning and using the formal method, with a good rate indicating that

Table 1. Comparison of Candidate Formal Methods [12]

Name	OO Support	Tool Support	Usability	Scalable	Concurrency	Formal Basis
X-machines	Yes	Very Poor	Poor	No	No	Yes (Formal Lang)
WSCCS	Yes	Poor	Poor	Limited	Yes	Yes (Process Algor.)
B	Yes	Average	Good	Yes	No	Yes (Set theory)
Z variants	Yes	Average to Good	Average	Yes	No	Yes (Set t./Pred. C.)
CSP	Yes	Good	Average	Yes	Yes	Yes (Algebraic)
Petri Nets	Yes	Average	Good	No	Yes	Yes
VDM++	Yes	Good	Good	Yes	Yes	Yes (Set theory)
UML	Yes	Good	Good	Yes	Yes	No

the methods syntax are similar to popular programming languages and easy to learn. The scalability criterion is the fourth important criterion that indicates whether the representation is scalable enough to support complex Cougaar systems. Formal basis criterion, the least important one, provides insights into the richness of the formal methods to describe the system completely and correctly.

As indicated in the Table 3.1, among formal methods, VDM++ appears to possess all of the important characteristics required by CMDA system. Some of the other prospective formal methods include CSP and Petri Nets. While CSP does support OO representations and has good tool support, the usability of CSP method is only average. As for Petri Nets, scalability of Petri net models is a major issue. Although VDM++ satisfies the criteria requirements of CMDA system, the time constraints imposed by the project schedule might not permit complete formalization of Cougaar system. Hence, the most apt implementation approach for CMDA system might be to combine the UML and VDM++ methods to exploit the advantages of both methods.

3.2   Vienna Development Method (VDM)

The Vienna Development Method [13] is a notation and set of techniques for formally specifying object-oriented systems (with concurrent and real-time behavior) including modeling the systems, analyzing those models and progressing to detailed design and coding. VDM has its origins in the work of the IBM Vienna Laboratory in the mid-1970s. VDM, one of the most popular and frequently used formal methods, is also one of the few that has ISO Standards for its specification language - VDM-SL, Meta-IV [14]. VDM++ is an extension of the VDM which support object oriented modeling. In this subsection, we outline VDM++ details on performance against the criteria for selection.

Advantages

The advantages of using VDM++ for this project include:

Usability: One key hindrance in using formal methods is the lack of support for programming language like semantics. VDM++ provides a programming language like semantics, thereby enhancing the usability of the method among developers. Further, VDM++ can be used in varying depths from specifying the requirements more correctly and completely, and to develop models for analysis and for implementing the system.

Applicability: Unlike most formal methods that evolved from academic world, VDM method was developed by the industry for solving real world problems. Hence VDM and its extension, VDM++, are used extensively and successfully to solve industrial problems.

OO Modeling Support: VDM++ is designed with OO modeling in mind. Hence the language can be used to model object oriented system, like Cougaar, without any modifications. The language also supports multiple inheritance and provides mechanisms to specify constraints on data and operations. The support for OO modeling is one of the biggest advantages for using VDM++.

Tool Support: VDM has extensive tool support. The class of tools available for VDM includes (1) VDM through Picture (VtP) by IDE: to input/edit formal specifications, to specify requirements using pictures or graphics (2) SpecBox: to print formal specifications captured automatically, to check specifications for grammatical correctness and for specifications completeness (3) Delft VDM SL: to check specifications for grammatical correctness and for specifications completeness (4) mural for proof support, (5) VDM domain Compiler for automated code generation and (6) transformation tools for converting UML models to VDM and vice versa [15, 16]. Further, IFAD VDM++ Toolbox is a set of tools designed to support VDM++. The toolbox provides a number of features that include checker to validate syntax and type, test coverage and statistics tool, and C++, Java code generators. Further, the toolbox provides APIs that allow programs to access and modify the running instance of VDM++ models inside the toolbox. This helps easier interfacing with the Eclipse IDE.

Disadvantages

Mathematical Foundation: VDM++ is based on mathematical notations. Therefore, many domain experts and system developers may not like to encode system specifications using VDM++ language semantics. The disadvantage can be mitigated by developing wrappers that will hide the complexity of VDM++ semantics.

Time Constraints: Even for formal methods experts, large system development with VDM++ would be a lengthy endeavor. The modeling of GCAM components in VDM++ will be time consuming and difficult. Hence, modeling the entire Cougaar system using VDM++ has to be avoided.

3.3   VDM++ Toolbox and CMDA

The VDM++ Toolbox, developed by IFAD, is a set of tools that supports the object-oriented VDM++ extension of VDM-SL. The toolbox, which is part of the VDMTools, differs from most other CASE tools for formal methods in the way the functional aspects of a specification are analyzed. Some of the features of the VDMTools are Specification Manager, Pretty Printer, Syntax Checker, Test Coverage and Statistics Tool, Type Checker, Dependency Browser, Interpreter and Debugger, Dynamic Link Facility, Couplings to Third-party Tools, and Java Code Generator.

The features of the VDM Tools planned to be used for the CMDA system include Rose-VDM++ link to convert UML into VDM++, VDM++ to Java code generator, Syntax and Type checker and Test Coverage and Statistics Tool.

In the next section, we discuss the use of VDM++ and UML in the CMDA approach emphasizing the transformation implications.

4   CMDA Transformation Approach

The transformation challenges detailed above entails using multiple representations to represent the CDMA system components. The representation that we believe, best addresses the challenges is a combination of UML and VDM++. The CMDA project intends to build a developer environment that will offer developers components, which can be aggregated to represent the system in workflow, GDAM and GCAM levels. Each of the components named as Workflow Beans, GDAM Beans and Cougaar Beans respectively (in synonym with Java beans concept) will contain sections of software artifacts and related information pertaining to that bean. Some example sections of the software artifacts that beans contain include:

1. Requirements model from which the transformer gleans the partial set of requirements,
2. Design model from which the systems design model is assembled by the transformer,
3. References to the lower level beans or links to Java code which can implement the bean. These references are traversed by the transformer while assembling the systems code and
4. Test case fragments that contain information on how to assembly the unit test cases for the beans.

Further, the bean contains documentation information such as description about the bean, and constraints pertaining to data, operation and connections with other beans. The constraints may be divided into two groups: (1) Port constraints, detailing constraints on input ports of the bean, and (2) Role constraints, detailing the restrictions the bean has on the roles or services the bean provides or supports.

The contents and size of the sections and information in a bean are influenced by the abstract layer to which the bean belongs. For example, a GDAM beans requirement section will be larger than the requirement section of the Cougaar bean, while the code section of a GDAM bean might be pointer to the Cougaar beans or code that can implement the GDAM bean in Cougaar. The models in the design model section of each bean will be represented using UML while the VDM++ representation will be used to delineating connector and other constraint information. The code section will contain links to Java code libraries at GCAM level and pointers to lower levels in rest of the abstraction layers. The requirements might be a combination of XPDL, text and UML diagram while the constraints also contain mapping (or connection) information that are mostly rule based with some formalizations applied.

The workflow of the CMDA system starts with developer assembling the system by picking the right workflow bean components and connecting them to represent the workflow. The constraints pertaining to connection are encoded in the beans and developers are shown a detailed error message when they try to connect two dissimilar components. Once the workflow of the system is build, it could be verified for consistency. The developer is then shown a list of GDAM beans that can be chosen to map a particular workflow bean. The expert system will list only related GDAM beans based on the constraints specified by the developer at the workflow level. The rationale to allow developers chose the right component is to allow developers make design decisions with the system assisting them (by showing a list of possible solutions and patterns).

The GDAM beans are mapped into Cougaar beans in a similar fashion. In all layers, as and when required, the developer will input necessary information to satisfy the completeness and correctness of bean component. The usability of the system can be improved by developing wrappers that would mask the semantics complexities of the representation language. Once the models are built, the transformation engine will traverse through the beans at each level and generate the software artifacts based on predefined transformation rules.

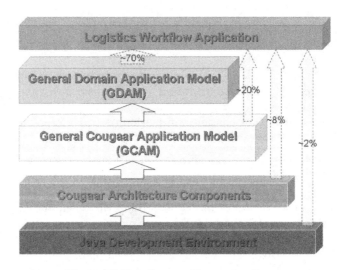

Fig. 3. CMDA System Abstraction Layers

Figure 3 delineates all the abstract layers that lie above the Java code. The GCAM layer, which has the largest number of components. The width of the boxes represents the extent to which the application can be represented by the layer. The ability to capture and/or implement the intended application's requirements increases as one progress through the layers, with the Java layer having the capability to implement all the requirements. The workflow is to be described using XPDL standard, defined by the Workflow Management Coalition, which provides a formal model for expressing executable processes that addresses all aspects of enterprise business processes. XPDL was chosen because the language focuses on issues relevant to the distribution of work and workflow processes than defining web services as in other standards such as BPML and BPEL.

The solid arrows moving upwards from the Java layer through the GDAM layer represent the composition of more concrete components to satisfy the domain level abstraction specified by the user. The dashed/transparent arrows pointing up to the domain application layer from the other layers depict the alternative components that can be obtained when a suitable GDAM component is not available. The values on the dashed/transparent arrow indicate the projected amounts of components from the various alternatives in the development environment.

5   Conclusions

Software development can be thought of as the evolution of abstract requirements into a concrete software system through a series of transformations and refinements. Even in moderately complex systems, this transformation is often too involved for fully automated means.

MDA provides a systematic way of capturing details during elaboration and refinement through the mapping from CIM to PIM, PIM to PSM and ultimately rendered as an executable software system. MDA as currently defined appears to have utility if used in moderation. However, for CMDA, it is not a panacea by any stretch. It still requires considerable work and strategic application.

Cougaar is complex requiring considerable mappings and transforms. For this reason, we chose an assembly centric approach with simple formalisms to start. The CMDA approach has substantial transformation challenges in generating software artifacts such as requirements, design, code, and test cases automatically. The artifacts are generated from models assembled using components or beans belonging to two abstract layers namely GDAM (abstracts the domain and generic agent system) and GCAM (abstracts the Cougaar system). A bean will contain nuggets of requirement, design, code, test and documentation details pertaining to that component along with transformation information. The CDMA system combines assembly approach with transformations in small concept to generate the artifacts.

A comparison study of formal methods was conducted to identify the suitable language representation for the GCAM and GDAM components. The selection criteria for the comparative study included criteria such as object-oriented support, usability and tool support. The study concluded that the complexity of the system, coupled with the need for completeness and correctness compels using a hybrid language representation (combination of UML and VDM++) to achieve transformations. The transformation engine will generate the required software artifacts, from the GDAM and GCAM models assembled by the developers, by parsing the various sections and portions in the beans.

Acknowledgements

This work has been supported, in part, by the DARPA STTR grant "AMIIE Phase II — Cougaar Model Driven Architecture Project," (Cougaar Software, Inc.) subcontract number CSI-2003-01. We would like to acknowledge the efforts, ideas, and support that we received from our research team including Todd Carrico, Sandy Ronston, Tim Tschampel, and H. Lally Singh.

References

1. Russell, S., Norvig, P.: Artificial Intelligence: A Modern Approach. Second edn. Pearson Education, Inc., Upper Saddle River, New Jersey (2003)
2. Dogac, A., Cingil, I.: Agent technology. In: B2B e-Commerce Technology: Frameworks, Standards and Emerging Issues. Addison-Wesley (2004)
3. —: Grasshopper 2: The agent platform. IKV++ Technologies AG (2004) http://www.grasshopper.de.
4. —: Agent oriented software. The Agent Oriented Software Group (2004) http://www.agent-software.com.
5. —: Cognitive agent architecture (Cougaar). Cognitive Agent Architecture (Cougaar) Open Source Project (2004) http://www.cougaar.com.
6. —: Java agent development framework (JADE). JADE Board (2004) http://jade.tilab.com.
7. —: The foundation for intelligent physical agents (fipa). FIPA Secretariat (2004) http://www.fipa.org/resources/livesystems.html.
8. —: Cougaar architecture document. Technical report, BBN Technologies (2004) Version for Cougaar 11.2.
9. —: Cougaar developers guide. Technical report, BBN Technologies (2004) Version for Cougaar 11.2.
10. Bass, L., Clements, P., Kazman, R.: Software Architecture in Practice. Addison-Wesley Publishing Co. (1998)
11. Dorfman, M., Thayer, R.: A Review of Formal Methods. Computer Society Press (1996)
12. Rouff, C., Vanderbilt, A., Truszkowski, W., Rash, J., Hinchey, M.: Verification of nasa emergent systems. In: Proceedings of the Ninth IEEE International Conference on Engineering Complex Computer Systems, Florence, Italy (2004)
13. —: Information on VDM. The Center for Software Reliability (2004) http://www.csr.ncl.ac.uk/vdm/.
14. Plat, N., Gorm-Larsen, P.: An overview of the ISO/VDM-SL standard. ACM SIGPLAN Notices (1992)
15. McGibbon, T.: An analysis of two formal methods: VDM and Z. Technical report, Data & Analysis Center for Software, Rome, NY 13441-4909 (1997) http://www.dacs.dtic.mil/techs/2fmethods/title.shtml.
16. —: The toolbox newsletter. IFAD (2000) http://shinsahara.com/www/vdm/NewsLetter/issue5.doc.

Facilitating the Specification Capture and Transformation Process in the Development of Multi-agent Systems

Aluízio Haendchen Filho[1], Nuno Caminada[2], Edward Hermann Haeusler[1],
and Arndt von Staa[1]

[1] PUC-Rio– Pontifícia Universidade Católica do Rio de Janeiro,
Departamento de Informática, Rua Marquês de São Vicente 225,
CEP 22453-900, Rio de Janeiro, RJ, Brasil
{aluizio, hermann, arndt}@inf.puc-rio.br
[2] UniverCidade – Centro Universitário da Cidade do Rio de Janeiro (UniverCidade/NUPAC)
caminada@atividade.com.br

Abstract. To support the development of flexible and reusable MAS, we have built a framework designated MAS-CF. MAS-CF is a *component framework* that implements a layered architecture based on contextual composition. Interaction rules, controlled by architecture mechanisms, ensure very low coupling, making possible the sharing of distributed services in a transparent, dynamic and independent way. These properties propitiate large-scale reuse, since organizational abstractions can be reused and propagated to all instances created from a framework. The objective is to reduce complexity and development time of multi-agent systems through the reuse of generic organizational abstractions.

1 Introduction

The characteristics and expectations of new application domains surrounding distributed systems have lead to the development of dynamic and evolving structures. After the advent of the Internet and with the recent emergence of new technologies, the application domain of MASs is expanding and nowadays it is used in many areas, such as e-business, web-services, knowledge management and now enterprise information systems [Faulkner2001, Griss2003, Adam2004, Giorgini2004]. Agent technology represent an extraordinary opportunity for information systems and corporate applications, because agents must be capable of managing and organizing information, recognizing personal tastes and making increasingly important decisions on behalf of their owners.

Nevertheless, the development of multi-agent systems is not trivial. To avoid the task of designing each new system, we need tools to help in the MAS construction, and by extension it is desirable to also have tools for reusing previous designed architectures and their relationships. There is a considerable research effort towards the development of frameworks for agent-based systems [Sycara1999, Wooldridge2000, Evans2001, Bellifemine2001]. Each framework has different

M.G. Hinchey et al. (Eds.): FAABS 2004, LNAI 3228, pp. 72–91, 2005.

application specific particularities, such as social capabilities, reasoning, flexibility for dynamic compositions, interoperability and so on.

Most approaches, however, focus on the reuse of application-specific concepts at the analysis, design and implementation levels (roles, protocols, agent architectures). Little research is conducted towards generic (i.e, application-independent) models [Faulkner2001, Zambonelli2002, Holvoet2003, Griss2003]. There is a large potential of reusing generic "organizational abstractions" – such as structures and patterns – for generic (i.e, application-independent) models [Zambonelli2002]. Reuse of generic software is recognized within the object-oriented community and has lead to the concepts such as design patterns and frameworks [Pree1999, Fayad1999].

The main focus of our work is the reuse of abstractional organizations applied to the development of multi-agent systems. Reuse an abstract architecture allow us not only to reuse the design and the implementation of the architectural software, but also the reuse of important individual agent properties, such as interaction, adaptation and collaboration, which can be completely or partially resolved at the architectural level. On the other hand, by freeing the developer from the task of implementing these complex properties on the agent, the work becomes simpler and can be better focused on the maintenance of the knowledge structure and on the learning capabilities of the agent.

This paper is structured as follows: the next section briefly describes the state-of-the art regarding agents and multi-agent systems. Section 3 describes the abstract architectural model, the communication model and interface specification. Section 4 describes the interaction model, formalized by means of service ontology. Section 5 describes how the architecture behavior has been formalized and how the specifications are being stored and transformed into reliable code. Related works are discussed in Section 6 and Contributions are listed in Section 7.

2 Agent and Multi-agent Systems

We have examined and identified through the literature the essential aspects surrounding agent-based technology. This section briefly presents some important concepts that will be used on the course of this work, namely agents and multi-agent systems.

2.1 Agents

There is no universally accepted definition of the term agent. Part of the difficulty to define agent arise from the fact that for different domains of applications, the properties associated with the agent concept assumes different levels of importance. There are many types of software agents with different characteristics such as mobility, autonomy, collaboration, persistence and intelligence.

The behavior of an agent depends on, and is affected by, the incorporated agency properties: interaction, adaptation, autonomy, learning, mobility and collaboration. Such properties were based on previous studies [Kendall1999, OMG2000, Garcia2001]. We have use the properties as follows, based on [Garcia2001]:

- *Interaction*: an agent communicates with the environment and other agents by means of sensors and effectors. These are available via the agent's provided and required interfaces;
- *Adaptation*: an agent should adapt its state and behavior according to new environmental conditions;

- *Autonomy*: an agent has its own control thread and can accept or refuse a request; in other words, by autonomy we understand the capacity of the agent to execute its activities without human intervention;
- *Learning*: an agent can learn on previous experience while interacting with its environment;
- *Mobility*: an agent is able to transport itself from one environment to another to achieve its goals;
- *Collaboration*: an agent can cooperate with other agents in order to achieve its goals and the system goals.

According OMG [OMG2000], *autonomy*, *interaction* and *adaptation* can be considered as fundamental properties of software agents, while learning, mobility and collaboration are neither a necessary nor sufficient condition for agenthood. There are several types of software agents, including information agents, user agents, interface agents and mobile agents. Each agent type has different application specific capabilities and agency properties. In order to have autonomy, an agent must possess a certain degree of intelligence allowing it to survive in a dynamic and heterogeneous environment [Correa1994]. Therefore, there is general consensus that autonomy is one of the central properties to the notion of agent.

2.2 Multi-agent Systems

There are several different ways to organize multi agent systems. In any given case, the best way depends on the purpose and objectives of the system, thus there are several types of multi-agent systems, each with its own particularities such as social capabilities, reasoning, interoperability and so on. Jennings [Jennings1996] proposes a framework that provides a structure to analyze and classify the activities of multi-agent systems according to two different perspectives: (i) the agent perspective: focuses on the characteristics of the agent involved with the MAS, such as internal architecture, structure and maintenance of knowledge, and abilities of reasoning and learning; (ii) the group perspective: includes group aspects such as organization, coordination, interaction and negotiation.

In MESSAGE [Evans2000], MAS architecture is defined through an organizational model, focused on the structure of the organization and the relationship between the agents it contains. The organizational model also describes mechanisms for conflict resolution and rules that enable agent groups to function as a unit serving a common purpose. Agents are identified based on a goal-oriented model, where organizational goals are decomposed and associated with tasks. Goal decomposition is carried out recursively, until the tasks associated with the goal can be completely fulfilled by an isolated agent or in collaboration with other agents. Agents are connected by organizational relationships (such as superior-subordinate and client-provider), proceedings of control management, workflows and interactions. Internal architecture and maintenance of the knowledge structure applies an approach similar to BDI (Beliefs, Desires, Intentions).

On the design of interoperable agents, JADE [Bellifemine2001] is a framework focused on interoperability based on the standardization of the language of knowledge. JADE can be considered an agent middleware that implements a platform and a development framework. The interaction model is implemented according to FIPA [FIPA2000] protocols. FIPA provides a standard language of communication based on protocols, an ontology necessary for the interaction between the agents from the system and from other systems. JADE provides an API to organize the system starting with a set of generic system services and agents. Services are transported through an interface mechanism to send/receive messages to/from other agents.

RETSINA [Sycara1999] focuses the agent architecture in a software infrastructure that allows heterogeneous agents to interact on the Internet. The RETSINA framework provides an abstract basic agent architecture consisting of, and integrating with, reusable modules and each module of an agent operates asynchronously. The RETSINA definition of multi-agent systems is driven by the vision that heterogeneous agents that autonomously organize their own social structures should populate multi-agent societies.

The descriptions show different ways to organize MAS. Nevertheless, most approaches focus the reuse in specific application concepts and on the individual properties of the agent, such as protocols, roles and internal architecture. Little research on the domain of multi-agent systems has been conducted emphasizing the reuse of generic organizational abstractions [Faulkner2001, Zambonelli2002, Holvoet2003, Griss2003].

3   The Architectural Model

In this section we present the main models that compose the framework architecture, thus, the abstract model, the structural model, the interface model and the logic model are described and commented.

3.1 The Abstract Model

The architecture of a multi-agent system can naturally be viewed as an organized computational society of individuals. For this reason, organizational abstractions should play a central role in the analysis and design of such systems. Zambonelli and Wooldridge [Zambonelli2002] state that "the introduction of high-level organizational abstractions can lead to cleaner and more manageable and reusable MAS design." Also according to Zambonelli, the organizational abstractions facilitate the design process because it leads to a cleaner separation between the component level (i.e., intra-agent) and system-level (i.e., intra-system). Holvoet [Holvoet2003] argue that "programming in the large" for reactive MASs should imply a reuse method that allows two things: (i) to describe MASs in an abstract, application-independent way and (ii) to reuse such abstract multi-agent system through application-specific adoptions.

In order to address these necessities, a few basic requisites of the model must be introduced. First we define MAS from an organizational view as a set of autonomous agents (possibly pre-existent) which common objective is the solution of a given problem [Jennings1996]. Nevertheless, the designer does not have to be focused on the solution of a specific problem. New problems may arise in the context of the MAS, and the society must be able to solve these new problems in collaboration. This can be achieved through the inclusion of new agents building compositions with pre-existing agents or by replacing obsolete agents. Therefore, the abstract model must provide an architecture that facilitates the inclusion of new agents at any given moment as new problems arise.

During the analysis phase, an understanding of the system and its structure can be done. In our case, this understood is captured in the system's organization, via architectural model. We view a organization as a collection of agents that provide and perform services, and take part in systematic, institutionalized patterns of interactions with other agents regulated by the architecture. Departing from the goals of the organization, services can be identified and allocated to new agents or to pre-existing ones.

3.2 Proposed Architecture

Our architecture was designed supported by the basic concepts present in component frameworks [Szyperski2002]. A component framework is a set of interfaces and interaction rules that govern how components "plugged into" the framework may interact. In particular, a component framework forms a framework that composes instances not based on directly declared connections or derivations (such as inheritance of a class framework), but based on the creation of contexts and the placement of instances in appropriate contexts [Szyperski2002]. Beyond the similar names, almost identical visions and superficially similar construction principles, component frameworks are very different from class frameworks [Bosch1999, Fayad1999] since the inheritance implementation is not commonly used between a component framework and the interfaces it supports.

Figure 2 illustrates the two main parts that compose our structural model: *System* and *Infrastructure*. *System* defines a structural model for the domain-specific MASs. We define domain according to [Sodhi2000, Tracz1994] as the space of the problem for a family of applications with similar requirements. *Infrastructure* defines a part that contains components that provide generic services, such as database access, translation services, HTTP services, GUI builders and others.

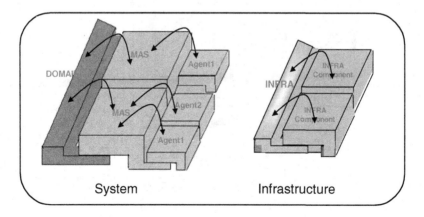

Fig. 2. The MAS-CF generic architecture

System can be seen in the left side of the Figure 2. It defines a three-tier architecture composed by the elements *Domain*, *MAS* and *Agent*. *Domain* is a component system, *MAS* is a component framework, and *Agent* is an abstract model for the instances plugged on the MAS. The Domain tier implements a set of rules of interaction that allows the communication and the sharing of services between different MAS and allows the communication between systems located in different domains. Different MAS located in a given domain can be plugged on the tier Domain. Note that tiers are described side by side with each other, while layers sit on top of each other. Traditional class framework merely structure individual components, independent of the placement in a tiered architecture. In the same way that MASs can be plugged on the Domain tier, agents can be plugged on the MAS tier.

Represented on the right side of the Figure 2, *Infrastructure* is a two-tier architecture where the Infra is a component framework and the generic Infra Components are instances of the Infra component framework. The communication between the System and Infrastructure is supported by an ontology, which describes the services and how they can be accessed. Details will be shown in the Section 4.

3.3 Communication Model

Based on fundamental principles present in component frameworks, we have defined the communication model considering that the exchange of information between agents will be implemented as connections between agents and the architecture. The objective is to allow the sharing and distribution of services in a transparent, independent and autonomous way. An agent or component is visible to the architecture and can communicate generating events, which trigger connections rules in the architecture. The communication is indirect, via a component framework that mediates and regulates component interactions. Figure 3 shows the communication model on the proposed architecture.

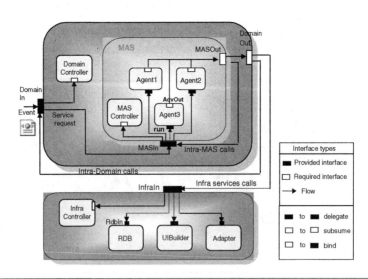

Fig. 3. The communication model

We use similar notation to SOFA [Plasil2002] to describe the communication between interfaces. Three different types of connections are distinguished: (i) *delegate*: a connection between a provided interface of a component and a provided interface of a subcomponent; (ii) *subsume*: a connection between a required interface of a subcomponent and a required interface of a component and (iii) *bind*: a connection between a required-interface and a provided-interface between two subcomponents. We have considered that the information flow between connections in bi-directional. The Java Virtual Machine places call returns in a stack. After the execution of an event, the system returns to the caller.

Services requests arrive from the environment through the interface *DomainIn*. These requests are decoded by the *DomainController* — which acts as an *abstract factory* [Gamma1995] — and are sent by the service to the responsible agent. Just as the *DomainController*, *MASController* and *InfraController* work as *abstract factories*. They encapsulate knowledge about which concrete classes are used for the system, and conceal the way that the instances of these classes are created and joined. It permits the configuration of the system with agents "product" that can vary widely in structure and functionality. As seen in the previous subsection, the concept of component framework can be applied in such a way that component frameworks are themselves components "plugged" into higher-tier component frameworks. Thus, by construction, a component framework accepts the insertion of instances at run-time. Agents and Infra Components can be dynamically registered and plugged on the framework.

3.4 Interface Model

One of the main ideas underlying frameworks is that semi finished components can be represented by abstract classes. Their purpose is to standardize the *class interface* for all instances or subclasses. Subclasses and instances can only augment the interface, and not change the names and parameters of methods defined in a superclass [Pree1999]. The term contract [Pree1999, Szyperski2002] is used for this standardization property: instances of subclasses of a class *A* support the same contract as supported by instances of *A*. A contract is a specification attached to an interface that mutually binds the client and the providers (implementers) of that interface. Thus, the semi-finished or ready-to-use components and agents of our framework can be implemented based on the contract of the abstract class.

On the lowest level tiers, the abstract class *Agent* provides two interfaces: a provided interface designated *AgentIn* and a required interface designated *AgentOut*. *AgentIn* provides a channel of communication through which agents can absorb events and is a flexible hot-spot [Pree1999]. The *AgentOut* interface establishes a communication channel from where services from other systems, agents or components may be requested. To this end, it is only necessary to agree to the contract established by the interface. The *AgentOut* interface is a frozen-spot. Note that *Agent* here represents a generic term. In practice, the interface assumes as prefix the name of the agent and as suffix the expressions *In* and *Out*. The two interfaces are encapsulated into the semi-finished abstract class *Agent* when instanced through the framework. The basic syntax of the contract is as follows:

```
public void AgentIn(String service, Vector in, Vector out)   → sensors
public void AgentOut(String service, Vector in, Vector out)  → effectors
```

The parameter *service* (String) defines the name of the requested service. The parameters possess semantic meaning similar to IDL CORBA. They can be of type *in* (flow from client to object) or *out* (flow from object to client). The operation result, whenever there is one, is essentially a distinguished *out* parameter. The specification of highly structured messages introduces a level of complexity, since the parameters frequently represent complex types or data structures, such as vectors of objects. The type Vector used on the *in* and *out* parameters make possible to use heterogeneous types of fields, such as Objects, arrays, Strings, and so on.

For the components of the Infra tier, only the provided-interface is instanced. Contrary to agents, components do not communicate among each other. As independent processing units, they do not request external services from other components or agents.

3.5 Logical Model

The UML provides the package mechanism [Larman1997] for the purpose of illustrating groups of elements or subsystems. Such a diagram may be called an architecture package design. A package defines a nested name space, so elements with the same name may be duplicated within different packages. Graphically, a package is shown as a tabbed folder; subordinate packages or classes may be within it. Figure 4 illustrates a more detailed breakdown of common packages in the architecture of the framework.

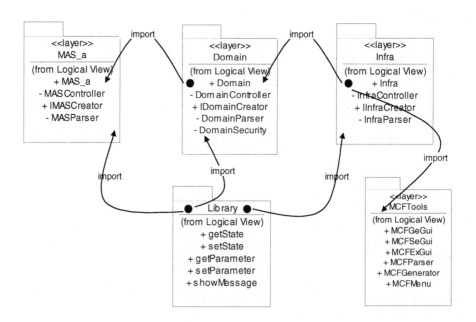

Fig. 4. Architectural units expressed in terms of UML packages

The framework contains a set of five packages: Domain, MAS, Infra, Library and MCFTools. Inside each package the encapsulated classes are listed. The three packages shown on the top represent the main tiers of the framework: Domain, MAS and Infra. Note that the three packages contain classes with the suffixes *Controller*, *Creator* and *Parser*. As seen on previous sections, the classes sporting the suffix *Controller* represent *abstract factories*, responsible for the dynamic creation of instances. The *Creator* interfaces (starting with the letter *I*) define a standard signature for the instances that can be created dynamically, establishing a plug-and-play structure. The classes sporting the *Parser* suffix implement programs that parse service catalogs (detailed in the next section) to retrieve the specification of the agent or component responsible for the execution of the service. When the agent is retrieved, it is delivered in the form of a *String* from the *Parser* class to the *Controller* class, which implements a *factory method* [Gamma1995] for the dynamic creation of instances.

The two packages shown bellow on Figure 4, *Library* and *MCFTools,* supply generic support services to the main packages of the framework. Library contains some classes that supply important generic services to the programs that control the interaction flux and the synchronism between processes. The classes *setState* and *getState* are responsible for the synchronism between processes. Class *setState* (producer) stores in a hashtable the next state for the action to be executed during the transition. The data is indexed based on a ID created for each instance, and associated to the state and corresponding action. Class *getState* (consumer) whenever called upon, retrieves the state stored in the hashtable and delivers to the process the instance and the action to be executed.

The MCFTools package provides a public interface to support the tasks of instancing the architecture and the elements, along with the necessary support for the specification of the service catalog. To this end, it makes a set of GUI classes available, such as *MCFMenu, MCFGeGui, MCFSeGui. MCFMenu* is the class that provides a common interface to a group of other components of the package and system, implementing a pattern *facade* [Gamma1995, Larman1997]. The disparate elements may be the classes in a package, a framework or a subsystem (local or remote). Along with the GUI classes, the package maintains a class called *MCFParser* that captures (when the architectural elements are instanced) the specifications described by the GUIs and stores it in the XML file. Finally, the *MCFGenerator* class is responsible for code generation, working inside the standards established by the standard code structure used by the framework (as per Section 5.2)

4   Interoperability

Consider the high level component Infra. New components, which implement generic services, can be plugged at run time; new services must be available to agents at run time. How to make new services available to the agents? How to allow agents to interact with each other without knowing in advance which services are available? The representations of the architecture were not sufficient to serve as a listing of all services provided. When a new agent is registered or instantiated by the framework, its services are registered in a XML ontology in the form of a services catalog.

The use of ontology serves us as a formal specification of the catalog of services provided. Every agent/component operating within the System or Infra part must abide to the specifications dictated by the services ontology. The same is true for components. Figure 5 shows how services registered on the catalog may be accessed through the controller components present on the layers. Different components access specific sections of the catalog and obtain information such as component instances, location of services and descriptions of the communication protocols.

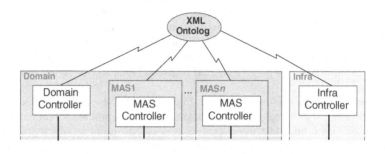

Fig. 5. Relationship between components and XML ontology

List 1 shows an example of how a services catalog can be structured in the form of an ontology. The tags *name* and *description* supply basic information about services provided by agents or by components. The *initiator* tag indicated the agent responsible for the execution of the service and the path tag indicates the physical location of the agent. It may be a physical address or a URL. The *type* tag indicates the type of protocol being used by the agent to deliver the message, initiate a conversation or supply a service.

```
-  <Services>
  -  <service>
        <name>Advising Receive</name>
        <description>...</description>
        <initiator>Advisor</initiator>
        <type>MAS-CF</type>
        <path>D\AcademicApplication\Advisement</path>
        <domain>Academic Applications</domain>
        <mas>Electronic Advisement</mas>
        <message>Contract MAS-CF</message>
     </service>
  </Services>
```

List 1 - XML specification of the catalog of services

The *Initiator* is the agent responsible for starting the execution of the service. The *Type* indicates the type of protocol used to deliver the message and to supply a speech act or a service. In this case, all tags are automatically retrieved from the specification and stored in XML format. Also present are the *name* and *description* tags, which supply basic information about the service. The XML catalog is critical to the system and during use a working copy is made to ensure system reliability. If the working copy fails a new copy is reconstituted from the original. Besides, the information contained on the XML catalog can be reconstituted from the interfaces on the original XML system specification through the use of special tools.

Semantic heterogeneity is one of the chief focus of any multi-agent system, this heterogeneity expresses the issue that any two interoperating agents must be certain when using a vocabulary of terms, or translations thereof, that they are using the same concepts with the same relevant inferences of relations as the other communicating agent [Sycara2003]. Two heterogeneous interoperating agents must be certain when using a vocabulary of terms or translations (FIPA to MAS-MF, for example) that they are using the same concepts with the same relevant inferences of relations as the other communicating agent. We argue that ontology, commonly defined in the literature as a *specification of a conceptualization*, is the representation that will provide this requirement [Gruber1998].

A conceptualization can be concretely implemented, for example, in a software component. Different types of ACL (Agent Communication Language) can be identified via *Type* tag and services are provided by adapter components to translate the MAS-CF messages to/from KQML [Finin1997], FIPA, UCL [Montesco2001] and other ACLs. It decodes the calls that arrive from the environment and identifies the language spoken by the agent, for example KQML or FIPA. These components can be registered and plugged into the Infra tier.

5   Describing and Transforming the Specifications

In this section we describe how the behavior of the framework is formalized through the use of FTS (Finite Transition System) [Arnold1994]. In the sequence, we show how the specification is described and transformed into reliable code.

5.1 The Behavior of the Framework

Most work on the semantics of parallel, communicating, concurrent or interacting processes is based on the concept of automaton. More generally, a finite state automaton formed of states and labeled transitions between those states, can describe a system whose state evolves over time [Arnold1994]. An agent is a computational entity handling sequences of events. To handle events, agents can emit events, absorb events, and process internal events [Plasil2002]. Method calls on interfaces turn into event, and the architecture's behavior is modeled via the event sequences (traces) on the architecture. The behavior of the architecture can be approximated and represented by FTS. A *transition system* consists of a set of possible states for the system and a set of transitions - or state changes – which the system can effect [Arnold1994].

The previously presented architecture (Figure 3) can be described as a concurrent FTS, as shown in Figure 6. The figure shows each tier represented as a FTS, working concurrently with other tiers. The label   indicates the target action or event, when the state triggers the transition. The set represented by the states $\{S_1, S_2\}$ encapsulate the provided- and required-interfaces *DomainIn* and *DomainOut* of the Domain tier, respectively. In a similar way, the set $\{S_4, S_5\}$, $\{S_{11}, S_{12}\}$ and $\{S_{81}\}$ compose the provided- and required-interfaces of the MAS, Agent and Infra tiers respectively. The states S_3, S_6, resp. S_{82} represent a set of nested states composed by the classes with the suffixes *Controller*, *Creator* and *Parser* of the Domain, MAS and Infra tiers, as seen on section 3.5.

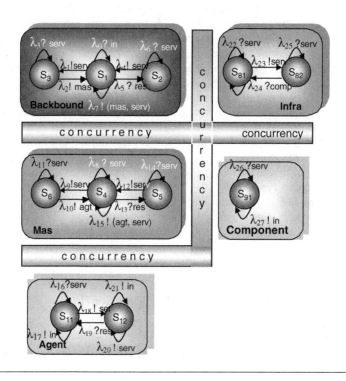

Fig. 6. The architectural model as FTS

Asynchronous behavior between states is represented through self-transition. A self-transition may represent a, asynchronous communication channel between two tiers ($(S_1$ to S_4, for example) or a recursive decomposition to nested states, as seen on S_3, S_6 e S_{11}. On the expressions that label the transitions, the character   represent the target action to be executed by the transition. The suffixes {!, ?} represent the action emitted or absorbed. Besides actions, variables are also described. Basically, the variables represent services (*serv*), instances (*mas*, *agt*, and *comp*) and results or data (*res*) modified by the states or processes.

In run-time, the program directs the flow via switch for the current state, evaluates the predicates and changes for the target state, performing the associated action. This can be seen in the code fragment presented on Figure 7 of the next subsection. ECA rules specifies how the architecture receives messages from the environment and from agents, how it verifies the service, direct services, sends messages and create instances of the architectural entities. The synchronism between tiers (considered as concurrent processes) is provided through CCS (Calculus for Communicating Systems) [Milner1985] expressions.

CCS expressions generate a set of *traces* over the architecture and the agents establish the restrictions, the sequence of execution and the synchronism between the concurrent tiers. The basic operators are the classic regular expressions *sequence*, *alternative* and *repetition*. The *enhanced operators* provide a notation to describe concurrency, using the known operators *or-parallel*, *and-parallel* and *restriction*. Several transitions can have the same source and target, i.e., the product mapping is not necessarily injective. The sequence of actions $S(c) = (t_1)(t_2)$ is called the trace of the path. Intuitively, the label of a transition indicates the action or the event, which triggers the transition.

5.2 Code Generation

When instancing MASs, agents or Infra components, the specifications captured and stored in XML file are transformed into reliable code using parser and generator programs. The parsers can read the specifications from the XML file using the standard XML document object model (DOM). DOM essentially maps every element of an XML document to an object. Such an object has methods to access the element's attributes, and DOM also supplies methods to navigate through documents and to locate the parent element and enumerate the child elements. After being parsed through the DOM, the information is supplied to the generator program, which transforms the parsed information into source code based on templates of MAS-CF entities.

During the implementation phase, code generation occurs at two separate times. First upon the instantiation of the architectural elements by the framework, when the code of the structural model is automatically generated. At this stage, the MAS (if it has not been instantiated), the agents and the internal layers of the agents can be instantiated. Afterwards, only the abstract method of semi-finished component can be implemented or plugged. Thus, the implementation of the internal architecture of the agent becomes independent from the framework. The internal implementation of the agents is free, and therefore any type of agent architecture or implementation model may be used.

In the design of rational agents, the role played by attitudes such as beliefs, desires (or goals) and intentions have been well recognized in the AI and agents literature. Systems and formalisms that give primary importance to intentions are often referred to as BDI (Belief, Desire, Intention) architectures. BDI-like architectures model the agent's behavior using a set of mental categories evolving in a mental cycle that allows the agent to make decisions and to act on the environment. These architectures raise from the process of deciding, moment by moment, which action to take towards its objectives.

Figure 7 shows a partial view of the generated Java code for the *Mas* (here *Mas* is an instance of the abstract model MAS) class. The interface *MasIn* (line 32), the parameters and the pre-condition (line 34) are supplied from the specification of the interface and the remaining items - states, transitions and actions - can be retrieved from the XML *service* specification. On line 36, the method *run()* of the library class *getState* retrieves the current state of this specific instance. Line 38 performs the transition via *switch* for the *case* that corresponds to the current state. Inside each *case*, the method *instanciaAgent()* of the abstract factory *MasController* is called and returns the instance responsible for forwarding or executing the requested service. On line 44, the target state is defined and stored using the method *run()* of our library class *setState* (line 45). On line 46, the agent returned in the *frame* instance performs the action associated with the transition.

```
24    private
25         final static int MasServiceReceive = 6;
26         final static int MasServiceRequest = 5;
27         final static int DomainServiceReceive = 1;
28         final static int DomainServiceRequest = 2;        ──▶  states
29         final static int AgentServiceReceive = 10;
30         final static int InfraServiceRequest = 20;
31
32    public void MasIn(String service, Vector in, Vector out)  ──▶  interface
33    {
34         if ((service.length()>0) &&(in!=null))         ──────▶  pre condition
35         {
36             int state = getState.run();                 ──────▶  current state
37
38             switch (state) {                            ──────▶  transition
39                 case MasServiceReceive:
40                     try
41                     {
42                         IFrameCreator frame = (IFrameCreator)
43                             MasController.instanciaAgente(service, in, out);
44                         state = AgentServiceReceive;   ──▶  target state
45                         setState.run(state);
46                         frame.run(service, in, out);   ──▶  action
47                         break;
48                     }
49                     catch (Exception e)
50                     {   showMessage.run(e);
51                         break;
52                     }
```

Fig. 7. Partial view of the generated code for the Mas class

The code of the Mas class presented above is almost completely frozen (except the name of the interface In - MasIn - on line 32, the name of the interface Out – MasOut - and the class name are hot-spots). It is completely generated when elements of the framework are instanced for the first time. The same happens for the classes Domain (through which different domains can be instanced) and Infra. The framework also generate the code for the abstract classes Agent and Component every time new agents or components are instanced. Specific implementation can be added on the hot-spots provided by the abstract classes of the last level.

We argue that the reuse of organizational abstractions, as well as the interaction facilities provided by the architecture reduces the complexity and facilitates the development of the cognitive capacities of the agents (learning and autonomy), since complex properties such as interaction, adaptation and collaboration can be addressed separately by the architecture. In this fashion, agent implementation can be better focused on the maintenance of its structures of knowledge gathering and on its mechanisms of learning.

6   Discussion and Related Works

The concept of connection as an architectural entity was established on the first ADLs, such as Darwin [Magee1997], UniCon [Shaw-Garlan1996], Wright [Allen1997] and ACME [Garlang1997] among others. The idea is to deal with aspects and system qualities in connectors, not in components. According to Szyperski [Szyperski2002], one of the problems with these approaches is that by introducing a pure connection-oriented approach, all components are restricted to only interact with other components if appropriately connected. On the other hand, a connector, when detailed, can easily heave substantial complexity and display a need to be partitioned into components itself. Thus, "connectors" turn into regular components and no special actions can be performed on the connections as such.

The concept of explicit connector has been loosing ground as time passes. Some ADLs, such as Rapide, have a very weak notion of connectors. Connections are specified with bindings between the provided service of a component and the required service of another component. Faulkner [Faulkner2001] proposed an ADL for multi agent systems using a similar concept. In his approach, Faulkner uses components, interfaces and services as architectural entities, without connectors. Connections are implemented as bindings between provided interfaces and services. Szyperski [Szyperski2002] states "contextual component frameworks can be used to reintroduce the intercepting behavior of connectors, but this time at the level of context boundaries." Contexts provide the generic-aspects, while components and/or agents provide the non-generic aspects of contexts by parametrizing generic contexts.

Our approach has a very weak notion of connector. The interaction rules are managed and performed by the architecture, resulting in calls to the other agents and services inside or outside of the organization. Its semantics consists of the rules defining the subtype (and supertype) relationship between tiers, and the services ontology providing the necessary mechanisms to interoperability support. Wooldridge [Wooldridge2000] states that agents are not built considering the existence of other specific agents; the idea is that interdependencies are likely to be reduced to make the system more flexible and reusable.

The preference for implicit connections, as opposed to explicit ones, is one of the key points in our approach, using a very weak notion of connector. Interaction rules are regulated and executed by the architecture, resulting in calls to other agents and components inside and outside the organization. The semantics consists of rules defining the relationship between superior and inferior layers and the ontology service providing support mechanisms necessary to interoperability. We share a concept introduced in [Wooldridge2000], whereas agents should not be built assuming the existence of other specific agents; the idea is that interdependencies may be reduced to make the systems more flexible and reusable.

Current frameworks for multi-agent systems such as JADE [Bellifemine2001], RETSINA [Sycara1999, Sycara2003], MESSAGE [Evans2002] and ZEUS [Azarmi2000] work with a structure much more focused on the individual properties of agents than on MAS architecture. These approaches provide an implementation that reinforces only partially the rules of interaction in the architecture. Unlike most frameworks for multi-agent systems, our framework focuses on the reuse of generic abstractional organizations instead on the individual agent properties such as roles, protocols and internal architecture.

7   Contribution and Practical Results

Our key contribution is to describe a MAS in an abstract and application-independent way, allowing large-scale reuse of the abstractional organizations. We were able to show, throughout the work, the support to architectural principles and the use of contextual compositions, allowing the reinforcement or solution at an architectural level, of some of the fundamental agency properties cited on Section 2 such as *interaction*, *adaptation* and *collaboration*. This makes the implementation of the agent much simpler since such aspects are addressed separately from the object's functional implementation. The following properties were directly or indirectly addressed at an architectural level:

- *interaction*: the rules of interaction established by the communication model forcing the instance of an agent to communicate via a control mechanism of the architecture makes possible the distribution and sharing of services in a transparent and independent way.
- *adaptation*: the abstract factories of the Domain, MAS and Infra tiers allow new agents or new version of agents replacing obsolete ones to be easily "plugged" in our framework, ensuring high flexibility and adaptability since the agents can easily adapt its state and behavior in run-time to new environment conditions.
- *collaboration*: the formalization of services through ontologies and catalogs communicate the semantics of the services provided by the agents and generic components, facilitating the assembly of composition and collaboration between agents via required- and provide-services. Forcing all agents to use a common vocabulary defined in one or more shared ontologies is an oversimplified solution especially when these agents are designed and deployed independently from each other.

Reusing an abstract architecture allows the reuse of not only architectural software design and implementation, but also of some agent properties that can be controlled via architecture mechanisms. Those benefits allow large-scale reuse reducing the time of system development and for system readiness.

We have instantiated a medical application for behavioral therapy using our framework. We were able to verify the facilities provided by the framework and at the same time evaluate certain non-functional requirements such as applicability, usability and performance among others. The system, called *MAS-CF Therapp* [Caminada2004] provides services for a larger application that uses Virtual Reality on the therapy of autistic children and children with a psychosis diagnosis. The system works in a distributed web environment, through the HTTP and TCP/IP protocols using *Java/JSP/Servlet* technology in conjunction with a *Java/Tomcat* server.

For the first time our MAS-CF framework could be evaluated in a real world application. From the viewpoint of practical applicability and use of the described techniques, the following could be evaluated:

- the contextual paradigm tiers of MAS-CF;
- the interaction model used by the framework;
- the viability of using MAS as well as the interaction with Virtual Reality techniques in such a way as to aid and support behavior therapy.

During the development process we could verify the advantages provided by the MAS-CF framework. The implementation of the agents was widely facilitated since the development was concentrated solely on the services provided and the relationships between layers necessary to providing these services. More concrete results will be obtained from future applications to be instantiated.

Acknowledgement

I would like to thank Professor Carlos Lucena for his contribution in this work.

The Brazilian Ministry of Science and Technology provides financial support to this research work through CNPq grants n° 140604/2001-4.

References

[Adam2004]   Adam E. and Mandiau R. "Design of a MAS into a Human Organization: Application to an Information Multi-Agent System." In Proc. 5th Agent-Oriented Information Systems, pages 1-15, Chicago, IL, USA, October 13, 2003. Springer Verlag, LNAI 3030, 2004.

[Allen1997]   R. Allen and G. Garlan. "Formalizing Architectural Connection." In Proc. 16th International Conference on Software Engineering, pages 71-80, Sorrento, Italy, May 1997.

[Amandi1997]   Amandi A.A.   "Programação de Agentes Orientada a Objetos". CPGCC UFRGS – Tese de Doutorado, Porto Alegre, 1997.

[Arnold94]   Arnold A. "Finite Transition Systems". PrenticeHall, Masson, Paris, 1994.

[Azarmi2000]   Azarmi N., Thompson S.   "ZEUS: A Toolkit for Building Multi-Agent Systems". Proceedings of Fifth Annual Embracing Complexity Conference, Paris April 2000.

[Baral997]   Baral, C., Lobo J., Trajcevski G. "Formalizing Workflows as Collections of Condition-Action Rules". Dept of Computer Science, University of Texas at El Paso. El Paso, Texas, USA,1997.

[Bellifemine2001]   Fabio Bellifemine, Agostino Poggi, Giovanni Rimassa: JADE: a FIPA2000 compliant agent development environment. Agents 2001: 216-217.

[Berners-Lee2001]   Berners-Lee, T.; Lassila, O. Hendler, J. – The Semantic Web – Scientific American - http://www.scientificamerican.com/2001/0501issue/0501berners-lee.html

[Bond1988]   Bond A.H. et al. "Readings in Distributed Artificial Intelligence." San Mateo, Morgan and Kaufmann, 1988.

[Bosch1999]   Bosch, J., Molin P., Mattsson M.; Bengtsson P.; Fayad M. "Framework problem and experiences" in M. Fayad, Building Application Frameworks, John Willey and Sons, p. 55–82, 1999.

[Breitman2004]   Breitman K, Haendchen Filho A., Haeusler E. H., Staa, A. V. "Using Ontologies to Formalize Service Specification in Multi-Agent Systems". Proceedings Of Third NASA/ IEEE Workshop on Formal Approaches to Agent Based Systems, Los Alamitos, California USA. To appear as LNCS, Springer-Verlag.

[Brussel1999]   Hendrik Van Brussel , Jo Wyns , Paul Valckenaers , Luc Bongaerts , Patrick Peeters, Reference architecture for holonic manufacturing systems: PROSA, Computers in Industry, v.37 n.3, p.255-274, Nov. 1998.

[Caminada2004]   Caminada, Nuno. "Uma Aplicação Terapêutica de Realidade Virtual Utilizando Tecnologia Baseada em Agentes de Software". Projeto Final de Conclusão do curso de Graduação em Ciência da Computação. UniverCidade – Unidade Ipanema. Junho 2004, Rio de Janeiro, Brasil.

[Connolly2000]   Connolly D. "Extensible Markup Language (XML)." February 2000. Available on-line: http://www.w3.org.XML/.

[Correa1994]   Correa Filho M. "A Arquitetura de Diálogos entre Agentes Cognitivos Distribuídos". COPPE da UFRJ. Tese de Doutorado, 1994.

[Dashofy2001]   Dashofy, E.M.,Hoek, A.v.d., and Taylor, R.N. "A Highly-Extensible, XML-Based Architecture Description Language." In Proceedings of the Working IEEE/IFIP Conference on Software Architecture (WICSA 2001). Amsterdam, The Netherlands, August 28-31, 2001.

[Evans2000]   Evans R. (Editor). "MESSAGE: Methodology for Engineering Systems of Software Agents". Deliverable 1, July 2000.

[Faulkner2001]   Faulkner S. "Towards an Agent Architectural Description Language for Information Systems". Technical Report, University of Louvain, Belgium, 2001.

[Fayad1999]   Fayad M.E. et al. "Building Application Frameworks". John Wiley & Sons, Inc. New York, 1999.

[Fensel2003]   Fensel, D.; Wahlster, W.; Berners-Lee, T.; editors – "Spinning the Semantic Web" – MIT Press, Cambridge Massachusetts, 2003.

[Finin1997]   Finin T. "KQML as an agent communication language." Proceedings of the Third International Conference on Information and Knowledge Management". CIKM-94, ACM Press, november 1994.

[FIPA1997]   Reference FIPA-OS V2.1.0. Nortel Networks Corporation, Ontario, Canada, 2000. FIPA-OS site http://www.emorphia.com/home.htm.

[Gamma1995]   Gamma E. et al. "Design patterns – elements of reusable object-oriented software." Addison-Wesley Longman, Inc., 1995.

[Garcia2003]   Garcia A., Lucena C.J.P. et al. (Eds.) "Software Engineering for Large-Scale Multi-Agent Systems". Lecture Notes in Computer Science – LNCS 2603, Springer Verlag, Germany, 2003.

[Garcia2001]   Garcia A., Torres V. In: "Sistemas Multi-Agentes". Editores: Carlos Lucena e Ruy Milidiú. Editora Papel Virtual, Rio de Janeiro, 2001.

[Garlan1997]   D. Garlan, R. T. Monroe, and D. Wile. "ACME: An architecture description interchange language." In Proc. CASCON'97, pages 169-183, Toronto, Canada, Nov. 1997.

[Gruber1998]   Gruber, T. – "A translation approach to portable ontology specifications." Knowledge Acquisition, 5(2):21--66, 1998.

[Griss2003]   Griss M.L., Kessler R.R. "Achieving the Promise of Reuse with Agent Component." Software Engineering for Large-Scale Multi-Agent Systems. Springer, LNCS 2603, Germany, 2003.

[Haendchen2004]   Haendchen Filho, A.; Staa, A.v.; Lucena, C.J.P. "A Component-Based Model for Building Reliable Multi-Agent Systems". Proceedings of 28th SEW - NASA/IEEE Software Engineering Workshop, Greenbelt, MD. IEEE Computer Society Press, Los Alamitos, CA, 2004, pg 41-50.

[Holvoet2003]   T. Holvoet and E. Steegmans. "Application-Specific Reuse of Agen Roles." Software Engineering for Large-Scale Multi-Agent Systems. Springer, LNCS 2603, Germany, 2003.

[Kendall1999]   Kendall E. et al. "A Framework for Agent Systems". In: Implementing Applications Frameworks – Object-Oriented Framewoks at Work. M.Fayadd et al. John Wiley & Soons, 1999.

[Kotak2003]   Kotak D. et al. "Agent-based holonic design and operations environment for distributed manufacturing". In: Computer in Industry, Volume 52, Issue 2, pg 95-108 – Elsevier Science Publishers B. V. Amsterdam, The Netherlands, 2003.

[Kwangyeol2003]   Kwangyeol R. et al. "Agent-Based Fractal Architecture and Modelling for Developing Distributed Manufacturing Systems." International Journal of Production Research, vol. 42, N0. 17 (2003) 4233-4255.

[Larman1997]   Craig Larman . "Applying UML and Patterns." Prentice Hall PTR, New Jersey, 1997.

[Magee1999]   Magee J., Kramer J. and Giannakopoulou D. "Behaviour Analysis of Software Architectures," presented at the 1st Working IFIP Conference on Software Architecture (WICSA1), San Antonio, TX, USA, 22-24 February 1999.

[Milner1985]   Milner R. "Lectures on a Calculus for Communicating Systems". Lectures Notes in Computer Science, Vol. 197 - Springer Verlag, 1985.

[Montesco2001]   Montesco C.A.E. et al. "UCL – Universal Communication Language". Universidade de São Paulo, Instituto de Ciências Matemáticas e da Computação. Technical Report, São Paulo, Brasil, 2002.

[Noriega1997]   P. Noriega. "Agent-mediated Auctions: THE Fiskmarked Metaphor." PHD Thesis, Universitat Autonoma de Barcelona, Barcelona, 1997.

[OMG2000]   Object Management Group. Agent Platform Special Interest Group. "Agent Technology – Green Paper", version 1.0, September 2000.

[Plasil2002]   Plasil F. et al. "Behavior Protocols for Software Components". IEEE Transactions on Software Engineering, Vol. 28, N. 11, November 2002.

[Pree99]   Pree, W. "Hot-spot-driven development" in M. Fayad, R. Johnson, D. Schmidt. Building Application Frameworks: Object-Oriented Foundations of Framework Design, John Willey & Sons, 1999.

[Sodhi2000]   Sodhi, J. et al. Software Reuse - Domain Analysis and Design Process. New York: McGraw Hill, 1998. 344 p.

[Srinivasan2001]   Srinivasan P. "An Introduction to Microsoft .NET Remoting Framework". Microsoft Corporation, July 2001.

[Sycara1999]   Sycara, K. "In-Context Information Management through Adaptative Collaboration of Intelligent Agents". Intelligent Information Agents. Edited by Matthias Klusch. Springer-Verlag, Berlin, 1999.

[Sycara2003]   Sycara K. et al. "The RETSINA MAS, a Case Study." Software Engineering for Large-Scale Multi-Agent Systems". Lecture Notes in Computer Science – LNCS 2603, Springer Verlag, Germany, 2003, p. 232-250.

[Szyperski2002]   Szyperski C. "Component Software – Beyond Object-Oriented Programming." Addison-Wesley and ACM Press, 2000.

[Todd2003]   Todd N., Szolkowski M. "Java Server Pages". Elsevier Ed., 2003.

[Tracz1994]   W. Tracz. Domain-Specific Software Architecture (DSSA) Frequently Asked Questions (FAQ). ACM Software Engineering Notes, 19(2):52-56, Apr. 1994.

[Vazquez2003]   Vazquez J. "The HARMONIA framework: the role of norms and electronic institutions in multi-agent systems applied to complex domains." Technical University of Catalonia, Barcelona, Spain. ISSN 0921-7126, IOS Press, 2003.

[Vitaglione2002]   Vitaglione G., Quarta F., Cortese E. "Scalability and Performance of JADE Message Transport System". Proceedings of AAMAS Workshop on AgentCities, Bologna, 16th July, 2002.

[**Zambonelli2002**]   Zambonelli, F., Jennings, N.R., Wooldridge M. Organisational Abstractions for the Analysis and Design of Multi-agent Systems. In P. Ciancarini, M.J. Wooldridge, AgentOriented Software Engineering, vol. 1957 LNCS, 235-251. Springer-Verlag: Berlin, Germany 2001.

[**Wooldridge2000**]   Wooldridge M., Jennings N. and Kinny D. "The Gaia Methodology for Agent-Oriented Analysis and Design". Proceedings of 3rd International Conference on Autonomous Agents, Seatle, WA, 1999.

Using Ontologies to Formalize Services Specifications in Multi-agent Systems

Karin Koogan Breitman, Aluízio Haendchen Filho, Edward Hermann Haeusler,
and Arndt von Staa

{karin, aluizio, hermann, arndt}@inf.puc-rio.br

Abstract. One key issue in multi-agent systems (MAS) is their ability to interact and exchange information autonomously across applications. To secure
agent interoperability, designers must rely on a communication protocol that allows software agents to exchange meaningful information. In this paper we propose using ontologies as such communication protocol. Ontologies capture the
semantics of the operations and services provided by agents, allowing interoperability and information exchange in a MAS. Ontologies are a formal, machine processable, representation that allows to capture the semantics of a domain and, to derive meaningful information by way of logical inference. In our
proposal we use a formal knowledge representation language (OWL) that translates into Description Logics (a subset of first order logic), thus eliminating
ambiguities and providing a solid base for machine based inference.The main
contribution of this approach is to make the requirements explicit, centralize the
specification in a single document (the ontology itself), at the same that it provides a formal, unambigous representation that can be processed by automated
inference machines.

1 Introduction

The anchor of our research is the multi agent architectural framework proposed in
[Haendchen03]. So far we have analyzed the architectures of several multi agent platforms, notably MESSAGE [Evans00], ZEUS [Azarmi00], JADE [Vitaglione02] and
proposed a framework whose innovative structural model overcomes most flexibility
shortcomings of other platforms at the same time that promotes large scale architectural reuse. The Agent Framework is described in detail in [Haendchen03, Haend-
chen04].

In the elaboration process of the Agent framework, we have identified the need for
a reference model that centralized the requirements for the services pro-
vided/requested by agents operating within our domain in a meaningful way. The
initial service specification was written in XML. The document was structured to
reflect the MAS architecture hierarchy, i.e., each section corresponded to one of its
architectural layers. Although highly structured, this document did not provide any
further semantics to aid either the understanding, verification or validation of the
specification. Agents could only interact if they shared the exact same specification.
No negotiation was possible, for the semantics of the services can not be fully
expressed in XML.

M.G. Hinchey et al. (Eds.): FAABS 2004, LNAI 3228, pp. 92–110, 2005.
© Springer-Verlag Berlin Heidelberg 2005

We decided to migrate to a more expressive representation. Ontologies were the natural choice, as they are becoming the standard for information interoperability on web [Goméz-Peréz04]. With the adoption of a ontological representation it was possible to formalize terms used in the previous XML service specification, i.e, services, objects, agents and components present in the architecture and the desired ways in which they should interact. In addition to the required syntax, the ontology specification was enriched with semantic content, thus allowing automatic verification, validation with users, and the possibility of negotiating with agents using different service specifications. Different ontologies can be negotiated through the processes of alignment, mapping or merging [McGuiness02, Bouquet03, Breitman03b]. This problem is defined as semantic coordination and can be described as the situation in which all parties have an interest in finding an agreement on how to map their models but given that there is more than one possibility, the right one (or a sufficiently good one) must be chosen [Bouquet03].

An ontology serves as the service specification of an agent operating in the domain, and will be used in making ontological commitments among other software agents [Fensel01]. An ontological commitment is an agreement to use a vocabulary in a way that is consistent with respect to the theory specified by the ontology, i.e., an agreement on what local models are about to achieve user goals [Bouquet03]. We build agents that commit to our ontology. Conversely we design ontologies in order to share knowledge with and among these agents [Gruber93]. The ontology concentrates the desired behaviors and service descriptions in a single document. It serves both as a specification and the reference model to which the agents operating in the domain should comply to.

The rest of the paper is divided as follows: in the next section we briefly introduce the ontology definition and representation language we adopted in the context of our research. In section 3 we describe the context of our MAS. In section 4 we show an example of our approach. In section 5 we briefly describe the lessons learned from this experience and, finally in section 6 we provide our conclusion remarks and future work.

2   Ontology

In order to secure interoperability among autonomous agents, a protocol in which to exchange the necessary information to support this process is required. We argue that ontology, commonly defined in the literature as a *specification of a conceptualization*, is the representation that will provide this requirement [Gruber98]. On one hand ontologies are expressive enough to capture the essential attributes present in MAS, in terms of their classes and relationships. On the other hand, ontologies provide the necessary formality in which to perform automated inference and model checking. According to Tim Berners Lee, ontologies will allow machines to process and integrate Web resources intelligently, enable quick and accurate web search, and facilitate communication between a multitude of heterogeneous web-accessible agents [Berners-Lee01].

We adopt the ontology structure O proposed by Maedche [Maedche02]. According to the author, an ontology can be described by a 5-tuple consisting of the core elements of an ontology, i.e., concepts, relations, hierarchy, a function that relates concepts non-taxonomically and a set of axioms. The elements are defined as follows:

$O := \{ C, \mathcal{R}, \mathcal{HC}, rel, \mathcal{A}^O \}$ consisting of :

- Two disjoint sets, C (concepts) and $\mathcal{R}$ (relations)
- A **concept hierarchy**, $\mathcal{HC}$: $\mathcal{HC}$ is a directed relation $\mathcal{HC} \subseteq C \times C$ which is called concept hierarchy or taxonomy. $\mathcal{HC}(C_1, C_2)$ means C_1 is a subconcept of C_2
- A **function** rel: $\mathcal{R} \rightarrow C \times C$ that relates the concepts non taxonomically
- A set of ontology **axioms** $\mathcal{A}^O$, expressed in appropriate logical language.

Most existing ontology representation languages can be mapped to this structure, e.g. RDF, Oil and DAML, but there seems to be a consensus to adopt OWL as the *de facto* language to represent ontologies. OWL is being developed by the W3 consortium as an evolution of the DAML standard [Hjem01, Hendler00, McGuiness03]. The OWL Web Ontology Language is designed for use by applications that need to process the content of information instead of just presenting information to humans. OWL facilitates greater machine interpretability of Web content than that supported by XML, RDF, and RDF Schema (RDF-S) by providing additional vocabulary along with a formal semantics. The OWL specification comprises three increasingly-expressive sublanguages: OWL Lite, OWL DL, and OWL Full. OWL Lite supports classification hierarchies and simple constraints, e.g., cardinality. It is intended as quick migration path from taxonomies and thesauri, i.e., that are free from axioms or sophisticated concept relationships. OWL DL supports *"expressiveness while retaining computational completeness (all conclusions are guaranteed to be computed) and decidability (all computations will finish in finite time)"* [McGuinees03]. DAML+OIL is equivalent, in terms of expressiveness, to OWL DL. Finally, OWL Full supports maximum expressiveness. According to the W3 consortium, it is unlikely that any reasoning software will be able to support complete reasoning for every feature of OWL Full.

The existence of a large repository of ontologies also influenced our decision to migrate to OWL as the ontology representation language used in our projects. In table I we show the mapping between the nomenclature used by the O ontology model and the one adopted by OWL.

OWL provides the modeling primitives used in frame based systems, i.e., concepts (or classes), the definition of its superclasses and attributes. Relations are also defined, but as independent entities, properties, instead of class attributes. The primitives provide expressive power and are well understood, allowing for automated inference. The formal semantics are provided by Description Logics (DL). DLs also known as terminological logics, form a class of logic based knowledge representation languages, based on the primitives above [Horrocks02]. DLs attempt to find a fragment of first order logic with high expressive power which still has a decidable and efficient inference procedure [Newell82, Heinsohn94]. FaCT is a working example of a system that provides reasoning support (i.e., consistency and subsumption checking) to OWL-encoded ontologies [Horrocks01].

Table 1. Terminology mapping between the O ontology structure and the ontology language OWL

O Ontology Structure		OWL
C	Concept	Class
R	Relation	Property
Hc	concept hierarchy	Subsumption relationship: SubClassOf
rel	function that relates the concepts non taxonomically	Restriction
AO	Axiom	Axiom

An OWL ontology is a sequence of axioms and facts, plus references to other ontologies, which are considered to be included in the ontology. OWL ontologies are web documents, and can be referenced by means of a URI. Ontologies also have a non-logical component that can be used to record authorship, and other non-logical information to be associated with an ontology [OWL, McGuiness03].

In the next section we present the MAS Framework we have been experimenting with and relate the construction process of its service ontology.

3   MAS Framework

Agent-oriented software engineering extends the conventional components' development approach, leading to the construction of more flexible and component-based MASs [Griss03], emphasizing reuse, low-coupling, high-cohesion and support for dynamic compositions. Rapid and problem-specific system construction can be attained through the use of model-driven development and reuse techniques in order to achieve a more flexible, adaptable, robust and self-managing application. These properties can be constituted by the combination of several technologies, such as component-based software engineering [Griss03,24,38], frameworks [Bosh99, Fayad99, Pree99, Roberts98], design patterns [Gamma95, Larman98], rule-based systems [Gelfond93, Paton95, Yu00] and now ontologies [Fensel03, Berners-Lee01, Hendler01]. The MAS Framework architecture comprises five layers: Domain, Multi Agent System (MAS), Agent, Module and Class. Figure 1 depicts the Framework architecture. Note that the Module and Class layers are located inside each agent, the modules are represented in Figure 1 by the circles labeled S30, S40, S50 and S51 (the classes are not represented in the Figure. They are internal parts of the modules). Note that there are two ontologies in the architecture, illustrated by circles S4, and S9. The first one, S4 is the upper ontology and contains the specification of shared domain services, i.e., infrastructure, interface and communication services that will always be instantiated by our Framework. This ontology was built by experts and is part of implementation of the Framework. The second ontology, located at the MAS level, illustrated by circle S9 in Figure 1, represents the agent specific ontology. It contains hot spots where particular application services are to be specified during the Framework instantiation process. As a consequence of the multi layered architecture of the

Framework, application services are specified under the domain level, i.e., as leaves of the upper ontology. For all practical purposes, the agent specific ontology is a composition of the upper ontology (top levels) with the addition of the specification of application specific services at the bottom levels (MAS and agent).

Each MAS centralizes its service specification in a single document (represented by the circle labelled S_9 in Figure1). In our architecture, agents preferentially receive services requirements through a single interface, instead of interacting directly with one another, using multiple interfaces. This communication is done using highly structured messages composed using the terminology formalized by the service ontology. This way, both the syntax required by the interface specification and the semantics associated to the terms used in the service request are now available. Providing clear semantics of of the terms in use, helps maintain clarity and transparency of the specification. It serves as an aid to the ontology validation process and also as a guide to non expert users in the processof including new service specifications at the agent layer.

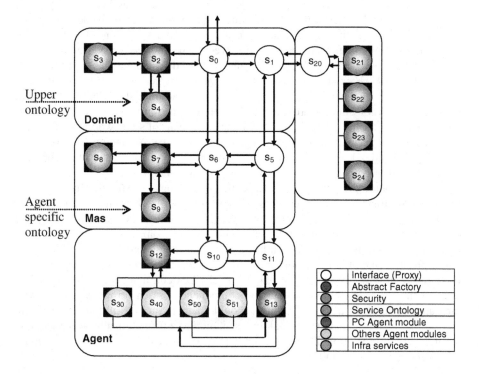

Fig. 1. MAS Framework Architecture

The syntax of the services provided by each agent, and how they can be accessed, is provided by the interface specification. Thus, an essential part of the process is defining a syntactic description of each interface and how the services can be accessed. The aim of the service specification ontology is to identify the services associated with each agent, specifying the main properties of these services. For each service that may be performed by an agent, it is necessary to document its properties. In particular we must identify inputs, outputs, pre-conditions, post-conditions, parameters, states, transitions and rules.

Initially we used an XML document to serve as the service specification. It contained descriptions of the services provided and interface parametrization. Although structurally sound, the XML document was found semantically weak and unfitting to describe some aspects of the service specification, e.g., rules and states. Migrating to an ontological respresentation was a natural move.

We had to question ourselves whether it was possible to express all the necessary information in the MAS service specification using the available ontology languages. As presented in section 2, the current W3C recommendation language for ontology modelling is OWL, the evolution of previous efforts in finding a standard ontology language. OWL comprises three different languages, the choice of which should be based in the level of expressiveness desired for the ontology in question. The first language, Lite OWL, was definetively not expressive enough to capture the necessary information present in the service specification. Our choice was between OWL-DL and Full OWL. The later, although allowing for maximum expressiveness, does not guarantee the possibility of automatic reasoning in computable time [OWL]. In our case, the use of inference to help verify overall specification consistency is very important, so we chose OWL-DL as the preferred language. The last ensures decidability and the existence of an efficient inference mechanism for the language [McGuiness03]. This choice, however, came with an additional modelling overhead. OWL-DL does not directly provide some modelling primitives, e.g., class attibutes and an-ary relationships. Those can be obtained by means of some workarounds . This is common practice in the mark up language community. Assuncíon Gómez-Pérez, Mariano Fernández-López and Oscar Corcho published a table of the most common workarounds (partially reproduced in Table II) [Gómez-Pérez04].

We build ontologies using the lexicon based ontology construction process proposed in [Breitman03]. This process is influenced by our background in requirements engineering and system specification and uses the Extended Lexicon of the Language (LEL) [Breitman03c, Leite93], referred to as Lexicon from here on, as the starting point. We initiate the process by building a Lexicon that captures the vocabulary of our application, i.e., the basic concepts and the relationships that bind them together in an informal way (using natural language). The Lexicon models a series of definitions of the services, objects, agents and components, present in the MAS architecture, and the desired ways in which they should interact. Such definitions evolve from an informal, natural language lexical representation to a formal, machine processable, ontological representation through the application of the lexicon-to-ontology mapping rules defined in [Breitman03].

The Lexicon represents domain information obtained with the help of well known elicitation techniques, e.g. questionnaires, observation, structured meetings. It captures both the denotation and connotation of important domain concepts. Differently from usual dictionaries, that capture the meaning (denotation) of an entry, the Lexicon also captures its connotation, i.e., the behavioral response or impacts that a lexicon entry might have in defining other entries [Leite93].

Table 2. Markup Language Workarounds [Goméz-Peréz04]

	RDF(S)	Daml+Oil	OWL
CONCEPTS			
Instance attributes	+	+	+
Class attributes	w	w	w
Facets			
Type constraints	+	+	+
Cardinality constraints	-	+	+
Procedural Knowledge	-	-	-
CONCEPT TAXONOMIES			
subclass-of	+	+	+
Disjoint decomposition	-	+	+
Exhaustive decomposition	-	w	w
Partition	-	+	w
RELATIONS			
Binary relations	+	+	+
N-ary relations	w	w	w

To build the service specification Lexicon we started with the elicitation of important domain[1] concepts. Those were present in the XML specification, but were not defined to satisfaction. To elicit their meaning, we applied questionnaires and structured interviews with domain experts, i.e., the software engineers involved in the construction of the first specificatio In Figure 2 we show an example of a lexicon entry. We depict the Advisor entry. The Lexicon elicitation and construction process is fully described in [Breitman03].

To generate the formal ontology we applied the process proposed by Breitman & Leite to the newly built Lexicon. This process consists of a set of rules that map Lexicon entries into the five ontological elements proposed by Maedche, described in section 2.

---

[1] Please note that we use the term domain in the broad sense, signifying the application domain as a whole. In this case, our domain is the entire multi agent framework, for which we intend to build a service specification, as opposed to its top layer that is incidentally named domain as well.

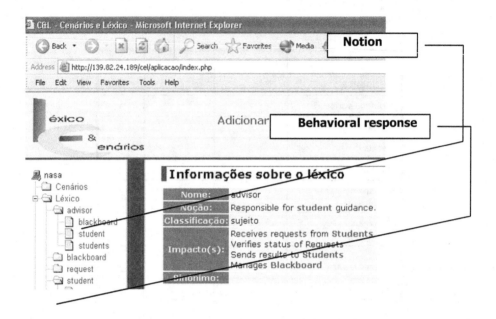

Fig. 2. Screen snapshot of the Lexicon entry Advisor in the C&L tool

Lexicon entries are typed in one of subject, object, verb or situation. Depending on the type a different set of rules is applied to the Lexicon entry and will result in its mapping to either an ontology concept or property. The notion of a Lexicon entry is mapped into the description of its correspondent ontology concept. Its behavioral responses serve as an aid in the identification of ontology properties, concept restrictions and non taxonomical relationships among ontology concepts. Axioms come from the identification of disjoint or generalization relationships held among Lexicon entries. The lexicon based ontology construction process is described in detail in [Breitman03]. This process is supported by C&L, an Open Source tool that automates great part of the lexicon to ontology mapping process. Some design decisions have to be taken by the software engineer and can not be fully automated [Breitman03c]. The tool also provides automated support for the creation and management of Lexicons [Felicíssimo04]. In Figure 3 we show the upper service ontology.

In this section we described the construction of the upper service ontology. Specific services provided by the agents are specificied in the application ontology, located at the MAS level, as shown in Figure 1. As mentioned before, it is a direct consequence of the multi layer architecture of the Framework that specific agent services are specified as leaves, i.e., placed under the lowest levels of the upper ontology. Evidently, those services are particular to each implementation and can not be provided by the upper ontology. Those specifications must be included by a software engineer, as part of the implementation of the MAS itself, and vary case by case. In the next section we exemplify our approach.

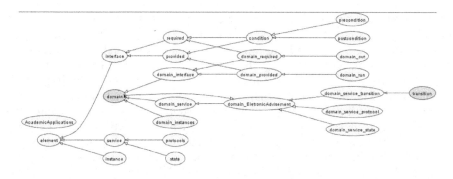

Fig. 3. Upper service ontology

4   Academic Control System: An example

To exemplify our approach we chose an academic control system MAS that tracks the undergraduate student advisement process. We focus on the services provided by the Advisor agent, as illustrated in Figure 4.

In the advising process, a student fills out a registration form with his/her name, student ID, the current semester and the details of the course he/she would like to take. After sending the request, the student receives the final results, either an enabling password or the justification for denying the request. The *Advisor* has the function of taking the student request and to conduct preprocessing, validating the student, verifying the syntactic aspects, checking the viability of the schedule, to direct the request result for the student or providing a request status.

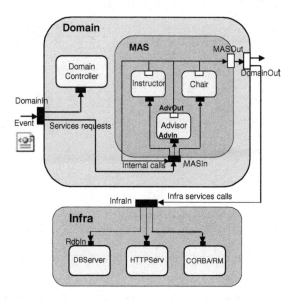

Fig. 4. System generic architecture as proposed in [Haendchen03], instantiated to the academic control MAS example

The agent *Chair* can make a slot available whenever the class is full, and the agent *Instructor* can dismiss pre-requisites for a course. The instructor and chair agents exchange messages with human agents through well-defined and well-structured e-mail messages. The advisor receives the request and verifies syntactic aspects, if the student has the prerequisites to the intended courses and checks to see if there are vacancies in the desired classes. If these conditions are met, the advisor authorizes the request by signing it and gives the student the registration password needed to register for the course. If these conditions are not met, the advisor directs the request according to the arguments of the event to the student, instructor or to the chair. While the process is under way, the student can ask the advisor for information about the progress of the request by e-mail. In any case, the advisor returns the request to the student via e-mail, specifying the result. Based in this information we modeled the Lexicon of the services provided by the system. In the academic control MAS case we used interviews and observation techniques to help elicit lexical information from the domain.

Through a series of refinements, the academic control Lexicon was mapped to its formal ontology. This process was semi automated, for some human input is necessary at specific decision points. The C&L open source tool automates this process and was used to support the construction of the academic system ontology [Silva03].

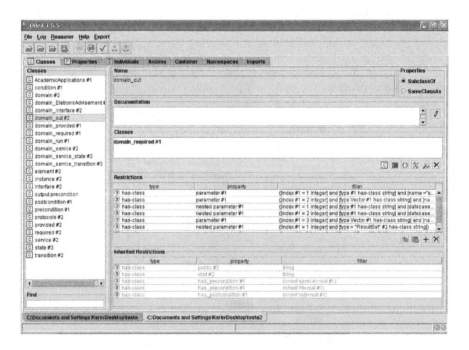

Fig. 5. The ontology concept domain_out implement using the OilEd tool

In Figure 5 we show a screen snapshot of the ontology of services provided by the academic control system. We focus on the the domain_out interface. Please note that, however some restrictions are defined at concept level, there is a great number of other restrictions inherited by its super classes (see the restriction box in the lowest right corner of Figure 5). The super class of class domain_out is indicated by Classes box, namely domain_required[2].) The ontology was implemented using the OilEd, a freeware tool for ontology editon developed at the University of Manchester, that exports to the chosen OWL format [Berchofer01].

We took special care to ensure overall model quality. We have validated the Lexicon with the users and verified using inspections [Kaplan00]. The ontology was verified using the FaCT (fast classification of terminologies) inference engine, publicly available at [FaCT04]. The reasoning services provided by this tool include inconsistency detection, determining subsumption and equivalence (among classes) relationships.

In Figure 6 we illustrate an inconsistency identified with the aid of of the inference mechanism. The ontology has axiom that states that the classes *MAS Security Checking* an *Domain Security Checking* are disjoint, i.e., their intersection is empty. This is is illustrated by the panel in left, that contains the list of axioms for the Academic Control ontology. In the right most panel we depict the ontology, as it was being built. In this process we specified a restriction in which a state would only be reached in the event that both *MAS Security Checking* an *Domain Security Checking* were activated.

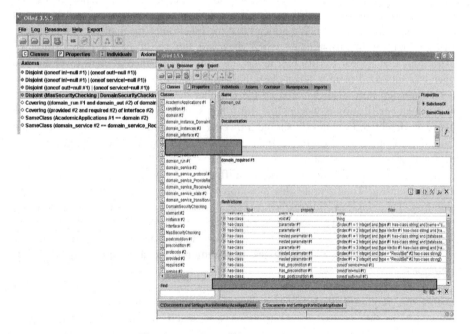

Fig. 6. Inconsistency in class domain_out

---

[2] The # symbol that appears as a suffix of the classes indicates the namespace of the class, i.e., the name of ontology where the specification of the class resides. OWL and similar mark up languages do not require that all concepts in the ontology are specified in the same document. By using the namespace mechanism, it is possible to reuse concepts defined in other ontologies, provided that a valid path to that document is given.

This situation is an impossibility, for the classes are forcibly (as explicited by the axiom in the left pane) disjoint. During the construction of the ontology this fact passed noticed by the designers. The consequences this error may bring to the implementation of the MAS are very serious, for that may cause the agent to halt or to enter a dead loop state. This fault was automatically detected with the use of the reasoner, as illustrated in Figure 7. We depict three panes; In the first one we show the interface to the FaCT reasoner. This tool is built in common Lisp and makes inferences over a description logic representation of the ontology. The ontology editor, OilEd translates the ontology to SHIQ (-----a description logic language dialect) and sends to the reasoner using a CORBA interface. On the second pane, middle one, we depict the log of the reasoning process. We enphasize that the class domain_out is unsatisfiable, but note that the reasoner also checks for errors in subsumption relationships and class instances. The third pane, rightmost, illustrates the graphical display of the inconsistency in the OilEd tool. Similarly to this case, the reasoner helped us detect other inconsistencies in the ontology. We also performed manual verification, using a process very similar to software walkthroughs: we gathered a group of three designers and revisited the material during a planned meeting. The chief designer of the ontology served as group mediator and conducted the meeting. The errors found we mostly sintactical, e.g., classes, properties and restrictions wrongly named or typos. A few inconsistencies such as the one illustrated in Figure 6 were also found. We noticed that the inheritance mechanism makes it very hard to identify inconsistencies when they are the result of a composition of restrictions that appear in different levels, i.e., one was defined at class level and the other was inherited from a super class. It is important to note that all of this type of inconsistencies were also detected by the reasoner in a later moment. We concluded that manual verification is worthwhile, for it helps identify problems that could not be otherwise detected. Practitioner should, however, focus in the terminology, usage and validation of ontological terms. Inconsistencies are more sistematically detected with the aid of an automatic reasoner.

The reasoner was also useful in the identification of a group of classes that partook a similar setting. To illustrate this situation we present the example of class *alert condition*. This class, as illustrated in Figure 7, is defined if two of its restrictions are true, namely **in!= null** and **security_check = 7**) . We defined this class in the ontology of the type SameClassAs, i.e., this is a necessary and sufficient condition to define any other class that possesses those requirements as a similar to class *alert condition*)

Class *domain_out* of the Academic Control ontology is an example of a class that fullfills this requirement. We depict this class and its restrictions in Figure 8 as follows. Note that one of the restrictions was specified among the class natural restrictions, the second came as an inherited restriction from its super class, domain_required. This mechanism is very interesting to help ensure that some conditons are met across the ontology.

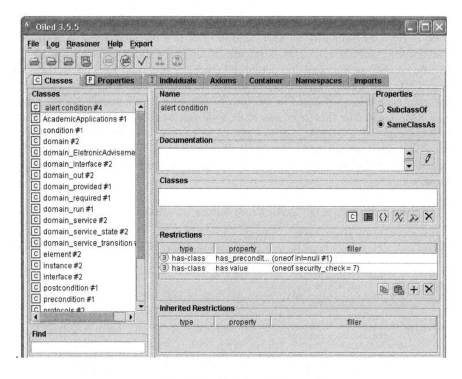

Fig. 7. Class alert condition

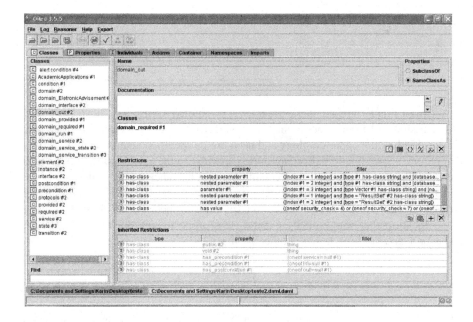

Fig. 8. Class domain_out, local and inherited restrictions

As an illustration we also show the OWL code for the *Domain_out* class in Table 3. Note the similarity to XML, and the fact that the language uses RDF constructors, e.g., subClassOf. This is intentional and is a direct consequence of the *"wedding cake"* architecture for ontology languages proposed by Tim Berners-Lee [Fensel03]. This model reflects the evolution of ontology mark up languages. Each new gain in semantics resulted in the construction of a new language layers, put on top a XML basis. The first layer was RDF, followed by RDF Schema. Because those were not expressive enough, a new wave of languages, including DAML, OIL and now OWL was proposed and put on top of the RDF layer. The result is that an OWL document contains OWL specific markup as well as primitives imported from layers below, e.g., rdfs: label.

Table 3. Example of OWL code for the domain_out class (partially represented)

```
<owl:Class rdf:about="file:/C:/Documents/AcademicAplications.owl#domain_out">
        <rdfs:label>domain_out</rdfs:label>
        <rdfs:comment><![CDATA[]]></rdfs:comment>
        <oiled:creationDate><![CDATA[2004-04-
18T22:20:56Z]]></oiled:creationDate>
        <oiled:creator><![CDATA[Karin]]></oiled:creator>
        <rdfs:subClassOf>
            <owl:Clas
rdf:about="file:/C:/OilED/ontologies/AcademicAplications.owl#domain_required"/>
        </rdfs:subClassOf>
        <rdfs:subClassOf>
            <owl:Restriction>
                <owl:onProperty
rdf:resource="file:/C:/Documents/Karin/AcademicAplications.owl#public"/>
                <owl:hasClass>
                    <owl:Thing/>
                </owl:hasClass>
            </owl:Restriction>
        </rdfs:subClassOf>
        <rdfs:subClassOf>
            <owl:Restriction>
                <owl:onProperty
rdf:resource="file:/C:/Documents/Karin/AcademicAplications.owl#void"/>
```

We must finally remark that a great level of ontology reuse is achieved as a result of our multi level Framework architecture. Generic services provided by every MAS are specified by the upper ontology and need not be specified again. The only services that require a specification effort are those particular to the agent in question. Even so, some of the inputs, pre and post conditions may be inherited from the super class under which the service is to be specified. The reuse of specifications not only reduces overall effort, but also serves to ensure quality because we are making use of a specification that was built by experts (less prone to mistakes), was verified by inspection, and has been tested in other applications[3].

---

[3] It is important to note that the upper ontology is continuously being refined as a result of reports from practitioners.

5   Lessons Learned

The evolution from the XML service specification to an OWL ontology was an over-all positive experience. Our initial concerns related to the power of the ontology representation to convey specification details were lifted as we were able to model every concept in the service specification in the ontology. During this process some work-arounds were needed, specially to formalize attributes such as function parameters and transitions.

The use of the FaCT reasoner helped verify the ontology and improve its overall quality. Automatic verification helped detect: inconsistencies (pre and post conditions, parameters, undesireable situations), errors, and some omission. Additional verification mechanisms will have to be used, as strings (e.g. Regular expressions) are processed as a block by the reasoner.

Our experience in building the service ontologies to support our MAS communication exchange has shown that this task is a very complex one. Despite the existence of methods to support ontology construction, it still remains more of a craft than a science [Fernandez-Lopez97, Gruninger95, Noy01-b, Sure03, Ushold96, Breitman03]. The decisions that have to be taken during this process, e.g., decide whether a concept should be mapped into a class or property, are very difficult and require expertise in concept modeling. By the same rule, the workarounds that have to be used in order to represent relevant specification concepts in the ontological representation are not trivial. It requires the ability to identify such concepts and to engender a workaround that maximizes the power of expression of the ontology.

Finally tool support for visualizing ontologies is still very poor. For ontology edition we have been using OilEd and Protégé [Berchofer01, Noy01]. Both tools fulfill our current editing requirements and have proven very reliable and easy to use. Our main concern today is the need for a tool that allows for a better visualization of the ontology, to help in the validation process[4].

6   Conclusion

In this paper we propose to using ontologies as a means to capture and publish the specifications of the services provided by the agents in a MAS. The ontology makes the requirements explicit, centralizes the specification in a single document, at the same that it provides a formal, unambigous representation that can be processed by automated inference machines [Sowa00]. The main contribution of this approach is to put in practice a standardized reference model when specifying new agents, components and object behavior in a MAS. We showed the feasibility of the approach by means of an example in which we constructed an ontology that specified the services provided within an academic control MAS.

The change from an XML representation to a OWL resulted in real quality gain for the service specification. The current ontological representation is more reliable, for it

---

[4] Both OilEd and Protégé provide visualization plug-ins. Those are static drawings of the ontology, usually too big and cumbersome. Neither plug in provides the necessary functionality required in the validation process.

can be automatically verified. Consistency is thus guaranteed by automatic inference. Furthermore Results from our analysis process (verification and validation) confirm that the OWL specification is more consistent and error free than the previous XML one.

The use of ontologies opens the possibility of interfacing with other MAS environments. As envisioned by James Hendler, the web of the future will be composed of a multitude of websites, network services and databases, each operating with its own local and contextualized ontology [Hendler01]. There is an ongoing effort to support the integration and alignment of different ontologies, in order to support communication and services exchange [Breitman03-d, Bouquet03, McGuiness02]. The ability to align different ontologies will make it possible to probe and request services in truly open ended environments, such as the web [Heflin01].

We are currently experimenting with semantic coordination of MAS ontologies.. We have developed a mechanism to align two different ontologies, CATO, that is publicly available in internet [Felicíssimo04]. We are using this mechanism to help integrate MAS operating in the health care domain. Our current experiment is trying to integrate services provided by a multi agent system used for the diagnoses and treatment of altistic children to similar health care multi agent systems. Our intention is to use the integration process to negotiate among different MAS thus providing new services that were not initially available, e.g., we are currently trying to align our MAS to the Retsina Calendar Agent as to provide appointment services.

The service specification ontology serves us in two ways. Externally of our Framework, the ontology communicates the semantics of the services provided by agents of our domain, thus allowing for exchanges among different MAS and interaction with other agents in Open Ended environments, such as the World Wide Web. Internally to our Framework structural, the ontology serves as a formal specification of the catalog of services provided. Every agent/component operating within our structural model must abide to the specifications dictated by the domain services ontology. The same is true to components and objects.

Future work includes the investigation of a visualization mechanism that would allow for the separation and display of services provided by each layer. The user interface of this mechanism will be inspired in the vision mechanisms of relational databases. At the same time we are considering the development of new plug ins that implement additional verification routines (e.g. lexical and syntactic analysers for strings – parameters, regular expressions), that are not currently covered by the inference mechanisms.

References

[Azarmi00]   Azarmi N., Thompson S. ZEUS: A Toolkit for Building Multi-Agent Systems. Proceedings of *Fifth Annual Embracing Complexity Conference*, Paris April 2000.

[Bechhofer01] Sean Bechhofer, Ian Horrocks, Carole Goble, Robert Stevens. OilEd: a Reasonable Ontology Editor for the Semantic Web. Proceedings of KI2001, Joint German/Austrian conference on Artificial Intelligence, September 19-21, Vienna. Springer-Verlag LNAI Vol. 2174, pp 396--408. 2001.

[Berners-Lee01] Berners-Lee, T.; Lassila, O. Hendler, J. – The Semantic Web – Scientific American - http://www.scientificamerican.com/2001/0501issue/0501berners-lee.html

[Bosh99] Bosch, J., Molin P., Mattsson M.; Bengtsson P.; Fayad M. Framework problem and experiences in M. Fayad, Building Application Frameworks, John Willey and Sons, p. 55–82, 1999.

[Bouquet03] Bouquet, P.; Serafini, L.; Zanobini, S.; - Semantic Coordination A new approach and an application – in Proceedings of the 2nd. International Semantic Web Conference – Florida, October 2003 – pp. 130-143.

[Breitman03] Breitman, K.K.; Leite, J.C.S.P - Lexicon Based Ontology Construction - Lecture Notes in Computer Science 2940- Springer-Verlag, 2003, pp.19-34.

[Breitman03b] Breitman, K.K.; Leite, J.C.S.P. - Semantic Interoperability by Aligning Ontologies - Proceedings of the Requirements Engineering and Open Systems (REOS)- Workshop at RE'03, Monterey, USA, September 2003.

[Breitman03c] Breitman, K.K., Leite, J.C.S.P.: Ontology as a Requirements Engineering Product, Proceedings of the International Conference on Requirements Engineering, IEEE Computer Society Press, 2003. pp. 309-319

[Evans00] Evans, R. "*MESSAGE*: Methodology for Engineering Systems of Software Agents". Deliverable 1, July 2000.[FaCT04] - http://www.cs.man.ac.uk/~horrocks/FaCT/

[Fayad99] Fayad M.E. et al. "Building Application Frameworks". John Wiley & Sons, Inc. New York, 1999.

[Felicíssimo04] Felicíssimo, C.H.; Leite, J.C.S.P., Breitman, K.K., Silva, L.F.S. - C&L: Um Ambiente para Edição e Visualização de Cenários e Léxicos - XVIIII Simpósio Brasileiro de Engenharia de Software (SBES) - Brasília - 18 a 22 de Outubro de 2004 - to appear.

[Fensel01] Fensel, D. – Ontologies: a silver bullet for knowledge management and electronic commerce – Springer, 2001– Fensel, D. – Ontologies: a silver bullet for knowledge management and electronic commerce – Springer, 2001

[Fensel03] Fensel, D.; Wahlster, W.; Berners-Lee, T.; editors – Spinning the Semantic Web – MIT Press, Cambridge Massachusetts, 2003.

[Fernandez-Lopez97] M. Fernandez, A. Gomez-Perez, and N. Juristo. METHONTOLOGY: From Ontological Arts Towards Ontological Engineering. In Proceedings of the AAAI97 Spring Symposium Series on Ontological Engineering, Stanford, USA, pages 33--40, March 1997.

[Gamma95] Gamma E. et al. "Design patterns – elements of reusable object-oriented software." Addison-Wesley Longman, Inc., 1995.

[Gelfond93] Gelfond M. "Representing Action and Change by Logic Programs". The Journal of Logic Programming. Elsevier Science Publishing Co, New York 1993.

[Goméz-Peréz04] Goméz-Pérez, A.; Fernadéz-Peréz, M.; Corcho, O., Ontological Engineering, Springer-Verlag, 2004.

[Gruber93] Gruber, T.R. – A translation approach to portable ontology specifications – Knowledge Acquisition – 5: 199-220

[Gruber98] Gruber, T. - A translation approach to portable ontology specifications. Knowledge Acquisition, 5(2):21--66, 1998.

[Gruninger95] Gruninger, M.; Fox, M. – Methodology for the Design and Evaluation of Ontologies: Proceedings of the Workshop on basic Ontological Issues in Knowledge Sharing, IJCAI-95 Canada, 1995.

[Haendchen03] Haendchen Filho, A.; Staa, A.v.; Lucena, C.J.P. "A Component-Based Model for Building Reliable Multi-Agent Systems". Proceedings of 28th SEW - NASA/IEEE Software Engineering Workshop, Greenbelt, MD. IEEE Computer Society Press, Los Alamitos, CA, 2004, pg 41-50.

[Haendchen04] Haendchen, A.; Caminada, N.; Haeusler, H.; von Staa, A. - Facilitating the Specification Capture and Transformation Process During the Formal Development of Multi-Agent Systems - Proceedings Of Third NASA/ IEEE Workshop on Formal Approaches to Agent Based Systems, Los Alamitos, California USA. To appear as LNCS, Springer-Verlag.

[Heflin01] Heflin, J.; Hendler, J. - A Portrait of the Semantic Web in Action - IEEE Intelligent Systems - March/April - 2001. pp.54-59.

[Heinsohn94] J. Heinsohn, D. Kudenko, B. Nebel, and H.-J. Profitlich. An empirical analysis of terminological representation systems. Artificial Intelligence, 68:367--397, 1994.

[Hendler00] Hendler, J.; McGuiness, D. - The DARPA agent Markup Language . IEEE Intelligent Systems. Vol 16 No 6, 2000. pp.67-73.

[Hendler01] Hendler, J. - Agents and the Semantic Web - IEEE Intelligent Systems - March/April - 2001. pp.30-37

[Hjem01] Hejem, J. - Creating the Semantic Web with RDF - Wiley, 2001

[Horrocks01] Ian Horrocks and U. Sattler. Ontology Reasoning for the Semantic Web. In In B. Nebel, editor, Proc. of the 17th Int. Joint Conf. on Artificial Intelligence (IJCAI'01), Morgan Kaufmann, pages 199--204, 2001.

[Horrocks02] I. Horrocks- Reasoning with expressive description logics: Theory and practice- A. Voronkov, editor, Proceedings of the 18th International Conference on Automated Deduction (CADE 2002)

[Kaplan00] Kaplan, G.; Hadad, G.; Doorn, J.; Leite, J.C.S.P. –Inspección del Lexico Extedido del Lenguaje– In Proceedings of the Workshop de Engenharia de Requisitos – WER'00 – Rio de Janeiro, Brazil – 2000.

[Larman98] Larman C. "Applying UML and Patterns". Prentice Hall PTR. Upper Saddle River, NJ, USA, 1998.

[Maedche02] Maedche, A. – Ontology Learning for the Sematic Web – Kluwer Academic Publishers – 2002.

[McGuiness02] McGuiness, D.; Fikes, R..; Rice, J.; Wilder, S. – An Environment for Merging and Testing Large Ontologies – Proceedings of the Seventh International Conference on Principles of Knowledge Representation and Reasoning (KR-2000), Brekenridge, Colorado, April 12-15, San Francisco: Morgan Kaufmann. 2002. pp.483-493.

[McGuiness03] McGuiness, D.; Harmelen, F. – OWL Web Ontology Overview – W3C Working Draft 31 March 2003

[Newell, 1982] Newell, A. (1982). The Knowledge Level. Artificial Intelligence, 18:87-127.

[Noy01-b] Noy, N.; McGuiness, D. – Ontology Development 101 – A guide to creating your first ontology – KSL Technical Report, Standford University, 2001.

[Noy01-b] Noy, N.; Sintek, M.; Decker, S.; Crubezy, R.; Fergerson, R.; Musen, A. – Creating Semantic Web Contents with Protégé 2000 – IEEE Intelligent Systems Vol. 16 No. 2, 2001. pp. 60-71

[OilEd] http://oiled.man.ac.uk/

[OWL] http://www.w3.org/TR/owl-ref/

[Paton95] Paton N.W. "Supporting Production Rules Using ECA-Rules in an Object-Oriented Context". Department of Computer Science. Technical Report. University of Manchester, UK, 1995.

[Pree99] Pree, W. Hot-spot-driven development in M. Fayad, R. Johnson, D. Schmidt. Building Application Frameworks: Object-Oriented Foundations of Framework Design, John Willey and Sons, p. 379–393, 1999.

[Roberts98] Roberts D., Johnson R. Evolving frameworks: A pattern language for developing object-oriented frameworks in Martin R.C., Riehle D., Buschmann F. Addison-Wesley,1998.

[Silva03] Silva, L.F.; Sayão, M., Leite, J.C.S.P.; Breitman, K.K. - Enriquecendo o Código com Cenários - 17 Simpósio de Engenharia de Software (SBES) - Manaus, AM - 2003 - ISBN 85-7401-126-6 - pp.161- 176.

[Sowa00] Sowa, J. F. – Knowledge Representation: Logical, Philosophical and Computational Foundations – Brooks/Cole Books, Pacific Grove, CA – 2000.

[Sure03] Sure, Y.; Studer, R. – A methodology for Ontology based knowledge management in Davies, J., Fensel, D.; Hamellen, F.V., editors – Towards the Semantic Web: Ontology Driven Knowledge management – Wiley and Sons – 2003. pp. 33-46.

[Ushold96] Ushold, M.; Gruninger, M. – Ontologies: Principles, Methods and Applications. Knowledge Engineering Review, Vol. 11 No. 2 – 1996. pp. 93-136

[Vitaglione02] Vitaglione G., Quarta F., Cortese E. Scalability and Performance of JADE Message Transport System. Proceedings of *AAMAS Workshop on AgentCities*, Bologna, 16th July, 2002.

[Yu00] Yu L. et al. "A Conceptual Framework for Agent Oriented and Role Based Workflow Modeling." Technical report, Institute for Media and Communications Management, University of St. Gallen, Switzerland, 2000.

Two Formal Gas Models for
Multi-agent Sweeping and Obstacle Avoidance

Wesley Kerr[1], Diana Spears[1], William Spears[1], and David Thayer[2]

[1] Department of Computer Science
[2] Department of Physics and Astronomy,
University of Wyoming, Laramie, WY 82071
wkerr@cs.uwyo.edu

Abstract. The task addressed here is a dynamic search through a bounded region, while avoiding multiple large obstacles, such as buildings. In the case of limited sensors and communication, maintaining spatial coverage – especially after passing the obstacles – is a challenging problem. Here, we investigate two physics-based approaches to solving this task with multiple simulated mobile robots, one based on artificial forces and the other based on the kinetic theory of gases. The desired behavior is achieved with both methods, and a comparison is made between them. Because both approaches are physics-based, formal assurances about the multi-robot behavior are straightforward, and are included in the paper.

1   The Sweeping and Obstacle Avoidance Task

The task being addressed is that of sweeping a large group of mobile robots through a long bounded region (a swath of land, a corridor in a building, a city sector, or an underground passageway/tunnel), to perform a search, i.e., surveillance. This requires maximum coverage. The robots (also called "agents") are assumed to lack any active communication capability (e.g., for stealth), and to have a limited sensing range for detecting other agents/objects. It is assumed that robots near the corridor boundaries can detect these boundaries, and that all robots can sense the global direction that they are to move. As they move, the robots need to avoid large obstacles (e.g., buildings). This search might be for enemy mines, survivors of a collapsed building or, alternatively, the robots might be patrolling the area. It is assumed that the robots need to keep moving, because there are not enough of them to view the entire length of the region at once. In other words, the robots begin scattered randomly at one end of the corridor and move to the opposite end (considered the goal direction). This is a "sweep." Once the robots get to the far end of the corridor, they reverse their goal direction and sweep again. Finally, if stealth is an issue then we would like the individual robot movements to be unpredictable to adversaries. It is

M.G. Hinchey et al. (Eds.): FAABS 2004, LNAI 3228, pp. 111–130, 2005.

conjectured that the behavior of a gas is most appropriate for solving this task, i.e., each robot is modeled as a gas particle.

2   Prior Approaches

There are many different methods for controlling groups of autonomous agents (swarms). Balch and Arkin [1] present a very popular approach – using behavior-based techniques. Behavior-based control uses a layered architecture based on arbitration between a suite of behaviors, such as avoidance, exploration, and planning. Although this technique has been successful in maintaining agent formations while going around obstacles, unfortunately it requires a lot of active communication and, typically, it requires small groups of heterogeneous agents that have prespecified roles. Fredslund and Matarić [2] present another behavior based technique using local interactions to create formations and avoid obstacles. This approach has already been ported to robots and experimental results show its successes at avoiding obstacles that are roughly the same size as the robots themselves. However, no solution is presented for the challenging case where the obstacle is the size of a city building.

Other research uses ethological models such as ants or bees to control the robots. In one such study [3], agents are modeled as individual ants in the colony. In this study, the robots leave long-term traces in the environment and require directed graphs to be imposed onto the terrain.

The approaches to swarm control that are of interest to us are rooted in physics. Spears and Gordon [4] have provided a technique called *physicomimetics* for controlling large groups of agents (modeled as particles), using virtual physics-based forces to move the agents into a desired formation, e.g., a hexagonal lattice. This technique scales well to large groups of agents and uses only local interactions. Using physicomimetics, agent swarms do a very nice job of staying in formation and avoiding obstacles, without the need for active communication, long-range sensing, or prespecified roles [5]. Nevertheless, a problem still exists when the agents are presented with a very *large* obstacle, e.g., a building in a city. As the agents move around the obstacle, they are unable to detect the agents that have chosen to move around the other side of the obstacle. Because of this, they are never able to regroup and leave an exposed and uncovered area downstream of the obstacle. The problem is that physicomimetics has traditionally been run in a mode that mimics the behavior of a crystalline solid. Yet solids are rigid and do not expand to fill/cover a region. This is the reason for investigating a gas approach to physicomimetics. The approach of Decuyper and Keymeulen [6] shows that a fluid metaphor works for solving arbitrarily complex mazes. The idea behind this research is that particles in a fluid automatically adapt to changes in the environment because of the fluid's dynamics. The research of Decuyper and Keymeulen has proven that the fluid metaphor is effective, but their approach requires a *global* grid in order to compute the fluid flow through the system. Our research, on the other hand, applies this same fluid metaphor, but using only *local* interactions.

3   Motivation for Gas Models

Both liquids and gases are considered fluids, but this paper focuses on gases. Gases offer excellent coverage, unpredictability of particle locations, and they can be bounded. In general, fluids (gases and liquids) are able to take the shape of their container and therefore are well suited to avoiding obstacles. Fluids are also capable of squeezing through narrow passages and then resuming full coverage when the passage expands. With gases, if we model a container, the gas will eventually diffuse throughout the container until it reaches an asymptotic state. Because gases have this property but liquids do not, gases are a more natural way to think of how to get particles around an obstacle, and why we chose to model a gas. Once the particles have moved around an obstacle, fluids have the ability to regroup. For example, consider releasing a gas from a container at the top of a room with obstacles. The gas inside the container is slightly heavier than the surrounding air. As the gas slowly falls to the ground, it separates around obstacles and expands back to cover areas under the obstacles.

Agents capable of mimicking fluid flow will be successful at avoiding obstacles and moving around them quickly. By mimicking gas flow in particular, the agents will be able to distribute themselves throughout the volume once they have navigated around the obstacle.

This article presents two formal gas models to solve the problem described above, and then compares them. The first approach is physicomimetics, also called *artificial physics (AP)*. The second is *kinetic theory (KT)*, which models virtual inter-particle and particle-wall collisions. Both of these approaches are amenable to straightforward physics analyses for providing behavioral assurances of the robot collective [7], [8].

4   The Physicomimetics Approach

Spears and Gordon [4] have created the artificial physics (AP) framework to control groups of autonomous agents. The goal of AP is one of reducing the potential energy of a system. Each agent in the system experiences a repulsive force from other agents that are too close, and an attractive force from other agents that are too far away. These forces, which are based on Newtonian physics, do not really exist in a physical sense, but the agents react to them as if they were real. Each agent can be described by a position vector x and a velocity vector v. Time is maintained with the scalar variable t. The simulation can be run in either 2D or 3D (to model swarms of micro-air vehicles). Agents in the system update their position, x, in discrete time steps, Δt. At each time step, each agent updates its velocity, v, based on the vector sum (resultant) of all forces exerted on it by the environment, which includes other agents within visibility range, as well as repulsive forces from obstacles and attractive forces from goals. This velocity, v, determines Δx, i.e., the next move of the agent. In particular, at each time step, the position of each particle undergoes a perturbation Δx. This perturbation depends on the current velocity, i.e., $\Delta x = v\Delta t$. The velocity of each particle at each time step also changes by Δv. The change in velocity is controlled by the force on the particle, i.e., $\Delta v = F\Delta t/m$, where m is the mass of that particle and F is the force on that particle. Note that this is the standard, Newtonian $F = ma$ equation.

By setting system parameters in AP, we can mimic solid, liquid, or gas states, as well as phase transitions between these states [7]. Traditionally, AP models a solid. To model a *gas* with AP, all agents experience purely repulsive forces from other agents as well as from obstacles and the side boundaries of the corridor.[1] Although AP was not designed to be an exact model of a gas, we have found that its behavior does a good job of mimicking a gas.

5   The Kinetic Theory Approach

There are two main methods for modeling fluids: the Eulerian approach, which models the fluid from the perspective of a finite volume fixed in space through which the fluid flows (typically the method of computational fluid dynamics), and the Lagrangian approach, in which the frame of reference moves with the fluid volume (typically the kinetic theory approach) [9]. Because we are constructing a model from the perspective of the agents, we choose the latter. Kinetic theory (KT) is typically applied to plasmas or gases, and here we model a gas. This overview of KT borrows heavily from Garcia [10].

When modeling a gas, the number of particles is problematic, i.e., in a gas at standard temperature and pressure there are 2.687×10^{19} particles in a cubic centimeter. A typical solution is to employ a stochastic model that calculates and updates the probabilities of where the particles are and what their velocities are. This is the basis of KT. One advantage of this model is that it enables us to make stochastic predictions, such as the average behavior of the ensemble. The second advantage is that with real robots, we can implement this with probabilistic robot actions, thereby avoiding predictability of the individual agent.

In KT, particles are treated as possessing no potential energy (i.e., an ideal gas), and collisions with other particles are modeled as purely elastic collisions that maintain conservation of momentum. Using some of the formulas for kinetic theory, we can obtain useful properties of the system. If we allow k to be Boltzmann's constant, such that $k = 1.38 \times 10^{-23}$ J/K, m to be the mass of the particle, and T to be the temperature of the system, then we can define the average speed of any given particle (in 3D) as,

$$\langle v \rangle = \int_0^\infty v f(v) dv = \frac{2\sqrt{2}}{\sqrt{\pi}} \sqrt{\frac{kT}{m}}$$

where $f(v)$ is the probability density function for speed.

Another property we can define for KT is the average kinetic energy of the particles:

---

[1] A frictional force is also included in the AP solid model, but is excluded in gas AP.

$$\langle K \rangle = \langle \frac{1}{2}mv^2 \rangle = \frac{3}{2}kT$$

Using KT, we are able to model different types of fluid flow. For our simulations, we modeled 2D Couette flow. The original code for this one-sided Couette flow is a translation of code from Garcia [10] to the Java programming language. Figure 1 shows a schematic for this one-sided Couette flow, where we have a fluid moving between two walls – one wall moving with velocity v_{wall}, and the other stationary. Because the fluid is a Newtonian fluid and has viscosity, we see a linear velocity profile across the system. Fluid deformation occurs because of the sheer stress τ, and wall velocity is transferred because of molecular friction on the particles that strike the wall. On the other hand, the particles that strike the non-moving wall will transfer some of their velocity to it. This does not cause the wall to move, since in a Couette flow the walls are assumed to have infinite length and therefore infinite mass. We chose a Couette flow so that we can introduce energy into the system and give the particles a direction to move. This effect is similar to AP modeling a goal force.

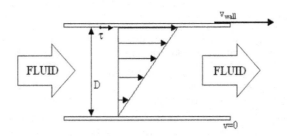

Fig. 1. Schematic for a Couette flow

The main differences between AP and KT are: (1) AP deals with forces. KT deals only with the resulting velocity vectors. (2) With the current force law used by AP, interactions are "soft collisions," i.e., repulsive forces cause small deviations in agent velocities. In KT, collisions cause radical, probabilistic changes in agent velocities. (3) For a given set of starting locations, AP is deterministic, whereas KT is stochastic.

6   Implementation

We created a 2D simulation world with a pair of corridor walls (which can be considered Couette walls), obstacles, and agents (modeled as gas particles). The fluid flow is unsteady with no turbulence, i.e., unsteady laminar flow.

First, we describe our AP gas approach, in which motion is due to attractive and repulsive forces. Recall that AP uses virtual Newtonian force laws. The force law used is:

$$F = |\boldsymbol{F}| = \frac{Gm_1m_2}{r^2} \qquad (1)$$

where G is a gravitational constant[2], m_1 and m_2 are the masses, and r is the distance between the agent and another object/agent. For a robotic implementation, there is a maximum possible force, F_{max}, i.e., $F \leq F_{max}$ always. The value of F_{max} used in our simulations is 1.5. The parameter G is set at initialization of the program. To maintain a desired distance, R, between agents in an AP solid, this force is repulsive if $r < R$ and attractive if $r > R$. For an AP gas, the force is always repulsive. Each agent has one sensor to detect the range and bearing to nearby agents, and one effector to move with velocity $\boldsymbol{v}$. To make the simulation a realistic model of robots, agents can only detect other agents/objects within a limited range, namely, $1.5R$. Our implementation assumes $R = 50$.

The corridor and obstacle wall forces are purely repulsive. For AP, the large-scale fluid motion is driven by an attractive goal force at one end of the corridor. Different force constants, G, are allowable for inter-agent forces and agent-wall forces. However, this paper assumes the same G, namely, 1,200. Note that if the forces for avoidance of an obstacle are equal to the attractive forces felt by the goal, the particles reach a stagnation point at the intersection with the obstacle – because all of the forces felt by the particle are in balance. To overcome this situation, when a particle experiences a repulsive force from an obstacle or wall that is the same in magnitude but in the opposite direction of the goal force, the particle translates this into a *tangential* repulsive force. When choosing an angle for the tangential force we must be careful to keep the particle from reaching a stagnation point and keep the particle from moving through the obstacle. Rotating the angle by 45° produces this result nicely. In particular, if the angle of the force is 180° then the angle for this force becomes 135° or 225°, depending on the direction chosen by the robot.

In parallel with the AP approach, we have also implemented the KT approach. Our KT approach models a modified (two-sided) Couette flow in which both Couette walls are moving in the same direction with the same speed. We invented this variant as a means of propelling all agents in a desired general direction, i.e., the large-scale fluid motion becomes that of the walls. Particle velocities start randomly and remain constant, unless collisions occur. (Note that with actual robots, collisions would be virtual, i.e., they would be considered to occur when the agents get too close. Wall motion would also be virtual.) The system updates the world in discrete time steps, Δt. We choose these time steps to occur on the order of the mean collision time for any given agent. Each agent can be described by a position vector $\boldsymbol{x}$ and a velocity vector $\boldsymbol{v}$. At each time step, the position of every agent is reset based on how far it could move in the given time step and its current velocity:

$$\boldsymbol{x} \leftarrow \boldsymbol{x} + \boldsymbol{v}\Delta t \ .$$

This is done for every agent in the system, and positions are updated regardless of walls and obstacles as well as other agents. Once the current agent's

---

[2] G is not related to actual gravity (which is purely attractive), but is a force constant used in the system.

position has been updated, a check is performed to see if that agent has moved through a wall (including an obstacle wall), in which case the position needs to be reset as if a collision occurred. If the agent strikes a moving wall, then some of the energy from the wall is transferred to the agent. This effect models the molecular friction of the fluid and speeds up the agent. The agent's position is reset as a biased Maxwellian distribution, based on where the agent strikes the wall and how far the agent would have been able to move if the wall were not there. On actual robots, wall collision detection will be done prior to moving. If the robot will intersect with the wall on its next move, then it determines its new position based on a collision, rather than actually colliding with the wall. Once all agents have moved and their positions have been reset based on collisions with the walls, inter-agent collisions are processed. The number of collisions in any given region is a stochastic function of the number of agents in that region (see [10] for details). This process continues indefinitely or until a desired state has been reached.

We have just described the KT approach to modeling Couette flow, modi-fied with a two-sided Couette. We next introduce obstacles into the world, and consider different methods for modeling interactions with obstacle walls.

For one, we could use a KT approach that treats the obstacle boundaries as stationary walls, and processes collisions the same as is done with Couette walls. Unfortunately, in the pure KT approach, agents do not perceive the location of an obstacle until they have collided with it. When colliding with an obstacle, the velocity of the particle off the obstacle is distributed Maxwellian in the goal direction and Gaussian in the lateral direction (i.e., orthogonal to the longitudi-nal goal direction). This produces excellent results when steady state is reached. A problem arises, however, since we are not modeling a steady state fluid flow. If we were given a steady flow, agents in the system would collide with other agents coming down the flow and through collisions would be pushed around the obstacle. Since flow is unsteady, one of the last agents in the system (i.e., upstream from all the other agents) could strike an obstacle and end up going in the opposite direction with no mechanism to turn it around.

The traditional AP (solid) approach to obstacle avoidance does extremely well at navigating around obstacles. Unfortunately, the AP solid approach does not maintain a good coverage of the environment once the particles have navigated around the obstacle. Figure 2 shows this in simulation. However, the AP *gas* approach (with repulsion only) is able to navigate around obstacles *and* retain good coverage, see Fig. 2. A question remains, nonetheless, as to whether we could do even better by combining AP and KT.

To address this question, we created a hybrid AP/KT algorithm, in which wall collisions generate large-scale motion, AP repulsive forces enable obstacle avoidance, and KT is responsible for agent-agent interactions. By treating the obstacle as a repulsive force, the agents *softly* bounce off the obstacle walls. This force causes the agent to turn, thereby allowing more particles to make it around the obstacle. Since the particles turn softly, they are more likely to hit one of the moving walls and continue in the direction of the flow until they have made it

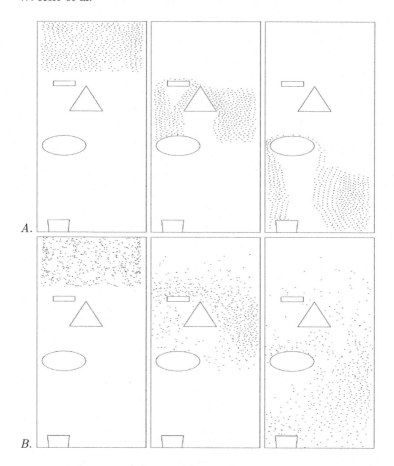

Fig. 2. AP controllers perform a sweep. *A.* AP solid *B.* AP gas

around the obstacle. We are able to achieve an even distribution of particles past the obstacle with this hybrid, as well as increase the number of particles that make it past in a shorter amount of time. Figure 3 shows the hybrid approach. Note that numerous alternative hybrids of AP and KT are possible; investigation of these others will be a topic for future research.

7   Experimental Results

To discover the strengths and weaknesses of each of our four methods (AP solid, AP gas, KT gas, and the AP/KT gas hybrid), we ran numerous empirical experiments with the simulator. Typical results are shown in Figures 2 and 3. In these figure, particles begin at the top and move to the bottom (which is the goal direction). The y-axis is vertical and the x-axis is horizontal. Our starting point was the AP solid approach to obstacle avoidance. Agent formations stayed

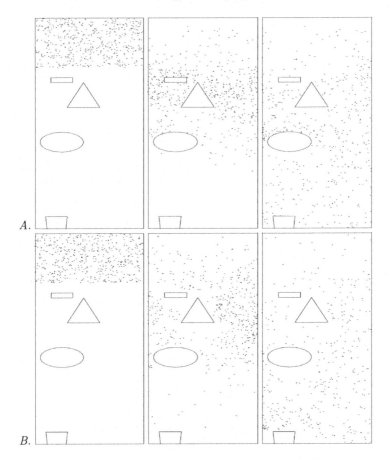

Fig. 3. KT controllers perform a sweep. *A.* KT *B.* AP/KT hybrid

intact with this approach, but coverage was very poor. AP gas yielded results far better than AP solid for coverage behind the obstacles (Fig. 2).

Like AP gas, pure KT has yielded excellent coverage. However, problems arose with KT because of the unsteady fluid flow, as discussed above. Furthermore, because of the unsteady nature of the flow, it typically took longer for the entire group of KT particles to get around all of the obstacles (if they were able to do so) than for AP particles to get around the obstacles.

Recall that the hybrid AP/KT approach avoids stagnation points. Other difficulties arise for the AP/KT method. One difficulty arises when two obstacles are very close together, i.e., sufficiently close that the forces exerted from them are able to dominate the goal forces and inter-particle forces. This leaves us with unexplored areas inside our corridor of obstacles (Fig. 4). All methods using force laws had problems dealing with this situation.

We have also encountered another potential problem for the KT approaches. The problem does not appear to be due to agent-agent interactions. Rather,

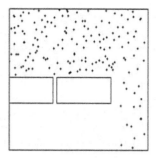

Fig. 4. Obstacle field that has a narrow corridor within. The force-based methods will be unable to explore this area

the problem arises when trying to address *both* the large-scale movement *and* avoidance of multiple obstacles. We notice this when the obstacle density is increased between the walls. Because the KT methods use collisions with Couette walls for propulsion in a goal direction, the width of the region between these walls determines the coverage of the world. In particular, if the walls contain a group of obstacles several layers abreast, we cannot guarantee that the central region of the Couette, far from the walls, will be covered by the agents. The pure AP models do not have this problem.

In summary, AP solid has very poor coverage, whereas all of the gas models produce excellent coverage, which reaffirms our motivation for choosing gas models. AP and AP/KT hybrid are better than KT for navigating around obstacles, although they have greater difficulty navigating through narrow corridors.

8   Theoretical Predictions

One of the key benefits of using a physics-based multi-agent system is that extensive theoretical (formal) analysis tools already exist for making predictions and guarantees about the behavior of the system. Furthermore, such analyses have the added benefit that their results can be used for setting system parameters for achieving desired multi-agent behavior. The advantages of this are enormous – one can transition directly from theory to a successful robot demo, without all the usual parameter tweaking. For an example of such a success (using AP solid), see [5]. To demonstrate the feasibility of applying physics-based analysis techniques to physics-based systems, we make predictions that support some of our claims regarding the suitability of gas models for our surveillance task.

Before describing the experiments, let us first present the metric used for measuring error between the theoretical predictions and the simulation results. *Relative error* is used, which is defined as:

$$\frac{|\,theoretical - actual\,|}{theoretical}$$

For each experiment, one parameter was perturbed (eight different values of the affected parameter were chosen). For each parameter value, 20 different runs through the simulator were executed, each with different random initial agent positions and velocities. The average relative error (over the 20 runs) and the standard deviation from the average were determined from this sample.

Next, consider the experiments. Recall that our objectives are to sweep a corridor and to avoid obstacles along the way. A third objective for the *swarm* of agents is that of coverage. We define two types of coverage: *longitudinal* (in the goal direction) and *lateral* (orthogonal to the goal direction). Longitudinal coverage can be achieved by movement of the swarm in the goal direction; lateral coverage can be achieved by a uniform spatial distribution of the robots between the side walls. The objective of the surveillance task is to maximize both longitudinal and lateral coverage in the minimum possible time. The number of particles, initial distribution of particles, and termination criterion are determined individually for each experiment, based on earlier studies.

To measure how well the robots achieve the task objective, we observe:

1. The distribution of velocities of all agents in the corridor. This is a measure of both sweep time and total coverage (i.e., a wide distribution typically implies greater coverage of the corridor length and width).
2. The degree to which the spatial distribution of the robots matches a uniform distribution. This is a measure of lateral coverage of the corridor
3. The average agent speed (averaged over all agents in the corridor). This is a measure of total coverage.

Measurement of each of these three aspects of the system (velocity distribution, spatial distribution, average speed) corresponds to each of our three experiments. Recall (above) that for each experiment, we vary the value of one parameter. The reason for varying such parameter values is to allow a system designer to optimize the design – by understanding the tradeoffs involved. In other words, we have observed that there is a tradeoff between the degrees of longitudinal coverage, lateral coverage, and sweep speed – greater satisfaction of one can lead to reduced satisfaction of the others, making this a Pareto-optimization task. By varying parameter values and showing the resulting velocity and spatial distributions and average speed, a system designer can choose the parameter values that yield desired system performance. Finally, why show *both* theory and simulation results for each experiment and each parameter value? Our rationale is that it is far easier for a system designer to work with the theory when deciding what parameter values to choose for the system. The designer can do this *if* the theory is predictive of the system. In our experimental results below, we show that the theory is indeed predictive of experimental results using our simulation.

For the sake of simplicity, in these experiments we use a subtask of our complete surveillance task. None of the experiments involve obstacles. For the first experiment, the agents are placed uniformly along the beginning of a long corridor and allowed to perform one sweep. In the second experiment, the agents are placed in a square container in an initially tight Gaussian distribution and allowed to diffuse to an asymptotic state. For the final experiment, the agents are placed at the beginning of a long corridor once again, and allowed to run for a predetermined number of time steps, after which the average speed is measured. In the second and third experiments, there is no goal force or wall movement, and therefore there is no directed bulk movement (transport) of the swarm.

8.1   Experiment 1: Velocity Distribution

The first theoretical prediction for our system is devoted to longitudinal coverage and sweep speed via movement. The theory predicts the velocity distribution for each of the approaches, AP and KT. It is assumed that fluid flow is in the y-direction (downward toward the goal), as in Fig. 2.

Recall that the AP approach is an implementation of $F = ma$. Assuming $F_y = g$, where g is the magnitude of the goal force, which is constant for all particles and is strictly in the goal direction, and assuming $m = 1$ (which is assumed throughout this paper), we have the following derivation (where v_y is the magnitude of the velocity in the y-direction, and v_x is assumed to be 0):

$$g = \frac{dv_y}{dt}$$

$$g \cdot dt = dv_y$$

$$g \int dt = \int dv_y$$

$$g \cdot t = v_y$$

$$v_y = gt$$

This shows that the velocity in the direction of the goal is just the force of the goal times the amount of time that has elapsed. We set up an experiment using this theoretical formula to determine the relative error for our experiments. The experiment placed 500 agents in the simulator and terminated in 100 time steps, since by this time the agents reach the maximum velocity that can be achieved on real robots. The parameter being varied is the goal force. The results are plotted in Fig. 5, and the relative error is roughly 1%.

For KT, a traditional one-sided Couette drives the bulk swarm movement. The complete derivation for the velocity profile of a Couette flow can be found in [9] (pages 417–420), but here we present a more concise version.

For steady, 2D flow with no external forces, there is a classical "Governing Equation" that predicts the y-direction momentum of the fluid. This Governing Equation is:

$$\rho v_y \frac{\partial v_y}{\partial y} + \rho v_x \frac{\partial v_y}{\partial x} = -\frac{\partial P}{\partial y} + \frac{\partial \tau_{yy}}{\partial y} + \frac{\partial \tau_{xy}}{\partial x}$$

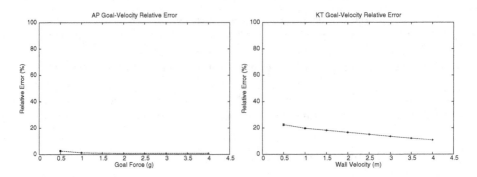

Fig. 5. Relative Error for Goal-Velocity (Prediction 1)

where ρ is the fluid density, v_x and v_y are the x- and y-components of velocity, P is the fluid pressure, and τ_{yy} and τ_{xy} are the normal and shear stresses, respectively. We can use this equation for momentum to derive the velocity. However, first we need to specialize the equation for our particular situation. For Couette flow, the equation becomes:

$$0 = \frac{\partial}{\partial x}\left(\mu \frac{\partial v_y}{\partial x}\right)$$

where μ is the fluid viscosity. Assuming an incompressible, constant temperature flow with constant viscosity, this becomes:

$$\frac{\partial^2 v_y}{\partial x^2} = 0 \tag{2}$$

Equation 2 is the Governing Equation for steady, 2D, incompressible, constant temperature Couette flow. Integrating twice with respect to x to find v_y, we get:

$$v_y = c_1 x + c_2 \tag{3}$$

We can solve for c_1 and c_2 from the boundary conditions. In particular, at the stationary Couette wall ($x = 0$), $v_y = 0$, which implies that $c_2 = 0$ from Equation 3. At the moving wall ($x = D$), $v_y = v_{wall}$, where D is the Couette width and v_{wall} is the velocity of the moving wall, which is in the y-direction (toward the goal). Then $c_1 = v_{wall}/D$ from Equation 3.

Substituting these values for c_1 and c_2 back into Equation 3, we get:

$$\frac{v_y}{v_{wall}} = \frac{x}{D}$$

This is a linear profile.

We set up an experiment to measure the relative error generated by our simulation, with each particle behaving as if it were part of a one-sided Couette flow. Each experiment contained 3,000 particles, and ran for 50,000 time steps. When determining the error, we divided the world into seven discrete cells. For each cell, we determined the average velocity of the particles located in that cell.

The relative error was averaged across all cells and plotted in Fig. 5 for eight different wall speeds. One can see that the error is below 20%, with a reduction in error for KT as the wall speed is increased. Note that the original algorithms from Garcia [10] also have error between theory and simulation that is slightly below 20%. Reasons for this discrepancy between theory and simulation are elaborated in the discussion section below. When determining the longitudinal coverage via swarm movement, we are able to predict very accurately for both algorithms in the simple scenario, except at slow wall speeds for KT.

8.2   Experiment 2: Spatial Distribution

For the second experiment, we predict the lateral coverage via the spatial distribution. For this experiment, there is neither a goal direction nor obstacles. The agents' task is to diffuse throughout the system. The theory for each approach in gas formation predicts a uniform distribution throughout the system. For the experimental setup, we measured the distance from the uniform distribution once the gas reached an asymptotic state. Therefore, we divided our system into discrete cells and counted the number of particles in each cell. Theory predicts that the number of particles in each cell should be n/c, where n is the total number of particles and c is the total number of grid cells that cover our system.

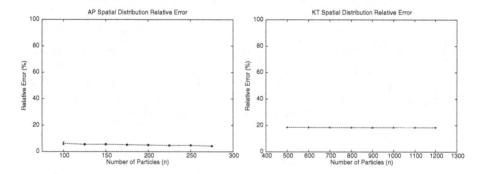

Fig. 6. Relative Error for Uniform Distribution (Prediction 2)

Our experimental system serves as a simple container to hold a gas. The gas should diffuse within the container until it reaches an asymptotic state and contains equal numbers of particles in each cell. We allowed the system several thousand time steps, starting from a tight Gaussian distribution about the center of the container, to reach this state and then measured the number of particles in each cell. This measurement was averaged over many time steps, since particles were still moving through the system and diffusion did not imply particles ceased to move. Both experiments were the same for AP and KT, and the results can be found in Fig. 6. In both cases, the parameter being varied is the number of particles. Once again, we are able to predict the spatial distribution with relative error less than 20%.

There is a noticeable downward trend for the relative error in the AP system as more particles are added to the system. Recall that in AP we use forces to affect other particles as well as forces from the walls to keep the particles inside the simulation. This requires that particles have a desired radius such that when another particle enters this radius, it is repelled away. As more particles are added to the simulation, the space is filled with particles that are constantly pushing each other away and moving into the only formation that will allow them all to fit, which is a uniform distribution.

8.3   Experiment 3: Average Speed

For the third experiment, we predict the average speed of the particles in the system. The average speed of the particles serves as a measure of how well the system will be able to achieve complete coverage, because higher speed implies greater coverage. The derivation for AP's prediction of average speed begins with a theoretical formula for AP system potential energy (PE) from [11]. This theory assumes that the particles start in a cluster of radius 0. There are two different situations, depending on the radial extent to which F_{max} dominates the force law $\boldsymbol{F} = m\boldsymbol{a}$. Recall that agents use F_{max} when $F > F_{max}$. This occurs when $\frac{G}{r^2} > F_{max}$ or, equivalently, $r \leq \sqrt{\frac{G}{F_{max}}} \equiv R'$. The first situation is when F_{max} is used only at close distances, i.e., when $0 \leq R' \leq 1.5R$. The second situation occurs when $R' > 1.5R$. Here we assume the first situation, i.e., a low value of G is used such that $G \leq F_{max}(1.5R)^2$, and F_{max} is only used at close distances. Because we are using AP *gas*, there is no friction and all forces are repulsive. We begin with a two-particle system. In this case, the formula is the sum of two integrals. The first represents the force felt by one particle as it approaches another, from a distance of $1.5R$ to R'. The second is the force F_{max} that is experienced when $0 \leq r \leq R'$. Then, using R' as defined above, with r the inter-agent distance, we have (V is the standard symbol for PE):

$$PE = V = \int_0^{R'} F_{max}dr + \int_{R'}^{1.5R} \frac{G}{r^2} dr$$

$$= F_{max}R' + G\int_{R'}^{1.5R} r^{-2}dr$$

$$= F_{max}R' + G(-r^{-1})|_{R'}^{1.5R}$$

$$= F_{max}R' + G\left(-\frac{1}{1.5R} + \frac{1}{R'}\right)$$

$$= \sqrt{GF_{max}} + G\left(\sqrt{\frac{F_{max}}{G}} - \frac{1}{1.5R}\right) \quad because \ R' = \sqrt{\frac{G}{F_{max}}}$$

$$= \sqrt{GF_{max}} + \sqrt{GF_{max}} - \left(\frac{G}{1.5R}\right)$$

$$= 2\sqrt{GF_{max}} - \left(\frac{G}{1.5R}\right)$$

Now we generalize V to N particles. V_N is our abbreviation for total potential energy, and

$$V_N = \sum_{i=0}^{N-1} iV = \frac{VN(N-1)}{2}$$

Note that all the potential energy transforms into kinetic energy (since there is no friction energy dissipation), i.e., $V_N \rightarrow KE$. Also, the total kinetic energy, KE, is equal to $\frac{1}{2}\sum_{i=1}^{N}(v(i))^2$, assuming $m=1$ and $v(i)$ is the speed of particle i. This formula for KE is equal to $\frac{N}{2}\langle v^2 \rangle$, where $\langle v^2 \rangle$ is the average of the particle speeds squared.

Setting $V_N = KE$, we get:

$$\frac{VN(N-1)}{2} = \frac{N}{2}\langle v^2 \rangle$$

$$V(N-1) = \langle v^2 \rangle$$

Substituting for V we get

$$\langle v^2 \rangle = V(N-1) = (N-1)\left[2\sqrt{GF_{max}} - \left(\frac{G}{1.5R}\right)\right]$$

From [12], we know that the relationship between $\langle v \rangle$ and $\langle v^2 \rangle$ is the following:

$$\langle v \rangle = \sqrt{\langle v^2 \rangle - \sigma^2}$$

where σ^2 is the variance of the velocity distribution. However, because the variance of the velocity distribution is not typically available when making a theoretical prediction, one approximation (which is an upper bound on the true theoretical formula because it assumes 0 variance) that we can use is:

$$\langle v \rangle \approx \sqrt{\langle v^2 \rangle} = \sqrt{(N-1)\left[2\sqrt{GF_{max}} - \left(\frac{G}{1.5R}\right)\right]}$$

Using this equation for AP, we ran through the experiments (starting with the particles in a tight cluster to match the theory), allowed the gas to reach an asymptotic state, and measured the relative error. For each experiment, there were 100 agents in the system. The total number of time steps required to reach this asymptotic state is different for each value of G since it requires that the agents are no longer interacting with each other. This terminating state can be found when all the agents have ceased to change their velocity. The parameter being varied is the gravitational force, G. As seen in Fig. 7, the error is less than 6%. Furthermore, if the system designer has *any* clue as to what variance to expect in speeds, the theoretical prediction will be greatly improved.

In addition to verifying the formula for $\langle v \rangle$, we also verified the predictiveness of the formula above for $\langle v^2 \rangle$, which is precise because it does not involve variance. The relative error in this case is less than 0.07% for all values of G, which is *extremely* low.

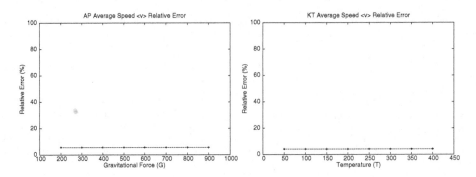

Fig. 7. Relative Error for Average Speed (Prediction 3)

We next show how we derive a KT formula for average speed by modifying the derivation for 3D $\langle v \rangle$ in [10] to a 2D formula for $\langle v \rangle$ (so it applies to our simulation). Assuming a system in thermodynamic equilibrium (since there is no bulk transport), with velocity components within the ranges $v_x + dv_x$ and $v_y + dv_y$, and k is Boltzmann's constant, m is the particle mass, v is the magnitude of the particle velocity (i.e., the particle speed), and T is the initial system temperature (a simple, settable system parameter), then the probability, $f(v_x, v_y)dv_xdv_y$, that a particle has velocity components in these ranges is proportional to $e^{(-mv^2/2kT)}dv_xdv_y$. In particular, we have:

$$f(v_x, v_y)dv_xdv_y = Ae^{(-mv^2/2kT)}dv_xdv_y = Ae^{(-mv_x^2/2kT)}e^{(-mv_y^2/2kT)}dv_xdv_y$$

because $v^2 = v_x^2 + v_y^2$, and A is a normalization constant that is fixed by the requirement that the integral of the probability over all possible states must be equal to 1, i.e.,

$$\int_0^\infty f(v_x, v_y)dv_xdv_y = 1$$

Therefore,

$$A = \frac{1}{\int_0^\infty e^{(-mv_x^2/2kT)}dv_x \int_0^\infty e^{(-mv_y^2/2kT)}dv_y}$$

To simplify the expression for A, we can use the fact (from pages 40-46 of [13]) that:

$$\int_0^\infty e^{(-mv_x^2/2kT)}dv_x = \sqrt{\frac{2\pi kT}{m}}$$

and then do likewise for v_y. Therefore:

$$f(v_x, v_y)dv_xdv_y = \left(\frac{m}{2\pi kT}\right)(e^{(-m(v_x^2+v_y^2)/2kT)})dv_xdv_y$$

where $\frac{m}{2\pi kT}$ is A.

Note, however, that $f(v_x, v_y)dv_x dv_y$ is a probability for a velocity *vector*, but we want average *speed*. To get average speed, the math is easier if we go from Cartesian to polar coordinates. In particular, to go from velocity to speed, we integrate over all angles.

In polar coordinates, $2\pi v dv$ is the area of extension (annulus) due to Δv. In other words, the area of an annulus whose inner radius is v and outer radius is $v + dv$ is $2\pi v dv$. Then the Maxwell-Boltzmann distribution of speeds, $f(v)dv$, is obtained by integrating the velocity distribution, $f(v_x, v_y)dv_x dv_y$, over all angles from 0 to 2π. This integration yields:

$$f(v)dv = 2\pi v \left(\frac{m}{2\pi kT}\right)(e^{(-mv^2/2kT)})dv$$

Canceling terms, the right-hand side becomes:

$$= v \left(\frac{m}{kT}\right)(e^{(-mv^2/2kT)})dv$$

Because $\langle v \rangle$ is an expected value,

$$\langle v \rangle = \int_0^\infty v f(v)dv = \frac{m}{kT}\int_0^\infty v^2 (e^{(-mv^2/2kT)})dv$$

From [14](page 609), we know that $\int_0^\infty e^{-ax^2}x^2 dx = \frac{1}{4}\sqrt{\pi}a^{-\frac{3}{2}}$. Substituting v for x and $\frac{m}{2kT}$ for a, we get:

$$\langle v \rangle = \left(\frac{m}{kT}\right)I(2) = \frac{1}{4}\sqrt{\frac{8\pi kT}{m}}$$

Once again, we set up an experiment to measure the actual average speed of the particles in the system. We allowed the system to converge to an asymptotic state for 50,000 time steps measuring the average speed. For each of the 500 particles in the system, we found the average speed, $\langle v \rangle$. This speed was used to find the relative error for the system. Since temperature drives changes in speed, we varied the temperature. Note that by setting T, a system designer can easily achieve desired behavior. The results can be found in Fig. 7 for the different temperatures. Our ability to predict the average speed of the particles is shown, by errors less than 10%.

9   Theoretical Predictions: Discussion

We are capable of predicting three different properties of the system, all of which affect coverage, with an accuracy of less than 20% error, and most with error less than 10%. A 10% error is low for a theoretical prediction.

By looking at the relative error graphs of both the AP and KT approaches, one notices that the AP error is always lower than that of KT (except in the case of $\langle v \rangle$, where the AP formula is a rough approximation). In fact, only KT gets 20% errors – AP errors are always substantially lower than 20%. Our rationale for AP having lower errors between theory and simulation is that AP uses a deterministic agent-positioning algorithm, whereas KT uses a stochastic algorithm for updating particle positions. Therefore, AP predictions are precise, whereas KT predictions are only approximate. Furthermore, as stated in [10], Monte Carlo simulations such as KT need very long runs and huge numbers of particles to acquire enough statistical data to produce accurate (theoretically predictable) results. We cannot guarantee this, since we are developing control algorithms for robotic swarms with a few to a few thousand robots. Therefore, our experiments show a higher error than desired for a Monte Carlo method but they are realistic for real-world swarms.

In conclusion, there appears to be a tradeoff. AP systems are more predictable – both on the macroscopic swarm level and on the level of individual agents. Therefore, if swarm predictability is a higher priority, then AP is preferable. On the other hand, if it is important that individual agents not be predictable (e.g., to an enemy), then KT is preferable.

10   Future Work

The next step is to develop a theory for the full surveillance task. Once this theory is complete, experiments need to be run to test all approaches: AP, KT, and various AP/KT hybrids. We plan to run numerous experiments to measure coverage versus time and determine which of the algorithms outperforms the others. Once that is complete, the next step is to port these approaches to our laboratory mobile robots. The solid AP approach has already been ported. Transitioning to AP *gas* will be straightforward. We will need to determine, using a more realistic robot simulator, how difficult (or easy) it will be to port KT to the actual robots.

References

1. Balch, T., Arkin, R.: Behavior-based formation control for multi-robot teams. IEEE Trans. on Robotics and Autom. 14 (1998) 1–15
2. Fredslund, J., Matarić, M.: A general algorithm for robot formations using local sensing and mimimal communication. In: IEEE Transactions on Robotics and Automation. (2002) 837–846
3. Koenig, S., Liu, Y.: Terrain coverage with ant robots: A simulation study. In: Agents'01. (2001) 600–607
4. Spears, W., Gordon, D.: Using artificial physics to control agents. In: IEEE International Conference on Information, Intelligence, and Systems. (1999) 281–288
5. Spears, W., Gordon-Spears, D., Hamann, J., Heil, R.: Distributed, physics-based control of swarms of vehicles. Autonomous Robots 17 (2004) 137–162
6. Decuyper, J., Keymeulen, D.: A reactive robot navigation system based on a fluid dynamics metaphor. In Schwefel, H.P., Männer, R., eds.: Parallel Problem Solving from Nature (PPSN I). Volume 496 of Lecture Notes in Computer Science., Springer-Verlag (1991) 356–362

7. Gordon-Spears, D., Spears, W.: Analysis of a phase transition in a physics-based multiagent system. In: Proceedings of FAABS II. (2002)
8. Jantz, S., Doty, K.: Kinetics of robotics: The development of universal metrics in robotic swarms. Technical report, Dept of Electrical Engineering, University of Florida (1997)
9. Anderson, J.: Computational Fluid Dynamics. McGraw-Hill (1995)
10. Garcia, A.: Numerical Methods for Physics. Second edn. Prentice Hall (2000)
11. Spears, W., Spears, D., Heil, R.: A formal analysis of potential energy in a multi-agent system. In: Proceedings of FAABS III. (2004)
12. Stark, H., Woods, J.: Probability, Random Processes, and Estimation Theory for Engineers. Prentice-Hall (1986)
13. Feynman, R., Leighton, R., Sands, M.: The Feynman Lectures on Physics. Addison-Wesley Publishing Company (1963)
14. Reif, F.: Fundamentals of Statistical and Thermal Physics. McGraw-Hill (1965)

A Formal Analysis of Potential Energy in a Multi-agent System

William M. Spears, Diana F. Spears, and Rodney Heil

University of Wyoming, Laramie WY 82071, USA
wspears@cs.uwyo.edu
http://www.cs.uwyo.edu/~wspears

Abstract. This paper summarizes a novel framework, called "physicomimetics," for the distributed control of large collections of mobile physical agents in sensor networks. The agents sense and react to virtual forces, which are motivated by natural physics laws. Thus, physicomimetics is founded upon solid scientific principles. Furthermore, this framework provides an effective basis for self-organization, fault-tolerance, and self-repair. Examples are shown of how this framework has been applied to construct regular geometric lattice configurations (distributed sensing grids). Analyses are provided that facilitate system understanding and predictability, including a quantitative analysis of potential energy that provides the capability of setting system parameters based on theoretical laws. Physicomimetics has been implemented both in simulation and on a team of seven mobile robots.

1 Introduction

The focus of our research is to build sensor network systems, specifically, to design rapidly deployable, scalable, adaptive, cost-effective, and robust networks (i.e., swarms, or large arrays) of autonomous distributed mobile sensing agents (e.g., robots). This combines sensing, computation and networking with mobility, thereby enabling deployment, assembly, reconfiguration, and disassembly of the multi-agent collective. Our objective is to provide a scientific, yet practical, approach to the design and analysis (behavioral assurance) of aggregate sensor systems.

Agent vehicles could vary widely in type, as well as size, e.g., from nanobots to micro-air vehicles (MAVs) and micro-satellites. Agents are assumed to have sensors and effectors. An agent's sensors perceive the world, including other agents, and an agent's effectors make changes to that agent and/or the world, including other agents. It is assumed that agents can only sense and affect nearby agents; thus, control rules must be "local." Desired global behavior emerges from local agent interactions.

This paper summarizes our *physicomimetics* framework for robot control. A theoretical analysis of potential energy is then provided, allowing us to properly set system parameters a priori. Finally, results of a multi-robot implementation are presented.

M.G. Hinchey et al. (Eds.): FAABS 2004, LNAI 3228, pp. 131–145, 2005.
© Springer-Verlag Berlin Heidelberg 2005

2   Relation to Alternative Approaches

System analysis enables both system design and behavioral assurance. Here, we adopt a physics-based approach to analysis. We consider this approach to fit under the category of "formal methods," not in the traditional sense of the term but rather in the broader sense, i.e., a formal method is a mathematical technique for designing and/or analyzing a system. The two main *traditional* formal methods used for this purpose are theorem proving and model checking. Why do we use a physics-based method instead of these more traditional methods? The gist of theorem proving (model checking) is to begin with a theorem (property) and prove (show) that it holds for the target system. But what if you don't know how to express the theorem or property in the first place? For example, suppose you visually observe a system behavior that you want to control, but you have no idea what causes it or how to express your property in concrete, logic-based or system-based terms? In particular, there may be a property/law relating various system parameters that enables you to predict or control the observed phenomenon, but you do not understand the system well enough to write down this law.

For such a situation, the traditional, logic-based formal methods are not directly applicable. One potentially applicable approach is empirical, e.g., machine discovery. We have chosen a theoretical (formal) physics-based approach because:

- Empirical techniques can tell you *what* happens, but not *why* it happens. Causal explanations are easier to understand, apply, build upon, and generalize.
- If a physics-based analysis technique is predictive of a system built on physics-based principles, then this analysis provides formal verification of the correctness of the system implementation. No such claims can be made for empirical results.
- *Finally, and most importantly, it is possible to go directly from theory to a successful robot demo, without the usual extensive parameter tweaking! We have already demonstrated such successes with our theories [1].*

3   The Physicomimetics Framework

In our physicomimetics framework, virtual physics forces drive a multi-agent system to a desired configuration or state. The desired configuration (state) is the one that minimizes overall system potential energy. We also refer to our framework as "artificial physics" or "AP".

At an abstract level, physicomimetics treats agents as physical particles. This enables the framework to be embodied in vehicles ranging in size all the way from nanobots to satellites. Particles exist in two or three dimensions and are considered to be point-masses. Each particle i has position $x = (x, y, z)$ and velocity $v = (v_x, v_y, v_z)$. We use a discrete-time approximation to the continuous behavior of the particles, with time-step Δt. At each time step, the position of each particle undergoes a perturbation Δx. The perturbation depends on the current velocity, i.e., $\Delta x = v \Delta t$. The velocity of each particle at each time step also changes by Δv. The change in velocity is controlled by the force on the particle, i.e., $\Delta v = F \Delta t / m$, where m is the mass of that particle and F is the force on that particle.[1] A frictional force is included, for self-stabilization. This force is modeled as a *viscous friction* term, i.e., the product of a viscosity coefficient and the agent's velocity (independently modeled in the same fashion by [2]).

The time step Δt is proportional to the amount of time the robots take to perform their sensor readings. A parameter F_{max} is added, which restricts the amount of acceleration a robot can achieve. A parameter V_{max} restricts the velocity of the particles. Collisions are not modeled, because AP repulsive forces tend to avoid collisions. Also, we do not model the low-level dynamics of the actual robot. We consider AP to be an algorithm that will determine "way points" for the actual physical platforms. Lower-level software can steer between way points.

Given a set of initial conditions and some desired global behavior, we define what sensors, effectors, and force F laws are required such that the desired behavior emerges.

4   Designing Lattice Formations

The example considered in this section was inspired by an application which required a swarm of MAVs to form a hexagonal lattice, thus creating an effective antenna [3].

Since MAVs (or other small agents such as nanobots) have simple sensors and primitive CPUs, our goal was to provide the simplest possible control rules requiring minimal sensors and effectors. Creating hexagons appears to be rather complicated, requiring sensors that can calculate range, the number of neighbors, their angles, etc. However, it turns out that only range and bearing information are required. To see this, recall that six circles of radius R can be drawn on the perimeter of a central circle of radius R. Figure 1 illustrates this construction. If the particles (shown as small circular spots) are deposited at the intersections of the circles, they form a hexagon with a particle in the middle.

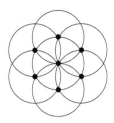

Fig. 1. How circles can create hexagons

---

[1] F and v denote the magnitude of vectors F and v.

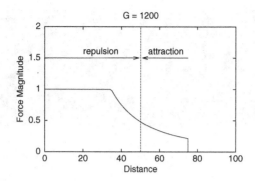

Fig. 2. The force law, when $R = 50$, $G = 1200$, $p = 2$ and $F_{max} = 1$

We see that hexagons can be created via overlapping circles of radius R. To map this into a force law, each particle repels other particles that are closer than R, while attracting particles that are further than R in distance. Thus each particle can be considered to have a circular "potential well" around itself at radius R – neighboring particles will want to be at distance R from each other. The intersection of these wells is a form of constructive interference that creates "nodes" of very low potential energy where the particles will be likely to reside. The particles serve to create the very potential energy surface to which they are responding![2]

With this in mind we defined a force law $F = Gm_im_j/r^p$, where $F \leq F_{max}$ is the magnitude of the force between two particles i and j, and r is the range between the two particles. The variable p represents a user-defined power, which can range from -5.0 to 5.0. When $p = 0.0$ the force law is constant for all ranges. Unless stated otherwise, we assume $p = 2.0$ and $F_{max} = 1$ in this paper. The "gravitational constant" G is set at initialization. The force is repulsive if $r < R$ and attractive if $r > R$. Each particle has one sensor that can detect the range and bearing to nearby particles. The only effector is to be able to move with velocity $v \leq V_{max}$. To ensure that the force laws are local, particles have a visual range of $1.5R$.

Figure 2 shows the magnitude of the force, when $R = 50$, $G = 1200$, $p = 2$, and $F_{max} = 1$ (the system defaults). There are three discontinuities in the force law. The first occurs where the force law transitions from F_{max} to $F = Gm_im_j/r^p$. The second occurs when the force law switches from repulsive to attractive at R. The third occurs when the force goes to 0.

The initial conditions are a tight cluster of robots, that propel outward (due to repulsive forces) until the desired geometric configuration is obtained. This is simulated by using a two dimensional Gaussian random variable to initialize the positions of all particles. Velocities of all particles are initialized to be 0.0, and masses are all 1.0 (although the framework does not require this).

---

[2] The potential energy surface is never actually computed by the robots. It is only computed in the simulation for visualization/analysis.

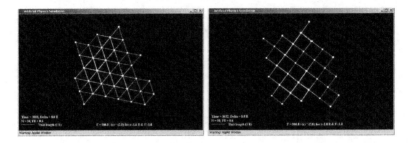

Fig. 3. Agents can form hexagonal and square lattices

Using this force law, AP successfully forms hexagonal lattices, with a small number of agents or hundreds. Square lattices are also easily obtained [4,1]. For a radius R of 50, a gravitational constant of approximately $G = 1200$ provides good results. The issue of how to set G, given other system parameters, is the focus of the analysis in this paper.

5   Energy Analysis

Because our force law is conservative (in the physics sense), the AP system should obey conservation of energy – if it is implemented correctly. Furthermore, as we shall see, the initial potential energy of the system in the starting configuration yields important information concerning the dynamics of the system.

First, we measured the potential energy (PE) of the system at every time step, using the path integral $V = -\int_s \mathbf{F} \bullet d\mathbf{s}$.[3] This can be thought of as the amount of work required to push each particle into position, one after another, for the current configuration of particles. Because the force is conservative, the order in which the particles are chosen is not relevant. Then we also measured the kinetic energy (KE) of the particles $(mv^2/2)$. Finally, since there is friction we also must take into account that energy as well, which we can consider to be heat energy. If there is no friction, the heat energy is zero.

Figure 4 illustrates an example of the energy dynamics of the AP system. As expected, the total energy remains constant over time. The system starts with only PE. Note that the graph illustrates one of the foundational principles of the AP system, namely, that the system continually evolves to lower PE, until a minimum is reached. This reflects a form of stability of the final aggregate system, requiring work to move the system away from desired configurations (thus increasing PE).

As the system evolves, the PE is converted into KE and heat, and the particles exhibit maximum motion, which is not very large (see Figure 4). Finally, however, the particles slow, and only heat remains. Note also that PE is negative after a certain point. This illustrates stability of individual particles (as well as the

---

[3] V is the traditional notation for potential energy.

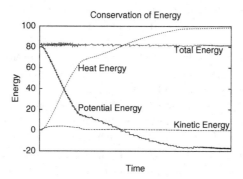

Fig. 4. Conservation of energy, showing how the total energy remains constant, although the amount of different forms of energy change over time

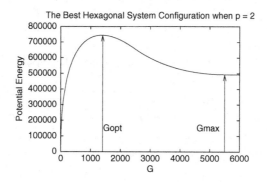

Fig. 5. The amount of potential energy of the initial configuration of the system is maximized for the G value that empirically yields the best results, which is roughly 1300. In this example $p = 2$. The arrows show the values of G_{opt} and G_{max}, respectively

collective) – it would require work to push individual particles out of these configurations. Hence this graph shows how the system would be resilient to moderate amounts of force acting to disrupt it, once stable configurations are achieved.

We have found that the initial configuration PE indicates important properties of the final evolved system, namely how well it evolves and the size of the formation. Intuitively, higher initial PE indicates that more work can be done by the system – and the creation of bigger formations requires more work. We have also observed that higher initial PE is correlated with better formations. Apparently there is more energy via momentum to push through local optima to global optima.

For example, consider Figure 5, which shows the PE of the initial configuration of a 200 particle system (when $p = 2$), for different values of G. In the figure, G_{opt} is the value of G at which the PE is maximized, and G_{max} is the largest useful setting of G (i.e., above G_{max} all forces are equal to F_{max}). Interestingly,

PE is maximized almost exactly at the range of values of G (around 1200 to 1400) that we have found empirically to yield the best structures.

We now compute a general expression for when PE is maximized. To find this expression for G_{opt}, we first need to calculate the potential energy, V. For simplicity, we begin by calculating the potential energy of a two particle system where the two particles are very close to each other.

It will be necessary to consider three different situations, depending on the radial extent to which F_{max} dominates the force law $F = G/r^p$. Recall that agents use F_{max} when $F \geq F_{max}$. This occurs when $G/r^p \geq F_{max}$ or, equivalently, when $r \leq (G/F_{max})^{1/p} \equiv R'$. The first situation occurs when F_{max} is used only when the other particle is at close range, i.e., when $0 \leq R' \leq R$. The second situation occurs when $R \leq R' \leq 1.5R$. The third situation occurs when F_{max} is always used, i.e., when $R' > 1.5R$. In this situation the force law is constant (F_{max}) and V remains constant with increasing G.

Let us now compute the PE for the first situation. It will be necessary to calculate three separate integrals for this situation. The first will represent the attractive force felt by one particle as it approaches the other, from a range of $1.5R$ to R. The second is the repulsive force of $F = G/r^p$ when $r < R$ and $F < F_{max}$. The third represents the range where the repulsive force is simply F_{max}. Then:

$$V = - \int_R^{1.5R} \frac{G}{r^p}\, dr + \int_{R'}^R \frac{G}{r^p}\, dr + \int_0^{R'} F_{max}\, dr$$

Note that the first term is negative because it deals with attraction, whereas the latter two terms are positive due to repulsion. Solving and substituting for R' yields:

$$V = \frac{(2R^{1-p} - (1.5R)^{1-p})G}{(1-p)} - \frac{pG^{1/p}}{(1-p)F_{max}^{(1-p)/p}}$$

The derivation of the second and third situations is similar (see Appendix for full derivations). The first situation occurs with low G, when $G \leq F_{max}R^p$. The second situation occurs with higher levels of G, when $F_{max}R^p \leq G \leq F_{max}(1.5R)^p$. The third situation occurs when $G \geq F_{max}(1.5R)^p$. In the third situation the PE of the system remains constant as G increases even further. Thus the maximum useful setting of G is $G_{max} = F_{max}(1.5R)^p$. We can see this in Figure 5 (which represent the full curves over all three situations), where $G_{max} = 5625$.

We next generalize to N particles for V (denoted V_N). Note that we can build our N particle system one particle at a time, in any order (because forces are conservative), resulting in an expression for the total initial PE:

$$V_N = \sum_{i=0}^{N-1} iV = \frac{VN(N-1)}{2}$$

with V defined above for the 2-particle system.

Now that we have a general expression for the potential energy, V_N, to find the expression for G_{opt} we need to find the value of G that maximizes V_N. First, we need to determine whether the maximum occurs in the first or second situation. It is easy to show that the slope of the PE equation for the second situation is strictly negative; thus the maximum must occur in the first situation. To find the maximum, we take the derivative of the V_N for the first situation with respect to G, set it to zero, and solve for G. The resulting maximum is at:

$$G_{opt}{}^\triangle = F_{max}R^p[2 - 1.5^{1-p}]^{p/(1-p)}$$

Note that the value of G_{opt} does not depend on the number of particles, which is a nice result. This simple formula is surprisingly predictive of the dynamics of a 200 particle system. For example, when $F_{max} = 1$, $R = 50$, and $p = 2$, $G_{opt} = 1406$, which is only about 7% higher than the value shown in Figure 5. Similarly, when $p = 3$, $G_{opt} = 64,429$, which is very close to observed values. The difference in values stems from the fact that in our simulation we have initial conditions specified by a two-dimensional Gaussian random variable with a small variance σ^2, whereas our mathematical analysis assumes a variance of zero. Despite this difference, the equation for G_{opt} works quite well.

As described in [4], we have also had success in creating square lattices. Performing a similar potential energy analysis yields a G_{opt} of:

$$G_{opt}{}^\square = F_{max}R^p\left[\frac{\sqrt{2}(N-1)(2-1.3^{1-p}) + N(2-1.7^{1-p})}{\sqrt{2}(N-1)+N}\right]^{p/(1-p)}$$

Note that G_{opt} actually depends on the number of particles N, which is the first time we have seen such a dependency. It occurs because we use two "species" of particles to create square lattices, which have different sensor ranges. However, because this difference is not large, the dependency on N is also not large. For example, with $R = 50$ and $F_{max} = 1$, then when $p = 2$ and there are 200 total particles, $G_{opt} = 1466$. With only 20 particles $G_{opt} = 1456$. Similarly, when $p = 3$ we obtain values of $G_{opt} = 67,330$ and $G_{opt} = 66,960$ respectively (for 200 and 20 particle systems).

6   Experiments with a Team of Robots

For our experiments, we used seven robots from the KISS Institute for Practical Robotics. For detecting neighboring robots, Sharp GP2D12 IR sensors are mounted, providing a 360 degree field of view, from which object detection is performed. The output is a list that gives the bearing and range to all neighboring robots. Once sensing and object detection are complete, the AP algorithm computes the virtual force felt by that robot. In response, the robot will turn and move to some position. This cycle of sensing, computation and motion continues until we shut down the robots or they lose power. The AP code is simple to implement. It takes a robot neighbor list as input, and outputs a turn and distance to move.

The goal of the first experiment was to form a hexagon with seven robots. Each robot ran the same software. The desired distance R between robots was 23 inches. Using the theory we chose a G of 270 ($p = 2$ and $F_{max} = 1$). The beginning configuration was random. The final configuration was a hexagon. The results are consistent, achieving the same formation ten times in a row with the same starting conditions and taking approximately seven cycles on average. For all runs the robots were separated by 20.5 to 26 inches in the final formation, which is only slightly more error than the sensor error.

For our second experiment we placed four photo-diode light sensors on each robot, one per side. These produced an additional force vector, moving the robots towards a light source (a window).[4] The results are shown in Figure 6, and were consistent over ten runs, achieving an accuracy comparable to the formation experiment above. The robots moved about one foot in 13 cycles of the AP algorithm.

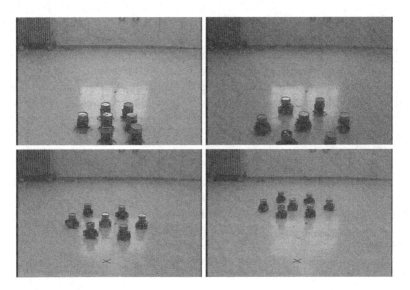

Fig. 6. Seven robots get into formation, and move toward the light

7   Summary, Related and Future Work

This paper presents a novel analysis of AP, focusing on potential energy. This analysis provides us with a predictive technique for setting important parameters in the system, thus enabling a system user to create (with good assurance) large formations. This static analysis combines many important parameters of the system, such as G, R, p, F_{max}, and sensor range. It also includes the geometry of the formations in a natural fashion. The parameter N was included as well,

---

[4] The reflection of the window on the floor is not noticed by the robots and is not the light source.

but it turns out to be of little relevance for our most important results. This is a nice feature, since our original motivation for the AP approach was that we wished it to scale easily to large numbers of agents. To include the other relevant dynamic parameters such as Δt, V_{max} and friction will require a more dynamic analysis.

The work that is most related consists of other theoretical analyses of swarm systems. Our comparisons are in terms of the goal and method of analysis. There are generally two goals: stability and convergence/correctness. Under stability is the work by [5, 6, 7]. Convergence/correctness work includes [5]. Other goals of theoretical analyses include time complexity [8], synthesis [9], prediction of movement cohesion [5], coalition size [6], number of instigators to switch strategies [10], and collision frequency [11].

Methods of analysis are also diverse. Here we focus only on physics-based analyses of physics-based swarm robotics systems. We know of four methods. The first is the Lyapunov analysis by [7]. The second is the kinetic gas theory by [11]. The third is the minimum energy analysis by [9]. The fourth develops macro-level equations describing flocking as a fluid-like movement [12].

The capability of being able to set system parameters based on theoretical laws has enormous practical value. To the best of our knowledge, the only analyses mentioned above that can be used to set system parameters are those of [6,10,12]. The first two analyses are of behavior-based systems, while the latter is of a "velocity matching" particle system.

In the long run, we'd like to design and analyze virtual worlds based on AP. The theoretical results being developed here would formalize the multi-robot motions in such a virtual world, which would then influence the coordination of actual robots.

References

1. Spears, W., Gordon-Spears, D., Heil, R.: Distributed, physics-based control of swarms of vehicles. Autonomous Robots, Issue on Swarm Robotics **17** (2004) 137–162

2. Howard, A., Matarić, M., Sukhatme, G.: Mobile sensor network deployment using potential fields: A distributed, scalable solution to the area coverage problem. In: Sixth Int'l Symposium on Distributed Autonomous Robotics Systems. (2002)

3. Kellogg, J., Bovais, C., Foch, R., McFarlane, H., Sullivan, C., Dahlburg, J., Gardner, J., Ramamurti, R., Gordon-Spears, D., Hartley, R., Kamgar-Parsi, B., Pipitone, F., Spears, W., Sciambi, A., Srull, D.: The NRL micro tactical expendable (MITE) air vehicle. The Aeronautical Journal **106** (2002)

4. Spears, W., Gordon, D.: Using artificial physics to control agents. In: IEEE International Conference on Information, Intelligence, and Systems. (1999) 281–288

5. Liu, Y., Passino, K., Polycarpou, M.: Stability analysis of m-dimensional asynchronous swarms with a fixed communication topology. In: IEEE Transactions on Automatic Control. Volume 48. (2003) 76–95

6. Lerman, K., Galstyan, A.: A general methodology for mathematical analysis of multi-agent systems. Technical Report ISI-TR-529, USC Information Sciences (2001)
7. Olfati-Saber, R., Murray, R.: Distributed cooperative control of multiple vehicle formations using structural potential functions. In: IFAC World Congress. (2002)
8. O. Shehory, S.K., Yadgar, O.: Emergent cooperative goal-satisfaction in large-scale automated-agent systems. Artificial Intelligence **110** (1999) 1–55
9. Reif, J., Wang, H.: Social potential fields: A distributed behavioral control for autonomous robots. In: Workshop on the Algorithmic Foundations of Robotics. (1998)
10. Numaoka, C.: Phase transitions in instigated collective decision making. Adaptive Behavior **3** (1995) 185–222
11. Jantz, S., Doty, K., Bagnell, J., Zapata, I.: Kinetics of robotics: The development of universal metrics in robotic swarms. In: Florida Conference on Recent Advances in Robotics. (1997)
12. Toner, J., Tu, Y.: Flocks, herds, and schools: A quantitative theory of flocking. Physical Review E **58** (1998) 4828–4858

Appendix: Derivation of Potential Energy Analysis

Hexagonal Formations

In this appendix are details for computing the general expression for PE, and where it is maximized. For simplicity, we begin by calculating the potential energy of a two particle system where the two particles are very close to each other.

It will be necessary to consider three different situations, depending on the radial extent to which F_{max} dominates the force law $F = G/r^p$. Recall that agents use F_{max} when $F \geq F_{max}$. This occurs when $G/r^p \geq F_{max}$ or, equivalently, when $r \leq (G/F_{max})^{1/p} \equiv R'$. The first situation occurs when F_{max} is used only when the other particle is at close range, i.e., when $0 \leq R' \leq R$. The second situation occurs when $R \leq R' \leq 1.5R$. The third situation occurs when F_{max} is always used, i.e., when $R' > 1.5R$. In this situation the force law is constant (F_{max}) and V remains constant with increasing G.

First Situation: Let us now compute the PE for the first situation. It will be necessary to calculate three separate integrals for this situation. The first will represent the attractive force felt by one particle as it approaches the other, from a range of $1.5R$ to R. The second is the repulsive force of $F = G/r^p$ when $r < R$ and $F < F_{max}$. The third represents the range where the repulsive force is simply F_{max}.[5] Then:

$$V = -\int_{R}^{1.5R} \frac{G}{r^p}\, dr \;+\; \int_{R'}^{R} \frac{G}{r^p}\, dr \;+\; \int_{0}^{R'} F_{max}\, dr$$

---

[5] Throughout our theoretical results we assume that $p \neq 1.0$, which is reasonable since we typically do not run AP with that setting.

Note that the first term is negative because it deals with attraction, whereas the latter two terms are positive due to repulsion. Then:

$$V = -\frac{Gr^{1-p}}{(1-p)}\bigg|_R^{1.5R} + \frac{Gr^{1-p}}{(1-p)}\bigg|_{R'}^R + F_{max}r\big|_0^{R'}$$

Expanding yields:

$$V = -\frac{G(1.5R)^{1-p}}{(1-p)} + \frac{GR^{1-p}}{(1-p)} + \frac{GR^{1-p}}{(1-p)} - \frac{G(R')^{1-p}}{(1-p)} + F_{max}R'$$

Substituting for R' yields:

$$V = \frac{G}{(1-p)}\left[2R^{1-p} - (1.5R)^{1-p} - \left(\frac{G}{F_{max}}\right)^{\frac{1-p}{p}}\right] + F_{max}\left(\frac{G}{F_{max}}\right)^{\frac{1}{p}}$$

Finally, simplification yields:

$$V = \frac{(2R^{1-p} - (1.5R)^{1-p})G}{(1-p)} - \frac{pG^{1/p}}{(1-p)F_{max}^{(1-p)/p}} \tag{1}$$

Second Situation: The derivation of the second and third situations is similar. For the second situation:

$$V = -\int_{R'}^{1.5R}\frac{G}{r^p}\,dr - \int_R^{R'}F_{max}\,dr + \int_0^R F_{max}\,dr$$

Then:

$$V = -\frac{Gr^{1-p}}{(1-p)}\bigg|_{R'}^{1.5R} - F_{max}r\big|_R^{R'} + F_{max}r\big|_0^R$$

Expanding yields:

$$V = -\frac{G(1.5R)^{1-p}}{(1-p)} + \frac{G(R')^{1-p}}{(1-p)} - F_{max}R' + F_{max}R + F_{max}R$$

Substituting for R' and simplifying yields:

$$V = \frac{G}{(1-p)}\left[\left(\frac{G}{F_{max}}\right)^{\frac{1-p}{p}} - (1.5R)^{1-p}\right] + F_{max}\left[2R - \left(\frac{G}{F_{max}}\right)^{\frac{1}{p}}\right] \tag{2}$$

Third Situation: For the third situation:

$$V = -\int_R^{1.5R}F_{max}\,dr + \int_0^R F_{max}\,dr$$

Then:

$$V = -F_{max}r\big|_R^{1.5R} + F_{max}r\big|_0^R$$
$$V = -F_{max}(1.5R) + F_{max}R + F_{max}R$$

$$V = \frac{F_{max}R}{2} \tag{3}$$

Generalization to N Particles: We next generalize to N particles for V (denoted V_N). Note that we can build our N particle system one particle at a time, in any order (because forces are conservative), resulting in an expression for the total initial PE:

$$V_N = \sum_{i=0}^{N-1} iV = \frac{VN(N-1)}{2}$$

with V defined above for the 2-particle system.

Optimum Value for G: Now that we have a general expression for the potential energy, V_N, we need to find the value of G that maximizes V_N. First, we need to determine whether the maximum occurs in the first or second situation. It is trivial to show that the slope of the PE equation for the second situation is strictly negative; thus the maximum must occur in the first situation. To find the maximum, we take the derivative of V_N for the first situation with respect to G, set it to zero, and solve for G. The constant $N(N-1)/2$ does not effect this computation:

$$\frac{dV_N}{dG} = \frac{(2R^{1-p} - (1.5R)^{1-p})}{(1-p)} - \frac{G^{(1-p)/p}}{(1-p)F_{max}{}^{(1-p)/p}} = 0$$

Hence:

$$\frac{(2R^{1-p} - (1.5R)^{1-p})}{(1-p)} = \frac{G^{(1-p)/p}}{(1-p)F_{max}{}^{(1-p)/p}}$$

Solving for G yields:

$$G_{opt} \overset{\triangle}{=} G = F_{max}R^p[2 - 1.5^{1-p}]^{p/(1-p)} \tag{4}$$

Note that the value of G_{opt} does not depend on the number of particles.

Square Formations

As described in [4], we have also had success in creating square lattices. The success of the hexagonal lattice hinged upon the fact that nearest neighbors are R in distance. This is not true for squares, since if the distance between particles along an edge is R, the distance along the diagonal is $\sqrt{2}R$. Particles have no way of knowing whether their relationship to neighbors is along an edge or along a diagonal.

Suppose each particle is given another attribute, called "spin". Half of the particles are initialized to be spin "up", whereas the other half are spin "down". Consider the square depicted in Figure 7. Particles that are spin-up are open circles, while particles that are spin-down are filled circles. Particles of unlike spin are distance R from each other, whereas particles of like spin are distance $\sqrt{2}R$ from each other. This "coloring" of particles extends to square lattices, with alternating spins along the edges of squares, and same spins along the diagonals.

Fig. 7. Square lattices can be formed by using particles of two "spins". Unlike spins are R apart while like spins are $\sqrt{2}R$ apart

We use the same force law as before: $F = Gm_im_j/r^p$. However, r is renormalized to $r/\sqrt{2}$ if two particles have the same spin. Once again the force is repulsive if $r < R$ and attractive if $r > R$. To ensure that the force law is local, particles cannot see other particles that are further than cR, where $c = 1.3$ if particles have like spin and 1.7 otherwise.

A similar potential energy analysis can be performed if one views the process as occurring in three stages: (1) compute the PE of clustering N spin-up particles together, (2) compute the PE of clustering N spin-down particles, and (3) compute the PE of combining both clusters.

Again, as with the hexagon formations, three situations can arise. Since the maximum PE again occurs with the first situation, we focus only on this situation for the remainder of the analysis.

First Situation for Spin-Up Particles: First, compute the PE of the initial configuration of two spin-up particles. When particles of like spin interact, r is renormalized by $\sqrt{2}$, and their sensor range is $1.3R$. Thus:

$$V = -\int_{\sqrt{2}R}^{1.3\sqrt{2}R} \frac{G}{(r/\sqrt{2})^p}\, dr + \int_{\sqrt{2}R'}^{\sqrt{2}R} \frac{G}{(r/\sqrt{2})^p}\, dr + \int_0^{\sqrt{2}R'} F_{max}\, dr$$

Then:

$$V = -(\sqrt{2})^p \frac{Gr^{1-p}}{(1-p)}\Big|_{\sqrt{2}R}^{1.3\sqrt{2}R} + (\sqrt{2})^p \frac{Gr^{1-p}}{(1-p)}\Big|_{\sqrt{2}R'}^{\sqrt{2}R} + F_{max}r\Big|_0^{\sqrt{2}R'}$$

Expanding yields:

$$V = -(\sqrt{2})^p \frac{G(1.3\sqrt{2}R)^{1-p}}{(1-p)} + 2(\sqrt{2})^p \frac{G(\sqrt{2}R)^{1-p}}{(1-p)}$$
$$- (\sqrt{2})^p \frac{G(\sqrt{2}R')^{1-p}}{(1-p)} + F_{max}\sqrt{2}R'$$

Substituting for R' and simplification yields:

$$V = \sqrt{2}\left[\frac{(2R^{1-p} - (1.3R)^{1-p})G}{(1-p)} - \frac{pG^{1/p}}{(1-p)F_{max}^{(1-p)/p}}\right]$$

Generalization to N Particles: The computation for V is very similar to that for the hexagonal lattice, differing only by a constant factor of $\sqrt{2}$ and the sensor range. We now generalize to N spin-up particles:

$$V_N = \frac{VN(N-1)}{2}$$

Aggregating all Particles: The computation for spin-down particles is identical. We now combine the two clusters of N spin-up and N spin-down particles:

$$V_{N+N} = V_N + V_N - \int_R^{1.7R} \frac{GN^2}{r^p} \, dr + \int_{R'}^{R} \frac{GN^2}{r^p} \, dr + \int_0^{R'} F_{max} N^2 \, dr$$

Then:

$$V_{N+N} = V(N-1)N - \left. \frac{GN^2 r^{1-p}}{(1-p)} \right|_R^{1.7R} + \left. \frac{GN^2 r^{1-p}}{(1-p)} \right|_{R'}^{R} + \left. F_{max} N^2 r \right|_0^{R'}$$

Expanding yields:

$$V_{N+N} = V(N-1)N - \frac{GN^2(1.7R)^{1-p}}{(1-p)} + \frac{2GN^2 R^{1-p}}{(1-p)} - \frac{GN^2(R')^{1-p}}{(1-p)} + F_{max} N^2 R'$$

Simplifying and substituting for R' yields:

$$V_{N+N} = V(N-1)N + N^2 \left[\frac{(2R^{1-p} - (1.7R)^{1-p})G}{(1-p)} - \frac{pG^{1/p}}{(1-p)F_{max}^{(1-p)/p}} \right]$$

To determine the value of G for which PE is maximized, we take the derivative of V_{N+N} with respect to G, set it to zero, and solve for G:

$$\frac{dV_{N+N}}{dG} = (N-1)N\sqrt{2} \left[\frac{(2R^{1-p} - (1.3R)^{1-p})}{(1-p)} - \frac{G^{(1-p)/p}}{(1-p)F_{max}^{(1-p)/p}} \right] +$$

$$N^2 \left[\frac{(2R^{1-p} - (1.7R)^{1-p})}{(1-p)} - \frac{G^{(1-p)/p}}{(1-p)F_{max}^{(1-p)/p}} \right] = 0$$

Hence:

$$(N-1)N\sqrt{2} \left[\frac{(2R^{1-p} - (1.3R)^{1-p})}{(1-p)} - \frac{G^{(1-p)/p}}{(1-p)F_{max}^{(1-p)/p}} \right] =$$

$$-N^2 \left[\frac{(2R^{1-p} - (1.7R)^{1-p})}{(1-p)} - \frac{G^{(1-p)/p}}{(1-p)F_{max}^{(1-p)/p}} \right]$$

Solving for G and simplifying yields:

$$G_{opt} \equiv G = F_{max} R^p \left[\frac{\sqrt{2}(N-1)(2 - 1.3^{1-p}) + N(2 - 1.7^{1-p})}{\sqrt{2}(N-1) + N} \right]^{p/(1-p)} \tag{5}$$

Note that in this case G_{opt} depends on the number of particles N. It occurs because of the weighted average of different inter-species and intra-species sensor ranges. However, because this difference is not large, the dependency on N is also not large (and approaches zero as N increases to infinity).

Agent-Based Chemical Plume Tracing Using Fluid Dynamics

Dimitri Zarzhitsky[1], Diana Spears[1], David Thayer[2], and William Spears[1]

[1] Department of Computer Science
[2] Department of Physics and Astronomy,
University of Wyoming, Laramie, WY 82071
dimzar@uwyo.edu

Abstract. This paper presents a rigorous evaluation of a novel, distributed chemical plume tracing algorithm. The algorithm is a combination of the best aspects of the two most popular predecessors for this task. Furthermore, it is based on solid, formal principles from the field of fluid mechanics. The algorithm is applied by a network of mobile sensing agents (e.g., robots or micro-air vehicles) that sense the ambient fluid velocity and chemical concentration, and calculate derivatives. The algorithm drives the robotic network to the source of the toxic plume, where measures can be taken to disable the source emitter. This work is part of a much larger effort in research and development of a physics-based approach to developing networks of mobile sensing agents for monitoring, tracking, reporting and responding to hazardous conditions.

1 Introduction

The objective of this research is the development of an effective, efficient, and robust distributed search algorithm for a team of robots that must locate an emitter that is releasing a toxic chemical gas. The basis for this algorithm is a physics-based framework for distributed multi-agent control [1]. This framework, called *physicomimetics* or *artificial physics (AP)*, assumes several to hundreds of simple, inexpensive mobile robotic agents with limited processing power and a small set of on-board sensors. Using AP, the agents will configure into geometric lattice formations that are preserved as the robots navigate around obstacles to a source location [2].

In this paper, we present a novel algorithm for chemical plume tracing (CPT) that is built upon the AP framework. The CPT task consists of finding the chemical, tracking the chemical to its source emitter and, finally, identifying the emitter. Here, we focus on the latter two subtasks. Our CPT algorithm combines the strengths of the two most popular chemical plume tracing techniques in use today. Furthermore, it is founded upon solid theoretical (formal) principles of fluid dynamics, which will make further analysis and improvement possible. Our algorithm assumes an AP-maintained lattice which acts as a distributed computational fluid dynamics (CFD) grid for calculating derivatives of *flow-field*

M.G. Hinchey et al. (Eds.): FAABS 2004, LNAI 3228, pp. 146–160, 2005.

variables, such as fluid velocity and chemical concentration. This paper consists of a formal study of the effectiveness of our novel algorithm, including comparisons with the two most popular alternatives on which it is built. To supplement the discussion of the underlying theory, we include results from software simulations that implement the theoretical scenarios we present, and include realistic elements of measurement discretization.

2   Motivation

The authors' goal is to design a search algorithm that scales well to a large number of robots, ranging perhaps from ten agents to a thousand and beyond. In order to achieve this goal, two things are necessary: a formal theory upon which the algorithm is based, and a suitable task that can be used to test the algorithm. The task of chemical plume tracing has posed problems for a number of years in a variety of manufacturing and military applications. In light of the current national concern with security and the possibility of a chemical terrorist attack, several private and government agencies have expressed interest in updating current techniques used to track hazardous plumes, and improving the search strategies used to locate the toxin emitter [3, 4, 5, 6].

Because the physicomimetics framework relies on application of virtual forces to construct and maintain the robotic lattice, physics is the natural choice for the theoretical foundation of our work. In particular, the well-studied field of fluid physics and mechanics is well-suited for the development and validation of our algorithms.

There is another advantage of using a physics-based foundation. Computational fluid mechanics requires computational meshes for sampling and processing of flow-field variable values. The lattice arrangements that emerge naturally from the physicomimetics framework can be used as computational meshes, thus forming a massively parallel system, capable of performing complex computations in real time, with the added benefit of resilience to failure, and ability to adjust when the environment characteristics change. The natural synergy between the different system components translates directly into an improved performance of the system. For instance, the construction of hexagonal formations requires the least amount of communication and sensor information within the agent control framework [7]; at the same time, a hexagonal lattice was shown [8] to have superior boundary characteristics for solving an important class of fluid mechanics problems.

3   Related Work

Current research in the field has been inspired by biological olfactory systems of lobsters and moths [9, 10, 11, 12]. The base requirement for any system that attempts to trace a chemical plume is of course the ability to sense the presence of the chemical agent, as well as its concentration. The best understood and most widely applied approach is that of *chemotaxis*, which consists of following a local gradient of the chemical concentration within a plume [13, 14, 11]. While chemotaxis is very simple to perform, it frequently leads to locations of high concentration in the plume that are not the source, such as a corner of a room. Furthermore, we have a proof, which we omit here due to space limitations, that a chemotaxis search strategy is likely to fail near the emitter's location, due to the fact that for a typical time-varying Gaussian plume density profile, the gradient goes to zero near the distribution's peak.

To overcome this problem, another common approach, called *anemotaxis*, has been developed. An anemotaxis-driven agent measures the direction of the fluid's velocity and navigates "upstream" within the plume [15, 14]. Such a strategy is successful in problems where the flow has no large-scale turbulence. In general, we do not have the luxury of assuming this type of airflow. On the contrary, the airflow could have large turbulent eddies that curl and circulate, thus creating a region where traveling upwind will result in a cycle, causing the anemotaxis technique to fail.

Early results from applying the solution of fluid dynamic problems to robotic systems are reported by Keymeulen and Decuyper [16, 17, 18]. In this work, a highly simplified model of fluid flow was used successfully in simulation to navigate a single robot in a semi-dynamic environment; the approach was inspired by the fact that fluid flow is a good model of the iterative, local-to-global route finding task optimization, since the local pressure fields that are responsible for the existence of the stable optimal path are void of local minima. In the development of their approach, Keymeulen and Decuyper relied on the concepts of a fluid *source* and *sink*, which they used to specify the robot's initial and goal locations. In the present work, we also base our method's development on these two concepts, and extensively utilize both mathematical and physical properties of these two entities in the verification of our algorithm.

Work by Balkovsky and Shraiman [19] on the subject of statistical analysis of the plume is also relevant. They develop a probability density function having a Gaussian form, and use it to develop a simplified model of the chemical plume, which is then traversed using an algorithm that takes the probability of the source's location into account. In the development of their algorithm, several assumptions were made regarding the type of the flow that the agent is expected to search. In our work we do not assume a particular flow-field, but rather establish several general categories of fluid flow and prove mathematically that our algorithm performs well in these broad and important categories.

Research by Parunak and Brueckner [11] makes a case for analysis of the self-organization property in multi-agent systems from the standpoint of entropy and the Second Law of Thermodynamics. They develop an analogy between entropy in the context of a system's energetic quality and informational disorder, and show how understanding and management of system entropy can be used to analyze a multi-agent system. They illustrate the idea by solving an agent coordination problem with the use of simulated randomly-diffusing pheromones. Our work complements their thermodynamic approach by looking at the conservation properties of matter, and improves it by providing a more realistic model of information flow within a system.

A promising approach to tracking and localizing a target with soft real-time constraints is discussed in Horling, et. al. [20]. The major contribution of their work is a radar network capable of operating under real-world conditions with realistic restrictions of noisy communication channels, limited sensory capabilities, and restricted computational power. The system however, only allows for fixed sensors and makes use of partially centralized sector and target manager agents, introducing local points of failure. In our approach, decisions are made in a fully decentralized manner, improving robustness of the entire system. In addition, our framework places no restriction on mobility of either the plume or the tracking agents.

Also of interest is the work of Polycarpou et al. [21], where the notion of artificial potential fields is used to find the goal object (an attractor) while avoiding obstacles (repellents). In order to apply potential fields, they create a map of the environment and the agents then are able to compute virtual forces based on the knowledge of the environment. However, such global maps are costly to build and mapping errors are a significant problem. The strategy we are proposing does not require environment mapping, and works well with the local information obtained in a highly distributed manner by the agents.

4   Computational Fluid Dynamics

Our approach makes use of the methods and concepts developed in the context of computational fluid dynamics (CFD), so a brief review of the relevant material will be useful. Flow of fluids is governed by three fundamental laws: the conservation of mass, conservation of momentum (Newton's Second Law), and the conservation of energy [22, 23]. There is also an equation that captures turbulent effects [24], but for simplicity we omit it here. Collectively, these equations are known as the Governing Equations. Equations that describe theoretical inviscid flows are also known as the Euler equations, while the more complex real viscous flows are described by the Navier-Stokes equations. These equations come in several forms, but we will focus on the conservation form, which is based on the time analysis of a differential volume spatially fixed in the flow field [23]. For instance, the simplest equation, the conservation of mass, is written as

$$-\frac{\partial \rho}{\partial t} = \nabla \cdot (\rho \mathbf{V}) \tag{1}$$

Here, ρ denotes the mass density of the chemical, $\mathbf{V}$ is the fluid's velocity (collectively, ρ and $\mathbf{V}$ are known as the *flow-field variables*), and t denotes time. For any real flow of practical interest, an analytical solution of the Governing Equations is impossible to obtain, due to the inherent non-linearity of the fluid dynamic systems. Thus, one CFD approach replaces the continuous partial derivatives with the corresponding discretized finite-difference approximations, and computes the unknown flow-field variables using a computational grid which

spans the region of interest. Our algorithm takes advantage of the lattice formations formed by our robotic agents to simulate the computational grid, thereby
allowing the agents to perform a sophisticated analysis of the flow and make
navigational decisions based on this analysis.

Other discretization methods, of which *finite-volume* and *finite-element* are
best known, are also applicable to the AP-driven robotic lattices. However, in
this paper, we only make use of the finite-difference discretization method because of its simple derivation from the Taylor-series expansion of partial derivatives [22]. For brevity and greater focus, we also ignore the interesting problem
of boundary conditions, and focus on a theoretically limitless domain. Since we
are interested in the problem of emitter localization, this simplification does not
have a significant impact on the solution, as long as the region in which plume
tracing is performed does not have walls nor obstacles. This limitation will be
addressed in the later stages of our research.

The work presented in the following sections deals with the development of
our physics-based solution to the chemical plume tracing task. It assumes a lattice of mobile agents with a limited, local view of the plume. The early theoretical
results have been verified in simulation, and more complex flow configurations
are currently being investigated.

5   Our Fluxotaxis Algorithm

The RHS of (1) represents the divergence of mass flux within the differential
volume. Divergence plays a key role in the proposed algorithm; it is therefore
helpful to briefly review the basics. Divergence is a convenient way to quantify
the change of a vector field in space. Although our approach is applicable to
3D geometries, for greater simplicity, we express the mass flux divergence in 2D
Cartesian coordinates as

$$\nabla \cdot (\rho \boldsymbol{V}) = u\frac{\partial \rho}{\partial x} + \rho\frac{\partial u}{\partial x} + v\frac{\partial \rho}{\partial y} + \rho\frac{\partial v}{\partial y} \tag{2}$$

where

$$\boldsymbol{V} = u\hat{\boldsymbol{i}} + v\hat{\boldsymbol{j}} \tag{3}$$

and $\hat{\boldsymbol{i}}$ and $\hat{\boldsymbol{j}}$ are unit vectors in the x and y coordinate directions, respectively.
If at some spatial point location P, $\nabla \cdot (\rho \boldsymbol{V}) > 0$, then it is said that point P
is a source of $\rho \boldsymbol{V}$, while $\nabla \cdot (\rho \boldsymbol{V}) < 0$ indicates a sink of $\rho \boldsymbol{V}$. It helps to point
out that the product $\rho \boldsymbol{V}$ is called the *mass flux* [23], and represents the time
rate of change of mass flow per unit area; dimensional analysis shows that $\rho \boldsymbol{V}$ is
simply mass/(area·time). The role of this quantity in the CPT task can be better
understood with the aid of the Divergence Theorem [25] from vector calculus:

$$\int_W \nabla \cdot (\rho \boldsymbol{V}) \mathrm{d}W = \oint_S (\rho \boldsymbol{V}) \cdot \mathrm{d}S \tag{4}$$

This equation, where W is the control volume and S is the bounding surface of
the volume, allows us to formally define the intuitive notion that a control volume

containing a source (e.g., emitter) will have a positive mass flux divergence, while a control volume containing a sink will have a negative mass flux divergence. This result serves as our basic criterion for theoretically identifying a chemical emitter. To the best of our knowledge, previous criteria for emitter identification are purely heuristic, e.g., [14]. Our method is the first with a solid theoretical basis.

Furthermore, this result is also the basis of our novel plume tracing algorithm, which we call *fluxotaxis*. With fluxotaxis, the robotic lattice will compute the local divergence of mass flux, and will follow its gradient (the direction of steepest increase). Mathematically, the gradient being followed is:

$$\nabla(\nabla \cdot \rho V) = \nabla(u\frac{\partial \rho}{\partial x} + \rho\frac{\partial u}{\partial x} + v\frac{\partial \rho}{\partial y} + \rho\frac{\partial v}{\partial y}) \qquad (5)$$

Each individual robot independently calculates this flux gradient (5). Due to the virtual cohesive forces holding the lattice together, the whole lattice will move in the flux gradient direction determined by the majority (with no explicit voting).

From (2) it is clear that the fluxotaxis algorithm combines information about both velocity and chemical density, and the fact that it also encapsulates the notion of mass flux, as demonstrated in (4), provides assurance that we will find the emitter as opposed to a local density maximum. The following section presents several formal proofs in support of this statement.

6   Fluxotaxis Theory

Our ultimate objective is to invent a foolproof mathematical formula that the robotic lattice can use to guide it to a chemical source. To date, the fluxotaxis formula is our best candidate, although it is not foolproof. With our objective in mind, we are currently beginning an in-depth study of the strengths and weaknesses of the fluxotaxis technique. Through such an analysis, we anticipate discovering a variant of the fluxotaxis method that will satisfy our objective.

In this section, we prove a sequence of lemmas that begin to elucidate the strengths of the fluxotaxis strategy as a local guide to the location of the chemical emitter. In subsequent papers, we will also explore and rectify its weaknesses. Here, we present initial versions of lemmas that have restrictive (albeit realistic) assumptions; future versions will relax these assumptions. We limit ourselves to lemmas because the final theorem is the complete navigation strategy that we intend to develop. Each of the following lemmas looks at a realistic scenario and demonstrates the performance of a fluxotaxis-managed, 1D robotic swarm.

All of these lemmas assume a local coordinate system shared by all of the robots in the robotic lattice. Such a shared coordinate system is achievable via local communication accompanied by coordinate transformations [2, 26]. The lemmas in this section assume a single coordinate axis for simplicity; generalization to 2D is expected to be straightforward, due to symmetries, and has already been verified in software simulations.

6.1   Fluxotaxis in Constant Velocity

Constant Velocity Lemma 1. Assume that the following conditions hold:

1. Chemical plume has a general Gaussian distribution $\rho(x) = \kappa e^{-(x-c)^2}$, centered at $x = c$.
2. Lattice position x_0 is such that $x_L < x_0 < x_R$, where x_L, x_R are solutions to $\partial^2 \rho(x)/\partial x^2 = 0$ (see Fig. 1); this implies that $\partial^2 \rho(x)/\partial x^2 < 0$ in the region of interest.
3. V is constant in magnitude throughout the flow, except right at the emitter $(x = c)$, and is an outward radial vector.

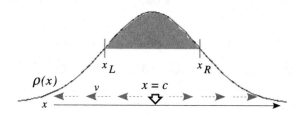

Fig. 1. The Gaussian chemical density distribution and the radial outflow velocity profile used in the Constant Velocity Lemma 1. The shaded area indicates the region where plume tracing is carried out by the fluxotaxis agents, and the arrow at $x = c$ marks the location of the chemical emitter

W.l.o.g., assume the existence of P_{emt} and P_{far} such that P_{emt} is closer to the emitter than P_{far}. Then execution of one step of the fluxotaxis algorithm implies that the agent lattice moves closer to the emitter, or equivalently

$$\left[u\frac{\partial \rho}{\partial x} + \rho\frac{\partial u}{\partial x} \right]_{\text{far}} < \left[u\frac{\partial \rho}{\partial x} + \rho\frac{\partial u}{\partial x} \right]_{\text{emt}} \tag{6}$$

Proof. The problem is symmetric with respect to the emitter's location $(x = c)$; thus it is sufficient to prove the case where $x_L < P_{\text{far}} < P_{\text{emt}} < c$. Because V is constant, $\partial u/\partial x = 0$, and (6) simplifies to

$$\left[u\frac{\partial \rho}{\partial x} \right]_{\text{far}} < \left[u\frac{\partial \rho}{\partial x} \right]_{\text{emt}}$$

Since u is a negative constant, the inequality can be simplified to

$$\left[\frac{\partial \rho}{\partial x} \right]_{\text{far}} > \left[\frac{\partial \rho}{\partial x} \right]_{\text{emt}}$$

Grouping like terms gives

$$0 > \left[\frac{\partial \rho}{\partial x} \right]_{\text{emt}} - \left[\frac{\partial \rho}{\partial x} \right]_{\text{far}}$$

Chemical Density (the highest density is in the middle, right at the emitter):

Fluid Velocity (uniform radial split at the emitter):

Lattice-Computed Divergence of Mass Flux (the maximum is near the emitter):

Two-Sided Lattice Trace (agents move inward, toward the emitter):

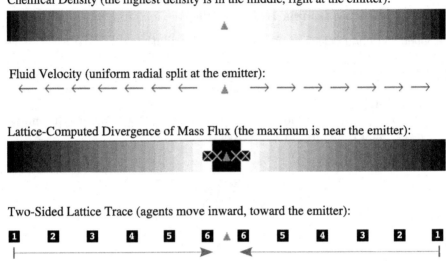

Fig. 2. Simulation results for the Constant Velocity Lemma 1. Individual agents are shown as black boxes with the white × in the middle, and the time trace of the two independent agent lattices is shown with boxed numbers indicating the location of the lattice at a given time step. The Lemma holds for any initial lattice configuration, and fluxotaxis successfully locates the chemical emitter

This is true because, by assumption 2,

$$0 > \frac{\partial^2 \rho}{\partial x^2} \qquad \qquad \square$$

Results of a software simulation for this lemma are shown in Fig. 2. In the figure, light-colored areas denote large values, and dark-colored areas correspond to small values. The location of the chemical emitter is marked by the triangle symbol. The initial positions of two separate agent lattices are at the outer edges of the environment, to the left and right of the emitter. During execution of the fluxotaxis algorithm, each agent (shown as a black box with a white × in the middle) computes the divergence of the mass flux using (2), with the partial derivatives replaced by the second-order accurate central difference approximation [27]. This value is recorded by the simulator for analysis purposes, and is displayed along with the final agent positions in the screen shot. Observe that the resulting divergence "landscape" has a global peak which coincides with the location of the emitter, and does not have any local maxima that could trap or mislead the agents. There is a small gap in the computed divergence plot near the emitter because the agents had terminated their search upon reaching the emitter. Each simulated agent (the black box) corresponds to one of the reference points (P_{emt} or P_{far}) in the Lemma's proof and, just as in the Lemma,

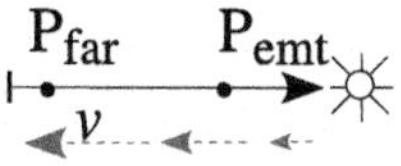

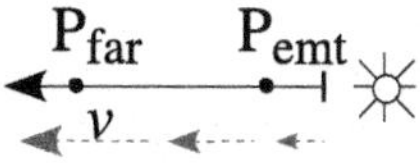

Fig. 3. Agent lattice coordinate axis orientation and the chemical source location in the Divergence Lemma 1

there are two agents per lattice. In this simulation, both agent lattices correctly moved toward the emitter in the center. In the proof of the Constant Velocity Lemma 1, we only considered the case where the lattice was to the left of the emitter; however, a similar proof can be given for the symmetric case, where the lattice starts out on the right side of the emitter, and the simulation in Fig. 2 demonstrates that the algorithm works regardless of the initial position of the agent lattice with respect to the chemical source.

6.2   Fluxotaxis at Source and Sink

Divergence Lemma 1. Fluxotaxis technique will advance the agent lattice toward a chemical source.

Proof. As before, assume a general Gaussian chemical plume distribution. W.l.o.g., assume the existence of two points P_{emt} and P_{far}, such that P_{emt} is closer to the source than P_{far} (see Fig. 3). Two cases result, based on the orientation of the lattice coordinate axis. (V is at the bottom of Fig. 3, below the axis.)

Case I assumes that the direction of the lattice coordinate axis is opposite to the direction of the fluid flow, and thus

1. $\partial^2 u/\partial x^2 \geq 0$
2. $\partial u/\partial x > 0$; thus $0 \geq u_{emt} > u_{far}$
3. $\partial^2 \rho/\partial x^2 \leq 0$
4. $\partial \rho/\partial x > 0$ and therefore $\rho_{emt} > \rho_{far}$

We need to prove that the agent will move toward the source, or

$$\left[u\frac{\partial \rho}{\partial x} + \rho\frac{\partial u}{\partial x} \right]_{far} < \left[u\frac{\partial \rho}{\partial x} + \rho\frac{\partial u}{\partial x} \right]_{emt} \tag{7}$$

Assumptions 1 and 3 imply

$$\left[\frac{\partial u}{\partial x} \right]_{far} \leq \left[\frac{\partial u}{\partial x} \right]_{emt} \quad \text{and} \quad \left[\frac{\partial \rho}{\partial x} \right]_{far} \geq \left[\frac{\partial \rho}{\partial x} \right]_{emt}$$

Together with assumptions 2, 4, and algebraic rules, Case I holds.   □

Case II is with the lattice coordinate axis in the same direction as the fluid flow, so that both u_{emt} and u_{far} are non-negative (see Fig. 3), and the previous assumptions become

1. $\partial^2 u / \partial x^2 \leq 0$
2. $\partial u / \partial x > 0$; thus $0 \leq u_{\text{emt}} < u_{\text{far}}$
3. $\partial^2 \rho / \partial x^2 \leq 0$
4. $\partial \rho / \partial x < 0$ and therefore $\rho_{\text{emt}} > \rho_{\text{far}}$

The agent will turn around and move toward the source if (7) holds. From assumption 1 we conclude

$$\left[\frac{\partial u}{\partial x}\right]_{\text{far}} < \left[\frac{\partial u}{\partial x}\right]_{\text{emt}}$$

Similarly, assumption 3 yields

$$\left[\frac{\partial \rho}{\partial x}\right]_{\text{far}} < \left[\frac{\partial \rho}{\partial x}\right]_{\text{emt}}$$

Algebraic application of the remaining assumptions shows that (7) holds.   □

Chemical Density (the highest density is in the middle, right at the emitter):

Fluid Velocity (radial flow speeds up away from the emitter):

Lattice-Computed Divergence of Mass Flux (the maximum is near the emitter):

Two-Sided Lattice Trace (agents move inward, toward the emitter):

Fig. 4. Simulation of a fluxotaxis-driven lattice (represented by black boxes) in the vicinity of a chemical source from the Divergence Lemma 1. The time trace, denoted by the numbered boxes, shows the location of each of the two different agent lattices at sequential time steps in the simulation. Both lattices correctly converge on the true location of the chemical emitter

Software simulation of this Lemma's configuration for both cases is shown in Fig. 4. As before, the fluxotaxis-driven lattice (represented by black boxes marked with the white × symbol) begins at the outer edges of the simulated world, and moves in toward the emitter, denoted by the triangle in the center.

(a) Case I   (b) Case II

Fig. 5. Location of the chemical sink and the two possible agent coordinate axis orientations in the Divergence Lemma 2

The direction of motion is determined by the gradient of the divergence of the mass flux, which is computed locally by each agent using a central difference approximation of the partial derivatives in (2), and as can been seen from the divergence plot, has the maximum value near the emitter's location. Similar to the previous simulation, the divergence value right at the emitter is not computed by the lattice, since the search terminates as soon as the emitter is found. Two fluxotaxis lattices are shown in the screen shot, and as expected, both of them successfully navigate toward the chemical source. As this figure illustrates, the initial position of a lattice with respect to the emitter does not impede the agents' ability to correctly localize the emitter.

Divergence Lemma 2. Fluxotaxis-controlled agents will move away from a chemical sink (see Fig. 5).

Proof. As before, assume a general Gaussian chemical plume distribution.W.l.o.g., assume the existence of two points P_{snk} and P_{far}, such that P_{snk} is closer to the sink than P_{far} (see Fig. 5). To prove that the agents will move away from the sink, we must show

$$\left[u\frac{\partial\rho}{\partial x} + \rho\frac{\partial u}{\partial x}\right]_{\text{snk}} < \left[u\frac{\partial\rho}{\partial x} + \rho\frac{\partial u}{\partial x}\right]_{\text{far}} \qquad (8)$$

Two cases result, based on the orientation of the lattice coordinate axis. (V is at the bottom of Fig. 5, below the axis.)

Case I occurs when the lattice coordinate axis points in the opposite direction to the fluid flow, so that both u_{snk} and u_{far} are negative (see Fig. 5). For this case, the assumptions are

1. $\partial^2 u/\partial x^2 \geq 0$
2. $\partial u/\partial x < 0$; thus $0 \geq u_{\text{snk}} > u_{\text{far}}$
3. $\partial^2 \rho/\partial x^2 \leq 0$
4. $\partial\rho/\partial x < 0$ and therefore $\rho_{\text{snk}} > \rho_{\text{far}}$

The agent will continue moving away from the sink if (8) is true. From assumption 1 we observe that

$$\left[\frac{\partial u}{\partial x}\right]_{\text{snk}} \leq \left[\frac{\partial u}{\partial x}\right]_{\text{far}}$$

Chemical Density (the highest density is in the center, but the emitter is absent):

Fluid Velocity (radial flow slows down near the center):

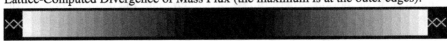

Lattice-Computed Divergence of Mass Flux (the maximum is at the outer edges):

Two-Sided Lattice Trace (agents move outward, away from the center of the sink):

Fig. 6. Simulated performance of the fluxotaxis algorithm within the chemical sink from the Divergence Lemma 2. As stated in the proof and visualized in the last time-step diagram, the robust fluxotaxis method forces the agent lattice out of the sink, even if the lattice starts out directly in the center of the sink, where the chemical concentration is at a local maximum. The robust physical foundation of the fluxotaxis algorithm allows it to outperform the simpler chemotaxis CPT strategy

Likewise, assumption 3 implies

$$\left[\frac{\partial \rho}{\partial x}\right]_{\text{snk}} \geq \left[\frac{\partial \rho}{\partial x}\right]_{\text{far}}$$

The remaining assumptions with algebraic simplification prove that (8) is true.

$\square$

Case II is when the direction of fluid flow and the lattice coordinate axis are the same, so that

1. $\partial^2 u/\partial x^2 \leq 0$
2. $\partial u/\partial x < 0$; thus $0 \leq u_{\text{snk}} < u_{\text{far}}$
3. $\partial^2 \rho/\partial x^2 \leq 0$
4. $\partial \rho/\partial x > 0$ and therefore $\rho_{\text{snk}} > \rho_{\text{far}}$

From assumptions 1 and 3 we conclude that

$$\left[\frac{\partial u}{\partial x}\right]_{\text{snk}} \leq \left[\frac{\partial u}{\partial x}\right]_{\text{far}} \quad \text{and} \quad \left[\frac{\partial \rho}{\partial x}\right]_{\text{snk}} \leq \left[\frac{\partial \rho}{\partial x}\right]_{\text{far}}$$

Algebraic simplification using assumptions 2 and 4 proves Case II.

$\square$

Simulation results for this lemma are presented in Fig. 6. Confirming the theoretical results just obtained, the high-density chemical build-up in the center of the environment does not fool the fluxotaxis algorithm, which correctly avoids the local spike in the density by directing the agents (again represented by black boxes) to the outer edge of the tracing region, where as can be seen from the divergence plot, the maximum mass flux divergence occurs. *The Divergence Lemma 2 proves that a fluxotaxis-driven agent lattice will escape from a sink.* However, a simple chemotaxis strategy is easily fooled by sinks, since by definition of a sink, $\partial\rho/\partial x > 0$ going into the sink. The fluxotaxis scheme is more robust in this case because it looks at the second order partial of ρ, and also takes the divergence of velocity into account. This simulation provides an example of how effectively the fluxotaxis technique merges the chemotaxis and anemotaxis CPT methods into a physically sound algorithm with valuable self-correcting properties.

7   Summary and Future Work

In this paper, we presented a new chemical plume tracing algorithm called fluxotaxis, that combines key strengths of chemotaxis and anemotaxis - the two most popular plume tracing methods. We showed that the fluxotaxis algorithm has been developed from the fundamental physical principles of fluid flow, and that it is able to overcome a major flaw of chemotaxis. We also built a formal mathematical tool set that we will employ to further improve the algorithm. In particular, we plan to soon extend the basic fluxotaxis approach outlined here to handle turbulent eddies, thus overcoming a major flaw of anemotaxis. To experimentally confirm our theoretical results, we will implement the algorithm on a massively distributed system of simple robotic agents currently under development for the task of toxic chemical plume emitter localization.

The most important contribution of our work is the development of a mobile robotic swarm control algorithm that can be analyzed with formal methods, such that the agents' behavior can now be mathematically predicted and guaranteed. Some of our work is relevant to the design and evaluation of artificial worlds, as it develops and refines methods for emulation of real-world physics in a simulated environment. The distributed nature of the CFD computations performed by the robotic swarm may also be of interest to the community. The contribution of this research is interdisciplinary and has a wealth of applications in domains other than the chemical plume tracing we discussed in this paper.

References

1. Spears, W., Gordon, D.: Using artificial physics to control agents. In: Proceedings of the IEEE Conference on Information, Intelligence, and Systems (ICIIS'99). (1999)
2. Spears, W., Spears, D., Hamann, J., Heil, R.: Distributed, physics-based control of swarms of vehicles. In: Autonomous Robots. Volume 17(2-3). (2004)

3. Board on Atmospheric Sciences and Climate: Tracking and Predicting the Atmospheric Dispersion of Hazardous Material Releases: Implications for Homeland Security. National Academies Press (2003)

4. Hsu, S.S.: Sensors may track terror's fallout. Washington Post (2003) A01

5. Cordesman, A.H.: Defending america: Asymmetric and terrorist attacks with chemical weapons (2001) report produced by the Center for Strategic and International Studies (CSIS).

6. Caldwell, S.L., D'Agostino, D.M., McGeary, R.A., Purdy, H.L., Schwartz, M.J., Weeter, G.K., Wyrsch, R.J.: Combating terrorism: Federal agencies' efforts to implement national policy and strategy (1997) Congressional report GAO/NSIAD-97-254, produced by the United States General Accounting Office.

7. Spears, W., Heil, R., Spears, D., Zarzhitsky, D.: Physicomimetics for mobile robot formations. In: Proceedings of the Third International Joint Conference on Autonomous Agents and Multi Agent Systems (AAMAS-04). Volume 3. (2004) 1528–1529

8. Carlson, E.S., Sun, H., Smith, D.H., Zhang, J.: Third Order Accuracy of the 4-Point Hexagonal Net Grid. Finite Difference Scheme for Solving the 2D Helmholtz Equation. (2003) Technical Report No. 379-03, Department of Computer Science, University of Kentucky.

9. Ball, P.: Odour-tracking trick sniffed out. Nature (2002)

10. Adam, D.: Wing scents. Nature (2000)

11. Parunak, H.V.D., Brueckner, S.: Entropy and self-organization in multi-agent systems. In: Proceedings of the International Conference on Autonomous Agents (AGENTS'01). (2001) 124–130

12. Koenig, S., Liu, Y.: Terrain coverage with ant robots: A simulation study. In: Proceedings of the International Conference on Autonomous Agents (AGENTS'01). (2001) 600–607

13. Sandini, G., Lucarini, G., Varoli, M.: Gradient driven self-organizing systems. In: Proceedings of the IEEE/RSJ International Conference on Intelligent Robots and Systems (IROS'93). (1993)

14. Chemical plume tracing. In Cowen, E., ed.: Environmental Fluid Mechanics. Volume 2. Kluwer (2002)

15. Hayes, A., Martinoli, A., Goodman, R.: Swarm robotic odor localization. In: Proceedings of the IEEE/RSJ International Conference on Intelligent Robots and Systems (IROS'01). (2001)

16. Decuyper, J., Keymeulen, D.: A reactive robot navigation system based on a fluid dynamics metaphor. In Schwefel, H.P., Männer, R., eds.: Parallel Problem Solving from Nature (PPSN I). Volume 496 of Lecture Notes in Computer Science., Springer-Verlag (1991) 356–362

17. Keymeulen, D., Decuyper, J.: The fluid dynamics applied to mobile robot motion: the stream field method. In: Proceedings of the 1994 International Conference on Robotics and Automation (ICRA'94). Volume 4., IEEE Computer Society Press (1994) 378–385

18. Keymeulen, D., Decuyper, J.: The stream field method applied to mobile robot navigation: a topological perspective. In Cohn, A.G., ed.: Proceedings of the Eleventh European Conference on Artificial Intelligence (ECAI'94), John Wiley and Sons (1994) 699–703

19. Balkovsky, E., Shraiman, B.: Olfactory search at high reynolds number. National Academies Press **99** (2002) 12589–12593

20. Horling, B., Vincent, R., Mailler, R., Shen, J., Becker, R., Rawlins, K., Lesser, V.: Distributed sensor network for real time tracking. In: Proceedings of the International Conference on Autonomous Agents (AGENTS'01). (2001) 417–424
21. Polycarpou, M.M., Yang, Y., Passino, K.M.: Cooperative control of distributed multi-agent systems. IEEE Control Systems Magazine (2001)
22. Tannehill, J.C., Anderson, D.A., Pletcher, R.H.: Computational Fluid Mechanics and Heat Transfer. Taylor and Francis (1997)
23. Anderson, J.D.: Computational Fluid Dynamics. McGraw-Hill, Inc. (1995)
24. Versteeg, H.K., Malalasekera, W.: An introduction to computational fluid dynamics: the finite volume method. Longman Scientific and Technical (1995)
25. Hughes-Hallett, D., Gleason, A.M., et al., W.M.: Calculus: Single and Multivariable. John Wiley and Sons (1998)
26. Craig, J.J.: Introduction to robotics: mechanics and control. Addison-Wesley Publishing Company, Inc. (1989)
27. Faires, J.D., Burden, R.: Numerical Methods. Brooks/Cole - Thomson Learning (2003)

Towards Timed Automata and Multi-agent Systems

G. Hutzler, H. Klaudel, and D.Y. Wang

LaMI, UMR 8042, Universit d'Evry-Val d'Essonne/CNRS
523, Place des Terrasses 91000 Evry, France
name@lami.univ-evry.fr

Abstract. The design of reactive systems must comply with logical correctness (the system does what it is supposed to do) and timeliness (the system has to satisfy a set of temporal constraints) criteria. In this paper, we propose a global approach for the design of adaptive reactive systems, i.e., systems that dynamically adapt their architecture depending on the context. We use the timed automata formalism for the design of the agents' behavior. This allows evaluating beforehand the properties of the system (regarding logical correctness and timeliness), thanks to model-checking and simulation techniques. This model is enhanced with tools that we developed for the automatic generation of code, allowing to produce very quickly a running multi-agent prototype satisfying the properties of the model.

Keywords: Agent oriented software engineering, formal models, agent oriented programming.

1 Introduction

Real-time reactive systems are defined through their capability to continuously react to the environment while respecting some time constraints. In a limited amount of time, the system has to acquire and process data and events that characterize its temporal evolution, make appropriate decisions and produce actions. Thus, the robustness of the system relies on its capability to present appropriate outputs (logical correctness) at an appropriate date (timeliness). Such applications are often critical. Their hardware and software architectures have to be specified, developed and validated with care. Then, they are set in order for the system to have a determinist and predictable behavior. The interest of multi-agent systems in this context may be considered as limited, especially because of autonomy and proactivity properties generally attributed to agents. In fact, the decision step in real-time systems is very often hidden and examples of usages of multi-agent paradigm in the real-time context [3, 18] exploit the distributed aspects of multi-agent systems much more than the autonomy aspects.

In this paper, we aim at addressing systems in which time constraints are neither critical (obtaining a response a little bit later than specified is acceptable) nor strict (when a normal delay of response is exceeded, the result is not

M.G. Hinchey et al. (Eds.): FAABS 2004, LNAI 3228, pp. 161–172, 2005.

immediately worthless but its value decreases more or less quickly with time). Another characteristic of such systems is the variability and unpredictability of treatments to process and their priority, but also of the availability of active entities (processors) in charge of processing. In such a context of dynamic scheduling in distributed systems, there is no solution yet capable to guarantee the respect of timing constraints. Our purpose is then to design this scheduling so as to optimize the compromise between the respect of logical correctness and timeliness, possibly by loosening some constraints when all of them cannot be satisfied simultaneously.

More precisely, rather than scheduling in its classical understanding, our concern here is the problem of adaptive reconfiguration of the processing chain during the execution. This reconfiguration can occur according to the available resources (sensors, processors, effectors), to the wished logical correctness, to the measured timeliness and to the events occurring in the environment. But, instead of doing this in a centralized manner, the agents will need to control the reconfiguration themselves, in addition to their normal activity of data processing.

Our objective here is to propose a complete approach, from a software engineering point of view, for the design of adaptive multi-agent systems. It covers all stages of software life cycle, from an abstract specification of the application architecture to a testable implementation, including formal verification of properties and simulation. The method is based on the formalism of timed automata [1], which allows to express systems as a set of concurrent processes satisfying some time constraints (section 3). We show that this formalism may be used in order to model a multi-agent system from the angle of data processing as well as that of dynamic treatment chain reconfiguration (section 4). Then, we show how model-checking and simulation may be used to verify selected properties of the system and analyze a priori its behavior (section 5). Finally, we address the problem of semi-automated translation from a timed automata specification to executable agents (section 6). But before giving more details about this work, it is necessary to give some words of explanation about our target application and its specificities.

2   Target Application and Objectives

The context in which we develop our approach is the project that we call *Dance with Machine* [12]. This project aims at staging a real-time dialogue between a human dancer-actor and a multimodal multimedia distributed cognitive system. The role of the latter is to achieve in real-time the captation and analysis of the performance of the dancer, and to build a multimedia answer to it. This answer may consist in visual animations projected on screens around the dancer, musical sequences, or actions by robots or other physical objects. We consider this application as a metaphorical transposition of the kind of interactions that we may forecast between human users and communicating objects. This is called *Ambient Cognitive Environments* (ACE), i.e., physical environments in which perception, processing and action devices have to organize dynamically and in a cooperative way in order to provide users with natural interaction and extended services.

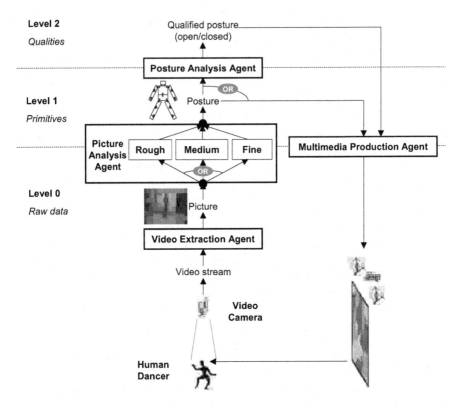

Fig. 1. Global architecture of the processing chain in the project "Dance with Machine"

The computerized setup is composed of a set of processors equipped with communication capabilities. They may also be connected to sensors (video cameras, biometric sensors, localization sensors, etc.) or effectors (screens, loudspeakers, engines, etc.). Each processor may run one or several agents, each of them being specialized for a specific kind of treatment. Data retrieved from the sensors must be handled by several agents before being converted into actions. Agents' work is to analyze, synthesize and transform the data that they get. Data produced by an agent are then transmitted to other agents in order to continue the processing. The data are finally used to generate pictures, sounds or actions, either when the analysis is precise enough, or when the available time is too limited. Figure 1 shows a very simplified view of this process. Only one perception modality is represented, which corresponds to a video camera.

The use of agents in this context is justified by the distributed nature of the application (captation, processing and action are distributed among several objects and processors). But the main reason why we use agents is to make the whole system adaptive in various contexts: when components are added or removed, when the global behavior of the system must change, or when time constraints are not met by the system. The main time constraint that the system should respect is the latency, i.e., the time between the acquisition of data by sensors, and the production of corresponding actions by the system, under one form or another. This latency should of course be kept as low as possible so that the reaction of the system seems instantaneous (at least very quick). On the other hand, the analysis of the dancer's performance should be kept as precise and thorough as possible. These two constraints are potentially contradictory since a precise and thorough analysis can take significantly more time than a rough and superficial one. The quality of an analysis can be measured along two complementary dimensions: the precision (for the measure of a parameter of the performance) and the thoroughness (when optional treatments are possible, a superficial processing will be limited to what is compulsory).

Our main purpose is to allow a very quick evaluation of various strategies in the control of the processing chain, in order to produce an efficient agent-based implementation of the system. We achieve it using a formal model of the system along with tools that we developed to automate the implementation of a functional prototype. Model-checking allows to verify that the systems complies to the specified constraints (latency, non-blocking, sequentiality of treatments, etc.). Simulation, for its part, allows to evaluate the quality of the compromise between logical correctness (is the quality of processing satisfactory?) and timeliness (does the system comply to time constraints?).

3   Introduction to Timed Automata

Real-time systems may be specified using numerous dedicated methods and formalisms. Most of them are graphical semi-formal notations allowing a state machine representation of the behavior of the system. Among the most popular formalisms, we may quote Grafcet [7], SA/RT [17], Statecharts [8], UML/RT [5]. Such visual representations do not enable to verify the properties of systems and it is necessary to associate a formal semantics to them, based in general on process algebras [9], Petri nets [6] or temporal logics [15]. Proposing a new formalism is not our intention here. On the contrary, we prefer to examine the potential benefit of real-time specification and verification techniques in the design and the programming of agent-based reactive systems. We chose for this purpose to use timed automata [1]. This formalism has the advantage to be relatively simple to manipulate and to possess adequate expressivity in order to model time constrained concurrent systems. Moreover, there exists for this model powerful implemented tools (e.g., UPPAAL [13]) allowing model-checking and simulation.

3.1   Standard Model

A timed automaton is a finite state automaton provided with a continuous time representation through real-valued variables, called *clocks*, allowing to express time constraints. Generally, a timed automaton is represented by an oriented graph, where the nodes correspond to states of the system while the arcs correspond to the transitions between these states. The time constraints are expressed

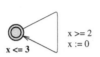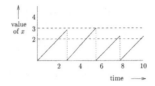

Fig. 2. Example of a timed automaton, where x is a clock. The guard $x \geq 2$ and the invariant $x \leq 3$ imply that the transition will fire after 2 and before 3 time units passed in the state

through *clock constraints* and may be attached to states as well as to transitions. A clock constraint is a conjunction of atomic constraints which compare the value of a clock x, belonging to a finite set of clocks, to a rational constant c. Each timed automaton has a finite number of *states* (locations), one of them being distinguished as *initial*. In each state, the time progression is expressed by a uniform growth of the clock values. In that way, in a state at each instant, the value of the clock x corresponds to time passed since the last reset of x. A clock constraint, called an *invariant*, is associated to each state. It has to be satisfied in order for the system to be allowed to stay in this state. Transitions between states are instantaneous. They are conditioned by clock constraints, called *guards*, and may also reset some clocks. They may also carry labels allowing synchronization. An example of timed automaton and a corresponding possible execution is shown in figure 2.

The behavior of a complex system may be represented by a single timed automaton being a product of a number of other timed automata. The set of states of this resulting automaton is the Cartesian product of states of the component automata, the set of clocks is the union of clocks, and similarly for the labels. Each invariant in the resulting automaton is the conjunction of the invariants of the states of the component automata, and the arcs correspond to the synchronization guided by the labels of the corresponding arcs.

3.2   Extensions in UPPAAL

We use UPPAAL for our modelling; a detailed presentation of this tool may be found in [13]. We remind here only the main characteristics and extensions with respect to the standard model [1]. In UPPAAL, a timed automaton is a finite structure handling, in addition to a finite set of clocks evolving synchronously with time, a finite set of integer-valued and Boolean variables. A model is composed of a set of timed automata, which communicate using binary synchronization through transition labels and a syntax of emission/reception. By convention, a label $k!$ indicates the emission of a signal on a channel k. It is supposed to be synchronized with the signal of reception, represented by a complementary label $k?$. Absence of synchronization labels indicates an internal action of the automaton. The execution of the model starts in the initial configuration (corresponding to the initial state of each automaton with all variable values set to zero), and is a succession of reachable configurations. The configuration change may occur for three reasons:

- by time progression corresponding to d time units in the states of the components, provided that all the state invariants are satisfied. In the new configuration, the clock values are increased by d and the integer variables do not change;
- by a synchronization if two complementary actions in two distinct components are possible, and if the corresponding guards are satisfied. In the new configuration, the corresponding states are changed and the values of clocks and of integer variables are modified according to the reset and update indications;
- by an internal action if such an action of a component is possible, it may be executed independently of the other components: the state and the variables of the component are modified as above.

Another peculiarity of UPPAAL, useful in expressing a kind of synchronicity of moves, is the notion of "committed" states, labelled in the figures by a special label C; see, for instance, the state *Choice* in the first automaton of figure 5. In such a state, no delay is permitted. This implies an immediate move of the concerned component. Thus, two consecutive transitions sharing a committed state are executed without any intermediate delay.

UPPAAL allows simulating systems specified in this way, detecting deadlocks and to verify, through model-checking, various reachability properties. Typically, it can answer the questions like "starting from its initial state, does the system reach a state where a given property is satisfied?", "starting from its initial state, is a given property always true?", or "starting from its initial state, can the system reach a given state in a given delay?".

4   Modelling a Decentralized Reactive System

As stated earlier, timed automata allow to model systems as a set of concurrent processes. We will detail gradually in the sequel the way they may be applied to our case study. The behavior of our agents consists in receiving and processing input data in order to generate and send new outputs. The processing has a duration, considered as fixed, and has to be performed repeatedly. The corresponding model is shown in figure 3.

Initially, the agent is waiting for new data in the state *Idle*. It starts processing on reception of the signal *WorkForAgentN* passing to the state *Processing*. It

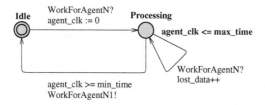

Fig. 3. A model of a simple agent

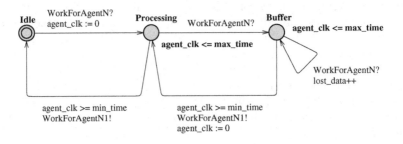

Fig. 4. A model of an agent with a buffer

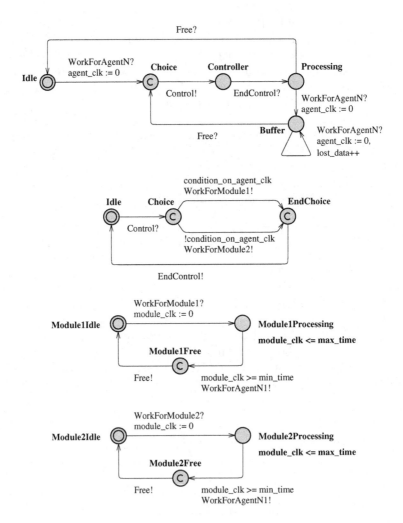

Fig. 5. A model composed of a generic agent, a controller module and two treatment modules

comes back to the state *Idle* at the end of its treatment, which takes a time comprised between *min_time* and *max_time*. The following agent is informed then (through the synchronisation on the channel *WorkForAgentN1*), that it can start processing.

This simple model presents however the following drawback: if a new treatment request comes to an agent when it is already processing, the corresponding data is lost. The number of such events is counted by incrementing the variable *lost_data*. Nevertheless, the loop at the state *Processing* is necessary to avoid deadlocks which may occur if the situation described above happens. A solution can be to introduce an additional state playing the role of a buffer (see figure 4).

Now, if a new request arrives to the agent while it is in the state *Processing*, it passes to the state *Buffer*. Then, it comes back to the state *Processing* at the end of the treatment, in order to start a next one. If a new request comes when the agent is already in the state *Buffer*, then the corresponding data is lost. At this stage, we shall still take into account the fact that a few modules (corresponding to various precisions of the processing) are available and may be used to analyze the dancer's posture. A first approach consists in duplicating the agent in charge of the corresponding treatment by associating to each copy a different duration constant. However, when a new data is available, it is transmitted to one of the agents chosen in a non-deterministic way. Thus, it is necessary to incorporate in the agent a controller responsible for choosing between different treatment modules. This solution is represented in figure 5.

When some data is ready to be processed, the controller module passes in the state *Choice*. The agent chooses to execute a treatment module depending on the value of the boolean expression *condition_on_agent_clk*. When the chosen module achieves processing, it informs about it the next agent in the processing chain, then it informs the controller by sending the signal *Free*.

5   Verification and Simulation

The controller presented in the previous section needs of course to be instantiated by fixing explicitly the criteria determining the choice between treatment modules. We present three different strategies that may be considered and address verification and simulation experiences which may be accomplished for some interesting properties. The particular context considered for this study is explained in figure 6.

The extraction agent produces an image every 50 ms, which has to be treated by the agent in charge of the analysis. This treatment should be performed either by a module capable to accomplish a complete analysis or by a module which can do only a partial one but taking less time ($t_{treatment_2} < t_{treatment_1}$). The controller has to be designed in such a way that it could be possible to conciliate two potentially contradictory criteria: analyzing all images or, in other words, avoiding loosing too many of them (timeliness) and performing a maximum of complete analyzes (logical correctness).

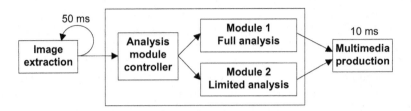

Fig. 6. A simplified scheme of the processing chain

5.1   Different Strategies of Choice

The first proposal is not really a strategy but we give it as a reference. It consists only in systematically alternating the two treatment modules.

In order to minimize the loss of images, the idea is to anticipate, when the agent performs the choice (t_{choice}), the date when the agent will receive a new image to analyze while it has already an image in its buffer and has not terminated its current analysis (t_{loss}). This is possible since the frequency of arrivals of new images is constant. Thus, in the second strategy, the module 1 will be chosen if and only if $t_{treatment_1} < t_{loss} - t_{choice}$.

In order to maximize the number of complete analyzes, one can loosen the previous constraint by allowing to use the module 1 even if its execution will necessarily entail a loss of an image. In the third strategy, the module 1 will be chosen if and only if $t_{treatment_1} < (t_{loss} - t_{choice}) * coef$, where $coef$ fixes the limits of allowance.

5.2   Results

For each strategy, it is possible to check with UPPAAL that the system satisfies certain properties. In particular, we checked that:

- there is no deadlock: A[ ] not deadlock;
- there is no image lost: A[ ] lost_data == 0;
- the ratio of the choice of module 1 is grater than a given threshold:
  A[ ] (nb1 * 100 / (nb1 + nb2 + lost_data)) > 50).

Moreover, it is possible to simulate the system during a given number of cycles and to check experimentally the ratio of lost images and images which could be analyzed completely versus treatment times $t_{treatment_1}$ and $t_{treatment_2}$, as shown in figure 7.

Model-checking techniques allow to verify formally and automatically if some properties of the system, considered as important, are satisfied in all possible system evolutions. On the other hand, simulation permits to obtain some empirical evaluation of performances of the system in terms of logical correctness and timeliness, depending on the characteristics of treatment modules and on the applied strategy. This allows also envisaging a supplementary control level for the agent in charge of the image analysis. This corresponds to a kind of "meta-strategy" which could adapt dynamically the strategy of choice depending on various constraints and fixed objectives.

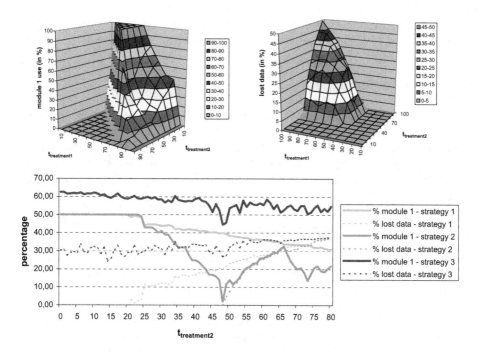

Fig. 7. The ratio of images analyzed with the module 1 (on the left) and the ratio of lost images (on the right), obtained for the second strategy and various values of time of treatment for modules 1 and 2. On the bottom, a comparison of the three strategies for $t_{treatment_1} = 80ms$ and $coef = 1.25$, for various values of $t_{treatment_2}$

6   Automated Code Generation

After having validated the model of the multi-agent system, both formally and experimentally, the next stage of development corresponds to translating it into an executable prototype. In order to do so, a naive idea could consist in implementing each timed automaton by a thread, since they are models of concurrent processes. Nevertheless, for a same agent modelled by several automata, it could involve several synchronization and lead to decline sensibly its performances, which could be awkward for a reactive system. Thus, a first step consists in performing first a synchronized product of all automata describing the same agent in order to transform it next into a skeleton of an application. The compiler that we developed produces this synchronized product by performing also a number of optimizations in order to minimize the size of the resulting automaton. Each agent is modelled consequently by a unique timed automaton, which can be translated into an executable form in several steps. First, only the finite state automaton aspects of the given timed automaton are considered. The states where it is necessary to let the time progress are assumed to correspond to some treatments. Our compiler translates it in terms of a state in which the agent does a break (which is supposed to be replaced by the corresponding treatment module when it is available). Finally, the synchronization signals between automata are associated to communications between the corresponding agents.

7   Conclusion

We presented in this paper a complete approach, from the software engineering point of view, for the modelling of adaptive real-time systems based on the multi-agent paradigm. The usage of timed automata specification and verification techniques played here a central and unifying role. We showed how this formalism, thanks to its capabilities to model concurrent processes having time constraints, can be adapted in order to represent multi-agent systems. Moreover, we demonstrated that it could be possible to model in a modular way an agent controller, capable to make decisions depending on some fixed objectives.

The advantage one can take from this formal specification is twofold: First, it is possible to check the model against various kinds of deadlock (or timelock) and more generally, against any property coming from a non-respect of time constraints, and avoid this way some problems at a very early stage of development. Second, it is worthwhile to take advantage of timed automata representation of the system in order to generate automatically application skeletons. To do so, we developed a specific compiler which, taking an XML representation of the timed automata specification, produces a skeleton based on the JADE multi-agent platform [4]. This prototype is finally used to validate choices made previously, during modelling and implementation, and to review and modify some of them if necessary.

Finally, the general purpose of this work consists in exploiting the approach described in this paper, the design patterns and the composition tools, in order to facilitate the design of an entire system. These design patterns could be coupled with machine learning techniques for the exploration of parameter spaces, in order to optimize agent behaviors when the model becomes more complex. Also, it would be interesting to develop an experimental protocol in order to validate, on the real prototype, the properties observed on the model. In this context, the presented work, even if it is at a preliminary stage, demonstrates however the feasibility of this approach and allows to foresee favorably the development of powerful and complete tools dedicated to the implementation of reactive multi-agent systems.

References

1. Alur R., Dill D. L., A Theory of Timed Automata, in Theoretical Computer Science, Vol. 126, No. 2, pp. 183-236, 1994.
2. Arai T., Stolzenburg F., Multiagent systems specification by UML statecharts aiming at intelligent manufacturing, in AAMAS'2002, pp. 11-18, 2002.
3. Attoui A., Les systmes multi-agents et le temps-rel, Eyrolles, 1997.
4. Belleifemine F., Caire G., Poggi A., Rimassa G., JADE - A White Paper, http://sharon.cselt.it/projects/jade/papers/WhitePaperJADEEXP.pdf, 2003.

5. Douglass B. P., Real-Time UML: Developing Efficient Objects for Embedded Systems, Addison-Wesley-Longman, Reading, MA, 1998.
6. Elmstrm R., Lintulampi R., Pezze M., Giving Semantics to SA/RT by Means of High-Level Timed Petri Nets, in RTSJ, Vol. 5, No. 2/3, pp. 249-271, 1993.
7. Groupe AFCET Systmes Logiques. Pour une reprsentation normalise du cahier des charges d'un automatisme logique, in RAII, Vol. 61 & 62, 1977.
8. Harel D., Statecharts : A Visual Formalism for Complex Systems, in Science of Computer Programming, Vol. 8, 1987.
9. Harel D., Pnueli A., Schmidt J. P., Sherman R., On the Formal Semantics of Statecharts, LICS 1987, pp. 54-64, 1987.
10. Hatley D. J. , Pirbhai I., Strategies for Real Time System Specification, Dover Press, 1987.
11. S. Horstmann and G. Cornell. Core Java 2, Vol. 1 & 2, Prentice Hall, 1999.
12. Hutzler G., Gortais B., Joly P., Orlarey Y., Zucker J.-D., J'ai dans avec machine ou comment repenser les rapports entre l'homme et son environnement, in JFI-ADSMA'2002, pp.147-150, Herms Science, 2002.
13. Larsen K. G., Pettersson P., Yi W., UPPAAL in a Nutshell, in Springer International Journal of Software Tools for Technology Transfer, 1(1-2), pp. 134-152, 1998.
14. Occello M., Demazeau Y., Baeijs C., Designing Organized Agents for Cooperation with Real-Time Constraints, in CRW'98, pp. 25-37, Springer-Verlag, 1998.
15. Manna Z., Pnueli A., The Temporal Logic of Reactive and Concurrent Systems: Specification, Springer-Verlag, 1991.
16. Soler J., Julian V., Rebollo M., Carrascosa C., Botti V., Towards a Real-Time Multi-Agent System Architecture, in COAS, AAMAS'2002, 2002.
17. Ward P., Mellor S., Structured Development for Real-Time Systems, Prentice-Hall, 1985.
18. Wolfe V. F., DiPippo L. C., Cooper G., Johnston R., Kortman P., Thuraisingham B., Real-Time CORBA, in IEEE TPDS, Vol. 11, no. 10, 2000.

An Approach to V&V of Embedded Adaptive Systems

Sampath Yerramalla[1], Yan Liu[1], Edgar Fuller[2], Bojan Cukic[1], and Srikanth Gururajan[3]

[1] Lane Department of Computer Science and Electrical Engineering
{yanliu, sampath, cukic}@csee.wvu.edu
[2] Department of Mathematics and Institute for Math Learning
ef@math.wvu.edu
[3] Mechanical and Aerospace Engineering Department,
West Virginia University, Morgantown WV 26506
srikanth@web.cemr.wvu.edu

Abstract. Rigorous Verification and Validation (V& V) techniques are essential for high assurance systems. Lately, the performance of some of these systems is enhanced by embedded adaptive components in order to cope with environmental changes. Although the ability of adapting is appealing, it actually poses a problem in terms of V&V. Since uncertainties induced by environmental changes have a significant impact on system behavior, the applicability of conventional V& V techniques is limited. In safety-critical applications such as flight control system, the mechanisms of change must be observed, diagnosed, accommodated and well understood prior to deployment.

In this paper, we propose a non-conventional V&V approach suitable for online adaptive systems. We apply our approach to an intelligent flight control system that employs a particular type of Neural Networks (NN) as the adaptive learning paradigm. Presented methodology consists of a novelty detection technique and online stability monitoring tools. The novelty detection technique is based on *Support Vector Data Description* that detects novel (abnormal) data patterns. The Online Stability Monitoring tools based on *Lyapunov's Stability Theory* detect unstable learning behavior in neural networks. Cases studies based on a high fidelity simulator of NASA's Intelligent Flight Control System demonstrate a successful application of the presented V&V methodology.

1 Introduction

The use of biologically inspired soft computing systems (neural network, fuzzy logic, AI planners) for online adaptation to provide adequate system functionality in changing environments has revolutionized the operation of realtime automation and control applications. In the instance of a safety-critical adaptive flight control system, these changes in the environment can be, for example, a stuck stabilator, broken aileron and/or rudder, sensor failure, etc. Stability and

M.G. Hinchey et al. (Eds.): FAABS 2004, LNAI 3228, pp. 173–188, 2005.

safety are two major concerns for such systems. In recent years, NASA conducted series of experiments evaluating adaptive computational paradigms for providing fault tolerance capabilities in flight control systems following sensor and/or actuator failures. Experimental success suggest significant potential for developing and deploying such fault tolerant controllers for futuristic airplanes [1, 2, 3, 4].

The non-probabilistic evolving functionality of realtime controllers, through judicious online learning, aid the adaptive system (aircraft) to recuperate from operational damage (sensor/actuator failure, changed aircraft dynamics: broken aileron or stabilator, etc). This adds an additional degree of complexity and system uncertainty. Since it is practically impossible to estimate and analyze beforehand all possible issues relative to adaptive system's safety and stability, these systems require a non-conventional, sophisticated V&V treatment. While adaptive systems in general are considered inherently difficult to V&V, system uncertainties coupled with other real time constraints make existing traditional V&V techniques practically useless for online adaptive systems and implementation of a non-conventional V&V technique a challenging task [5, 6]. This (in)ability to provide a theoretically valid and practically feasible verification and validation remains one of the critical factors limiting wider use of neural networks based flight controllers [5, 6, 7].

We propose a non-conventional V&V approach and derive a validation methodology suitable for online adaptive systems. We apply our approach to an adaptive flight control system that employs Neural Networks (NN) as the adaptive learning paradigm. Presented V&V methodology consists of an online novelty detection technique and online stability monitoring tools. The novelty detection technique is based on Support Vector Data Description (SVDD) in order to detect novel (abnormal) data patterns. As a one-class classifier, the support vector data description is able to form a decision boundary around the learned data domain with very little or even zero knowledge outside the boundary. The online stability monitoring tools based on Laypunov stability theory are designed to detect unstable (unusual) NN behavior. The underlying mathematics of the online monitoring tools is a rigorous mathematical stability verification technique. This technique emphasizes the need for a precise stability definition for adaptive systems and reasons about the self-stabilizing properties of the adaptive neural network within the control system's architecture.

1.1   Paper Overview

We propose a V&V framework that is suitable for online adaptive systems in Section 2. The presented validation approach requires an understanding of two complementary novelty detection and stability analysis techniques that are discussed in detail in Sections 3 and Section 4. In Section 5, test cases and simulation results describing the operational behavior of the online novelty and stability analysis are discussed in detail. We conclude the paper with a brief discussion on the prospects of the presented V&V approach for other online adaptive systems in Section 6.

2   A V&V Framework

One of the goals of our V&V and safety assurance approach is to ensure the correct diagnosis followed by blocking/permitting of novel (abnormal or unreliable) data from entering the online adaptive component, the neural network. We propose to use novelty detectors and safety monitors as online filters [8]. Figure 1 illustrates the V&V framework. The SVDD data analysis technique is capable of detecting anomalies in the neural network's inputs and outputs. Safety monitors disallow the propagation of unsafe controller gains (adjustments) from entering the controller. It is evident that such a device must require a wide range of system (aircraft) domain-knowledge. Therefore, we seek to define a control error adjustment and detection technique suitable for alerting from anomalous, unstable, and eventually unsafe aircraft behavior if the outputs from neural network adaptation were to enter the controller.

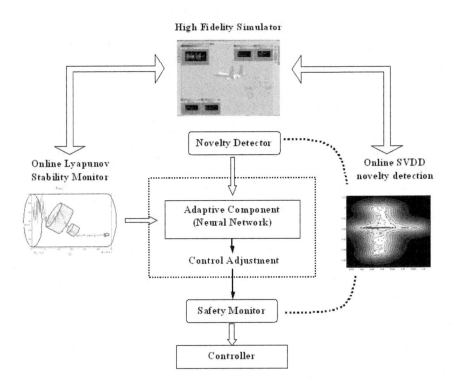

Fig. 1. Adaptive Flight Control System's V&V Framework

Another key step of the validation framework is the runtime stability monitor. Its goal is to determine whether, under given flight conditions, the neural network converges, i.e., if its state transition trajectories lead to a stationary state. The online monitor is complemented by mathematical stability proofs [9] that can define its engagement or disengagement. In other words, to preserve computational resources the online monitor may not be engaged in flight conditions that are considered to be *a priori* safe.

3   Novelty Detection Technique

In general, novelty detection techniques require beforehand knowledge of both nominal and off-nominal flight domains. However, for the validation of NN in online adaptive systems, it is impossible to anticipate all possible adverse environmental conditions and/or failure modes. Under flight failure scenarios, the performance of most regular classification models deteriorate due to restrictions in their generalization capabilities and low quality data. As a one-class classification tool, Support Vector Data Description (SVDD) technique is derived from Support Vector learning theory by Tax et. al. [10, 11]. Differing from general support vector classifiers that decide the maximum margin hyperplane to separate two classes, SVDD method tries to find an optimal decision boundary for a given data set. Thus, it provides the best possible representation of the target-class and offers inferences that can be used to detect the outliers from the nominal feature space. This, for our validation purposes, can be defined as the *"safe region"*, relating to nominal flight conditions.

SVDD is developed from the concept of finding a sphere with the minimal volume to contain all data [12, 13, 14]. Given a data set S consisting of N examples $x_i, i = 1, .., N$, the SVDD's task is to minimize an error function containing the volume of this sphere. With the constraint that all data points must be within the sphere, which is defined by its radius R and its center a, the objective function can be translated into the following form by applying Lagrangian multipliers,

$$L(R, a, \alpha_i) = R^2 - \sum_i \alpha_i \{R^2 - (x^2 - 2ax_i + a^2)\}$$

where $\alpha_i > 0$ is the Lagrange multiplier. L is to be minimized with respect to R and a and maximized with respect to α_i. By solving the partial derivatives of L, we also have:

$$\sum_i \alpha_i = 1;$$

and

$$a = \sum_i \alpha_i x_i,$$

which gives the Lagrangian with respect to α_i:

$$L = \sum_i \alpha_i (x_i \cdot x_i) - \sum_{i,j} \alpha_i \alpha_j (x_i \cdot x_j)$$

where $\alpha_i \geq 0$ and $\sum_i \alpha_i = 1$.

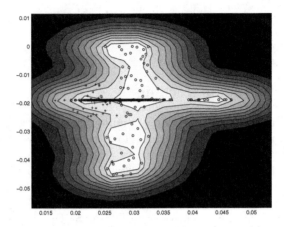

Fig. 2. SVDD with different distances from the center

In the solution that maximizes L, a large portion of α_i's become zero. The rest of α_i's are greater than zero and their corresponding objects are those called support objects. They lie on the boundary that forms a sphere that contains the data. Hence, object z is accepted by the description when:

$$\|z - a\|^2 = (z - \sum_i \alpha_i x_i)(z - \sum_i \alpha_i x_i) \leq R^2.$$

Real world systems usually produce multi-dimensional highly nonlinear data that are inseparable by a linear discriminant. This makes the data description harder to obtain. Similar to the Support Vector Machine (SVM) [10], by replacing some kernel function $K(x, y)$ with the product of (x, y) in the above equations, we are able to map our data from a high dimensional space onto a Hilbert space, which is also referred to as the "feature space". In the feature space, objects can be classified with lower complexity. Selecting the well-known Gaussian kernel function, where $K(x, y) = exp(-\|x - y\|^2/s^2)$, we now have:

$$L = 1 - \sum_i \alpha_i^2 - \sum_{i \neq j} \alpha_i \alpha_j K(x_i, x_j).$$

The formula of checking object z now becomes:

$$1 - 2\sum_i \alpha_i K(z, x_i) + \sum_{i,j} \alpha_i \alpha_j K(x_i, x_j) \leq R^2.$$

Since the SVDD is used as a one-class classifier, in practice, there are no actual outliers well defined other than those randomly drawn from the rest of the space outside the target class. However, by applying the SVDD method, we can obtain a relatively sound representation of the target class. To detect outliers, a more precise criteria should be inferred from empirical testing or pre-defined thresholds. By setting the boundaries to a certain distance from the center, Figure 2 illustrates the different boundaries with respect to different parameter settings. A rule of thumb here is that the greater the value of the distance from the center, the rougher the boundary. Therefore, the number of the outliers that can be detected decreases. In practice, a pre-defined threshold can be used as the furthest distance of a data point from the center, which the system can tolerate. Such pre-defined thresholds need sufficient testing within each specific data domain.

4   Online Monitoring

Self-organizing neural networks, introduced by Kohonen [15] and modified by several others [17, 18, 19] over the last twenty years, offer topology-preserving adaptive learning capabilities that can, in theory, respond and learn to abstract from a much wider variety of complex data-manifolds, the type of data encountered in an adaptive flight control system.

The adaption of neural networks can successfully model the topology and abstract the information from data patterns that have a predictable structure. However, during online adaptation, the data patterns may be presented to the network at a varying sampling rates. The presented data can exhibit pathological dimensional stratification, such as uniformity or functional discontinuities. It has been observed (experimentally) that under these circumstances, the neural network encounters difficulties in learning and abstracting information from the presented data, eventually leading to a deteriorating network performance. In such cases the neural network might fail in its primary goal *"to successfully learn and provide a better estimate of the learnt parameters to the flight controller"*. This degradation in the network's performance is depicted in a loss of its self-stabilizing properties. The goal of an online stability monitor is to capture and analyze the self-stabilizing properties of the network in the hope that it will be able to detect unstable neural network behavior and warn the pilot/system of the imminent threat to the controller.

The construction of an online stability monitor is based on rigorous mathematical stability analysis methodology, *Lyapunov's direct method* [16]. According to this method, a system is said to be stable near a given solution one can construct a Lyapunov function (scalar function) that identifies the regions of the state space over which such functions decrease along some smooth trajectories near the solution. In the discrete sense, Lyapunov stability can be defined as follows:

Definition 1. *Lyapunov Stability*
If there exists a Lyapunov function, $V : \mathbb{R}^O \to \mathbb{R}$, defined in a region of state space near a solution of a dynamical system such that

1. $V(0) = 0$
2. $V(x) > 0 : \forall x \in O, x \neq 0$
3. $V(x(t_{i+1})) - V(x(t_i)) = \Delta V(x) \leq 0 : \forall x \in O$

then the solution of the system is said to stable in the sense of Lyapunov.

$x = 0$ represents a solution of the dynamical systems and $\mathbb{R}^O$, O represent the output space and a region surrounding this solution of the system respectively.

According to the above definition a system is stable if all solutions of the state that start nearby end up nearby. A good distance measure of nearby must be defined by a Lyapunov function (V) over the states of the system. By constructing V, we can guarantee that all trajectories of the system converge to a stable state. The function V should be constructed keeping in mind that it needs be scalar $(V \in \mathbb{R})$ and should be non-increasing over the trajectories of the state space. This is required in order to ensure that all *limit points* of any trajectory are stationary.

Definition 2. *Asymptotic Stability (AS)*
If in addition to conditions 1 and 2 of Definition 1, the system has a negative-definite Lyapunov function

$$\Delta V(x) < 0 : \forall x \in O \tag{1}$$

then the system is Asymptotically Stable.

Asymptotic stability adds the property that in a region surrounding a solution of the dynamical system trajectories are approaching this given solution asymptotically.

Definition 3. *Global Asymptotic Stability (GAS)*
If in addition to conditions 1 and 2 of Definition 1, the Lyapunov function is constructed such that,

$$\lim_{t \to \infty} V(x) = 0 \tag{2}$$

*over the **entire** state space then the system is said to be Globally Asymptotically Stable.*

A notable difference between AS and GAS is the fact that GAS implies any trajectory beginning at *any* initial point will converge asymptotically to the given solution, as opposed to AS where only those trajectories beginning in the neighborhood of the solution approach the solution asymptotically. The types of stability defined above have increasing property strength.

Global Asymptotic Stability $\Longrightarrow$ Asymptotic Stability $\Longrightarrow$ Lyapunov Stability.

The reverse implication does not necessarily hold as indicated by the Venn diagram of Figure 3. Thus a strict Lyapunov function should force every trajectory to asymptotically approach an equilibrium state. Even for non-strict Lyapunov functions, it is possible to guarantee convergence by LaSalle's invariance principle. In mechanical systems a Lyapunov function is considered as an energy minimization term. In economic and finance evaluations it is considered as a cost-minimization term, and for computational purposes it can be considered as an error-minimization term. Figure 4 shows a Lyapunov function for

Fig. 3. Relative strengths of Stability

the NN operation where the decreasing cylinder radii indicate a converging, stable operation. The online stability monitor essentially computes Lyapunov and Lyapunov-like functions (similar to the one shown in Figure 4) based on the current states of the neural network learner and analyze each function to evaluate the overall network stability. Thus, online stability monitoring complements analytical stability analysis techniques by being being able to detect system states that deviate away from stable equilibria in real-time.

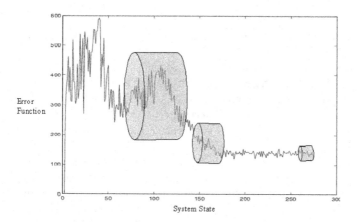

Fig. 4. A converging Lyapunov-like function

5   Case Study

The knowledge gained through the design and evaluation of new control schemes is of direct use in performance verification and validation. Proper experimentation is required to justify realism and applicability of the proposed techniques into actual practice.

5.1 The Intelligent Flight Control System

The Intelligent Flight Control System (IFCS) was primarily developed by NASA as a novel flight control system with the primary goal to *"flight evaluate control concepts that incorporate emerging soft computing algorithms (NN or AI techniques) to provide an extremely robust aircraft capable of handling multiple accident and/or an off-nominal flight scenario"* [1, 2, 7].

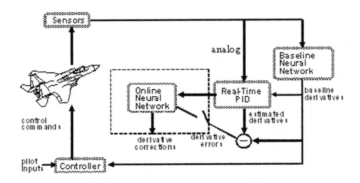

Fig. 5. The Intelligent Flight Control System

The diagram of Figure 5 shows the architecture of the IFCS using Dynamic Cell Structure (DCS) neural network, referred to as the Online Learning Neural Network (OLNN). The control concept can be briefly described as follows. Notable discrepancies from the outputs of the the Baseline (Pre-trained) Neural Network (PTNN) and the Real-time Parameter Identification (PID), either due to a change in the aircraft dynamics (loss of control surface, aileron, stabilator) or due to sensor noise/failure, are accounted by the Online Learning Neural Network. The primary goal of OLNN is to learn online and provide a better estimate for future use of these discrepancies, commonly known as Stability and Control Derivative errors. The critical role played by the online learning neural network in fine-tuning the control parameters and providing a smooth control adjustments is the motivation for the need for a practical, nonconventional validation methodology.

Major advances in the development of modern control laws have generated the need for developing very detailed and sophisticated simulation environments for R&D purposes. Novel techniques for adaptive flight control achieves maturity through extensive experimentation in simulated environments. Figure 6 shows the interface of the IFCS F-15 simulator developed by the WVU research team. The control framework of the simulator is based on the IFCS architecture, shown in Figure 5. Through the high fidelity simulator, we are able to collect valuable data representing nominal flight conditions as well as some failure scenarios.

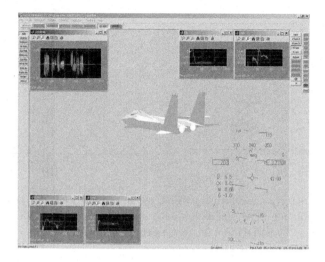

Fig. 6. NASA-WVU F-15 Simulator

5.2 Flight-Data Description

The simulation data depicts nominal and off-nominal flight conditions of approximately 10 seconds of flying time corresponding to 200 frames of data at the simulation rate of 20Hz. A data frame is a point in a seven-dimensional space corresponding to 4 sensor readings (independent variables) and 3 stability and control derivative errors from PID and PTNN (dependant variables). The NN tested here is the $DCS - C_z$ network, one of the five DCS-subnetworks of the IFCS. The independent variables are Mach number (the ratio of the speed of the aircraft to the local speed of sound), alpha (aircraft's angle of attack) and altitude of the aircraft. The dependent variable are three stability and control derivative errors generated by the difference between PID and PTNN.

In the following sections, we first present novelty detection results using SVDD on the NN training data. Online stability monitoring results for NN learning are described next. Both tools are tested on two failure mode data sets obtained from the simulator. The two specific types of failures induced in the IFCS simulator are control surface failures (stuck aileron, stabilator) and loss of control surface. A control surface failure (locked left stabilator, stuck at +3 Degree) is simulated into the system at the 100th data frame. In another simulation, a loss of control surface (50% missing surface of right aileron) failure is also induced at the 100th data frame.

5.3 Novelty Detection Using SVDD

We first simulate one run of nominal flight conditions of 40 seconds with a segment of 800 data points saved. After running SVDD on the nominal data, we obtain a sound data description of nominal flight conditions. The data description is then used to detect novel data that falls outside the boundary. The crosses

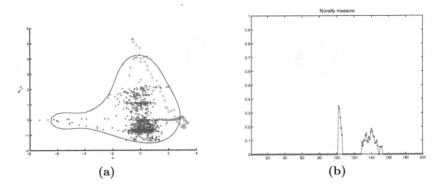

Fig. 7. Novelty detection results using SVDD on control surface failure simulation data. (a): SVDD of nominal flight simulation data is used to detect novelties. (b): Novelty measures returned by SVDD tool for each testing data point

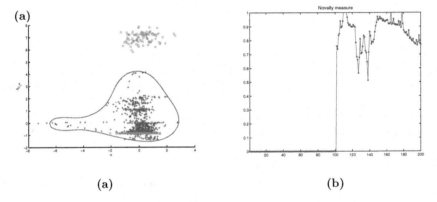

Fig. 8. Novelty detection results using SVDD on loss of control surface failure simulation data. (a): SVDD of nominal flight simulation data is used to detect novelties. (b): Novelty measures returned by SVDD tool for each testing data point

in Figure 7(a) and Figure 8(a) represent the nominal data points on which the boundary is found by SVDD.

We then use the boundary formed by SVDD to test on failure mode simulation data. Novelty detection results for control surface failure simulation data and loss of control surface failure simulation data are shown in Figure 7 and Figure 8 respectively. Circles in Figure 7(a) and Figure 8(a) represent failure mode simulation data. In the plot of Figure 8(a), depicting the loss of control surface failure, a large portion of failure mode data falls outside the boundary. The loss of control surface failure indicates a more substantial damage than the stuck-at type of failure. Consequently, the data points in Figure 8(a) fall further outside the nominal data boundary than the data points in Figure 7(a). The novelty measures shown in Figure 7(b) and Figure 8(b) are probability-like measures computed for each data point based on the distance from the SVDD boundary formed on the nominal flight condition data. Correspondingly, in plots of Figure 7(b) and Figure 8(b), we can see that the novelty measures of loss of control surface failure data after 100th data frame are larger than those of control surface failure data. In both figures, after the 100th point, when failures occurred, SVDD detects the abnormal changes and returns with the highest novelty measures. This demonstrates the reasonably effective and accurate detection capabilities of our SVDD detector.

5.4   Online Stability Monitoring

Described novelty detection mechanisms provide an independent approach to reliable failure detection, thus enhancing the ability of the system analyst to evaluate the mechanisms in charge of the activation adaptive component(s). Online stability monitors serve the purpose of evaluating whether adaptive subsystem provide adequate accommodation capabilities that address specific environmental conditions. In other words, the monitors track the adaptation process and continually evaluate the difference between the current state abstraction provided by the learning device (DCS neural network in our case study) and its desired goal.

Adaptive systems are associated with uncertainty, many degrees of freedom and high noise-level in real flight conditions. Due to their complexity, we may not always be able to check to see if each dimension of the input data is effectively abstracted and represented by the neural network. Lyapunov theory provides the tool to collapse the multidimensional evaluation criteria into one or a few meaningful bounded functions. The data sets being modeled in the case study represent short data sequences for one out of five neural networks in the intelligent flight control system. We constructed four Lyapunov-like functions to reduce the need for checking effective learning by each dimension. Rather than looking onto several dozen graphs, the adequacy (stability) of learning can be assessed from the analysis of these four graphs, representing the Lyapunov functions.

The four Lyapunov-like functions are specific for the DCS neural network of the Intelligent Flight Control System. Their formal description would require detailed presentation of the DCS learning algorithm, which is outside of the scope of this paper. In general terms, the DCS network is a so called self-organizing map. Self-organizing maps evolve their topology to reflect as closely as possible the topological characteristics of the data set being approximated. Therefore, by measuring euclidian distances within the evolving network and comparing them with actual distances in the training data set, we may derive the measure of the goodness of approximation. The four Lyapunov like functions were defined because they evaluate different aspects of DCS adaptation: the Kohonen's rule and the competitive Hebbian rule [19, 20]. Furthermore, we noticed that these four functions react with different intensities to different training data sets. Given that these data sets represent actual aircraft failure scenarios, selected Lyapunov-like functions complement each other.

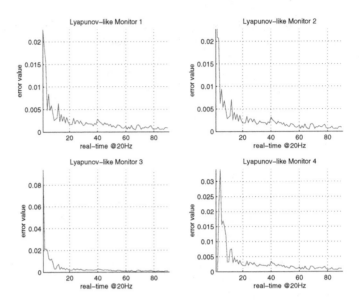

Fig. 9. Online Monitors: Pre-control Surface Failure

As the neural network starts to adapt to the presented failure mode data, the run-time monitor is engaged. It continually monitors the behavior of the neural network. Figure 9 shows the plots of the four Lyapunov-like monitors before a control surface failure (locked left stabilator, stuck at +3 Degree) is induced into the system, and before it propagates into the neural network. Figure 9 shows no predominant spikes in the individual monitors, indicating the lack of intense adaptation in nominal conditions. Because the neural network does not attempt to change the control input to the flight control system, its output bears very limited overall system risk during this period.

Figure 10 shows the plots of the four Lyapunov-like monitors after the control surface failure (locked left stabilator, stuck at +3 Degree) is induced into the system and after the failure propagates into the neural network. Figure 11 shows the plots of the four Lyapunov-like monitors after the loss of control surface (50% missing surface of right aileron) is simulated into the system and after the failure propagates into the neural network. The plots show a predominant spike at time frame 100 (the time of the failure). The spikes indicate the successful detection of the unusual (failed) environmental condition by monitoring the internals of the neural network. In the short term, the neural network undergoes a significant degree of adaptation. The high values of the Lyapunov-like functions indicate that the neural network needs additional time (and learning cycles) to faithfully represent its newly arrived (in real-time) input data set. During this period, the confidence on neural network's output diminishes drastically, i.e., the network is not providing the desirable failure accommodation. But, Within the next 50 or so frames in Figure 10, the values of Lyapunov-like monitors approach 0, indicating that the failure has been accommodated through adaptation. The failure

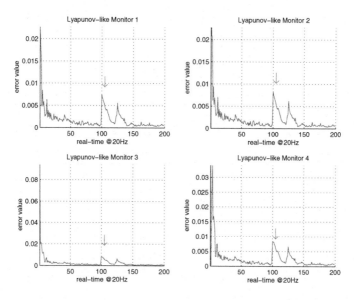

Fig. 10. Online Monitors: Post-control Surface Failure

accommodation delay is longer in Figure 11, an expected indication of the severe failure condition (the loss of a control surface). At this point, a verification and validation engineer needs to assess the adequacy of the failure accommodation mechanism with respect to the overall system safety requirements, evaluate alternative designs, and prepare suitable V&V recommendations to the safety board.

6   Conclusions

We developed a non-conventional approach for validating the performance adequacy of the neural network embedded in an online adaptive flight control system. The validation framework consists of

- Online filters (novelty detectors) that check the validity of inputs and control outputs, and
- Runtime stability monitors that examine the stability properties of the neural network adaptation.

Experimental results from the data collected on an F-15 aircraft flight simulator show that:

1. SVDD can be adopted for defining nominal performance regions for the given application domain. Our techniques provided successfully automated separation between faulty behaviors and normal system events in real-time operation.

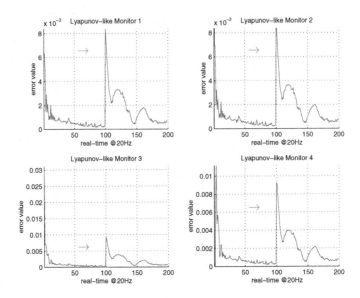

Fig. 11. Online Monitors: Post-Loss of Control Surface Failure

2. Based on the originally developed concept of Lyapunov-like functions applied for the first time to neural network learning, the online stability monitors have shown a successful realization of convergence tracking of adaptation error towards a stable (or unstable) and safe (or unsafe) state in the adaptive flight control system.

We conclude that the proposed methodology provides a good approach for validating online adaptive system's safety, stability and performance. The observed efficiency and scalability of both methods give us the expectation that the proposed V&V method can be successfully applied to other types of online adaptive learners.

Acknowledgment

This work was funded in part by grants to West Virginia University Research Corp. from the NASA Office of Safety and Mission Assurance (OSMA) Software Assurance Research Program (SARP) managed through the NASA Independent Verification and Validation (IV &V) Facility, Fairmont, West Virginia, NASA research grant No. NAG4-163; and from the NSF CAREER award CCR-0093315.

References

1. The Boeing Company, Intelligent flight control: advanced concept program, Technical report, 1999.

2. Charles C. Jorgensen. Feedback linearized aircraft control using dynamic cell structures, *World Automation Congress* (ISSCI), Alaska, 1991, 050.1-050.6.
3. M. Napolitano, C. D. Neppach, V. Casdorph, S. Naylor, M. Innocenti, and G Silvestri, A neural network-based scheme for sensor failure detection, identification and accomodation, *AIAA Journal of Control and Dynamics*, Vol. 18, No. 6, 1995, 12801286.
4. Institute of Software Reseach. Dynamic cell structure neural network report for the intelligent flight control system, Technical report, Document ID: IFC-DCSR-D002-UNCLASS-010401, January 2001.
5. Johann Schumann, and Stacy Nelson, Towards V&V of neural network based controllers. *Workshop on Self-Healing Systems*, 2002.
6. D. Mackall, S. Nelson, and J. Schumann. Verification and validation of neural networks of aerospace applications, Technical report, CR-211409, NASA, 2002.
7. M.A. Boyd, J, Schumann, G. Brat, D. Giannakopoulou, B. Cukic, and A. Mili, Validation and verification process guide for software and neural nets, Technical report, NASA Ames Research Center, September 2001.
8. Y. Liu, S. Yerramalla, E. Fuller, B. Cukic and S. Gururajan. Adaptive Control Software: Can We Guarantee Safety? *Proc. of the 28th International Computer Software and Applications Conference, workshop on Software Cybernetics*, Hong Kong, September 2004.
9. S. Yerramalla, E. Fuller, B. Cukic. Lyapunov Analysis of Neural Network Stability in an Adaptive Flight Control System, *6th Symposium on Self Stabilizing Systems (SSS-03)*, San Francisco, CA, June 2003.
10. V. N. Vapnik, Statistical learning theory, Wiley, 1998.
11. D.M.J. Tax, and R.P.W. Duin, Outliers and data descriptions, *Proc. ASCI 2001, 7th Annual Conf. of the Advanced School for Computing and Imaging* (Heijen, NL, May 30-June 1), ASCI, Delft, 2001, 234-241.
12. D.M.J. Tax, and R.P.W. Duin, Support vector domain description, *Pattern Recognition Letters*, Vol. 20, No. 11-13, 1999, 1191-1199.
13. D.M.J. Tax, and R.P.W. Duin, Data domain description using support vectors, *Proc. European Symposium on Artificial Neural Networks* (Bruges, April 21-23, 1999), D-Facto, Brussels, 1999, 251-257.
14. D.M.J. Tax, A. Ypma, and R.P.W. Duin, Support vector data description applied to machine vibration analysis, *Proc. 5th Annual Conference of the Advanced School for Computing and Imaging* (Heijen, NL, June 15-17), ASCI, Delft, 1999, 398-405.
15. Teuvo Kohonen, The self-organizing map, *Proc. of the IEEE*, Vol. 78, No. 9, September, 1990, 1464-1480.
16. V. I. Zubov, Methods of A. M. Lyapunov and their applications, U.S. Atomic Energy Commission, 1957.
17. Bernd Fritzke, Growing cell structures - a self-organizing network for unsupervised and supervised learning, *Neural Networks*, Vol. 7, No. 9, May 1993, 1441-1460.
18. Thomas Martinetz, and Klaus Schulten, Topology representing networks, *Neural Networks*, Vol. 7, No. 3, 1994, 507-522.
19. Jorg Bruske, and Gerald Sommer, Dynamic cell structure learns perfectly topology preserving map, *Neural Computation*, Vol. 7, No. 4, 1995, 845-865.
20. N.Rouche, P.Habets, and M.Laloy, Stability theory by Liapunov's direct method, Springer-Verlag, New York Inc., 1997.

Verifying Multi-agent Systems via Unbounded Model Checking*

M. Kacprzak[1], A. Lomuscio[2], T. Łasica[3], W. Penczek[3,4], and M. Szreter[3,**]

[1] Białystok University of Technology, Institute of Mathematics and Physics,
15-351 Białystok, ul. Wiejska 45A, Poland
mdkacprzak@wp.pl
[2] Department of Computer Science, King's College London,
London WC2R 2LS, United Kingdom
alessio@dcs.kcl.ac.uk
[3] Institute of Computer Science, PAS, 01-237 Warsaw, ul. Ordona 21, Poland
{tlasica, penczek, mszreter}@ipipan.waw.pl
[4] Podlasie Academy, Institute of Informatics, Siedlce, Poland

Abstract. We present an approach to the problem of verification of epistemic properties in multi-agent systems by means of symbolic model checking. In particular, it is shown how to extend the technique of unbounded model checking from a purely temporal setting to a temporal-epistemic one. In order to achieve this, we base our discussion on interpreted systems semantics, a popular semantics used in multi-agent systems literature. We give details of the technique and show how it can be applied to the well known train, gate and controller problem.

Keywords: Model checking, unbounded model checking, multi-agent systems.

1 Introduction

Verification of reactive systems by means of model-checking techniques [3] is now a well-established area of research. In this paradigm one typically models a system S in terms of automata (or by a similar transition-based formalism), builds an implementation P_S of the system by means of a model-checker friendly language such as the input for SMV or PROMELA, and finally uses a model-checker such as SMV or SPIN to verify some temporal property ϕ the system: $M_P \models \phi$, where M_P is a temporal model representing the executions of P_S. As it is well known, there are intrinsic difficulties with the naive approach of performing this operation on an explicit representation of the states, and refinements of symbolic techniques (based on OBDD's, and SAT [1] translations) are

---

* The authors acknowledge support from the Polish National Committee for Scientific Research (grant No 4T11C 01325, a special grant supporting ALFEBIITE), the Nuffield Foundation (grant NAL/00690/G), and EPSRC (GR/S49353/01).
** M. Szreter acknowledges support from the US Navy via grant N00014-04-1-4063 issued by the Office of Naval Research International Field Office.

M.G. Hinchey et al. (Eds.): FAABS 2004, LNAI 3228, pp. 189–212, 2005.

being investigated to overcome these hurdles. Formal results and corresponding applications now allow for the verification of complex systems that generate tens of thousands of states.

The field of multi-agent systems (MAS) has also recently become interested in the problem of verifying complex systems. In MAS the emphasis is on the autonomy, and rationality of the components, or agents [22]. In this area, modal logics representing concepts such as knowledge, beliefs, intentions, norms, and the temporal evolution of these are used to specify high level properties of the agents. Since these modalities are given interpretations that are different from the ones of the standard temporal operators, it is not straightforward to apply existing model checking tools developed for standard *Linear Temporal Logic* (LTL) (or *Computation Tree Logic*, CTL) temporal logic to the specification of MAS. One further problem is the fact that the modalities that are of interest are often not given a precise interpretation in terms of the computational states of the system, but simply interpreted on classes of Kripke models that guarantee (via frame-correspondence) that some intuitive properties of the system are preserved[1]. This makes it hard to use the semantics to model any actual computation performed by the system [21]. For the case of knowledge, the semantics of interpreted systems [8], popularized by Halpern and colleagues in the 90's, can be used to give an interpretation to the modalities that maintains the traditional S5 properties, while, at the same time, is appropriate for model checking [9]. Indeed, a considerable amount of literature now exists on the application of interpreted systems and epistemic logic to the application areas of security, modelling of synchronous, asynchronous systems, digital rights, etc. It is fair to say that this area constitutes the most thoroughly explored, and technically advanced sub-discipline among the formal studies of multi-agent systems available at the moment.

1.1   State of the Art and Related Literature

The recent developments in the area of model checking MAS can broadly be divided into streams: in the first category standard predicates are used to interpret the various intensional notions and these are paired with standard model checking techniques based on temporal logic. Following this line is for example [23] and related papers. In the other category we can place techniques that make a genuine attempt at extending the model checking techniques by adding other operators. Works along these lines include [19, 20, 12, 17, 16, 15, 14, 10].

In [19] local propositions are used to translate knowledge modalities on LTL structures. Once this process is done, the result can be fed into a SPIN model checker. Unfortunately, in this approach local propositions need to be computed by the user.

---

[1] For example, in epistemic logic it is customary to use equivalence models to interpret a knowledge modality K so that it inherits the properties of the logical systems S5 [2]; in particular axioms T, 4, and 5 (which are considered to be intuitively correct for knowledge) result valid.

These works were preceded by [12], where van der Meyden and Shilov presented theoretical properties of the model checking problems for epistemic linear temporal logics for interpreted systems with perfect recall. In particular, it was shown that the problem of checking a language that includes "until" and "common knowledge" on perfect recall systems is undecidable, and decidable fragments were identified.

In [17, 16, 15] an extension of standard temporal verification via model checking on obdd's to epistemic and deontic operators is presented and studied.

In [14, 10] an extension of the method of bounded model checking (one of the main SAT-based techniques) to CTLK a language comprising both CTL and knowledge operators, was defined, implemented, and evaluated. While preliminary results appear largely positive, any bounded model checking algorithm is mostly of use when the task is either to check whether a universal CTLK formula is actually false on a model, or to check that an existential CTLK formula is valid. This is a severe limitation in MAS as it turns out that many of the most interesting properties one is interested in checking actually involve universal formulas. For example, in a security setting one may want to check whether it is true that forever in the future a particular secret, perhaps a key, is mutually known by two participants.

1.2   Aim of This Paper

The aim of this paper is to contribute to the line of SAT-based techniques, by overcoming the intrinsic limitation of any bounded model checking algorithm, and provide a method for model checking the full language of CTLK. The SAT-based method we introduce and discuss here is an extension to knowledge and time of a technique introduced by McMillan [11] called *unbounded model checking (UMC)*. A byproduct of the work presented here is the definition of fixed point semantics for a logic CTL_pK, which extends CTLK by past operators.

Like any SAT-based method, UMC consists in translating the model checking problem of what is in this case a CTL_pK formula into the problem of satisfiability of a propositional formula. UMC exploits the characterization of the basic modalities in terms of *Quantified Boolean Formulas* (QBF), and the algorithms that translate QBF and fixed point equations over QBF into propositional formulas. In order to adapt UMC for checking CTL_pK, we use three algorithms. The first one, implemented by the procedure *forall* [11] (based on the Davis-Putnam-Logemann-Loveland approach [4]) eliminates the universal quantifier from a QBF formula representing a CTL_pK formula, and returns the result in *conjunctive normal form* (CNF). The remaining algorithms, implemented by the procedures *gfp* and *lfp* calculate the greatest and the least fixed points for the modal formulas in use here. Ultimately, the technique allows for a CTL_pK formula α to be translated into a propositional formula $[\alpha](w)$ [2] in CNF, which characterizes all the states of the model, where α holds.

---

[2] Note that w is a vector of propositional variables used to encode the states of the model.

For the case of CTL it was shown by McMillan [11] that model checking via UMC can be exponentially more efficient than approaches based on BDD's in two situations:

- whenever the resulting fixed points have compact representations in CNF, but not via BDD's;
- whenever the SAT-based image computation step proves to be faster than the BDD-based one.

Although we do not prove it here, we expect a similar increase in efficiency for model checking of CTL_pK over interpreted systems.

The rest of the paper is structured in the following manner. Section 2 introduces interpreted systems semantics, the semantics on which we ground our investigation. The logic CTL_pK is defined in Section 3. Section 4 summarize the basic definitions that we need for CNF and QBF formulas, and fixes the notation we use throughout the paper. A fixed point characterization of CTL_pK formulas is presented in Section 5. The main idea of symbolic model checking CTL_pK is described in section 6, where algorithms for computing propositional formulas equivalent to CTL_pK formulas are also given. Two examples on the use of the algorithms of this paper are given in Section 7. Preliminary experimental results are shown in Section 8, whereas conclusions are given in Section 9.

2   Interpreted Systems Semantics

Any transition-based semantics allows for the representation of temporal flows of time by means of the successor relation. For example, UMC for CTL uses plain Kripke models [11]. To work on a temporal epistemic language, we need to consider a semantics that allows for an automatic representation of the epistemic relations between computational states [21]. The mainstream semantics that allows to do so is the one of interpreted systems [8].

Interpreted systems can be succinctly defined as follows (we refer to [8] for more details). Assume a set of agents $A = \{1, \ldots, n\}$, a set of local states L_i and possible actions Act_i for each agent $i \in A$, and a set L_e and Act_e of local states and actions for the environment. The set of possible global states for the system is defined as $G = L_1 \times \ldots \times L_n \times L_e$, where each element $(l_1, \ldots, l_n, l_e)$ of G represents a computational state for the whole system (note that, as it will be clear below, some states in G may actually be never reached by any computation of the system). Further assume a set of protocols $P_i : L_i \to 2^{Act_i}$, for $i = 1, \ldots, n$, representing the functioning behaviour of every agent, and a function $P_e : L_e \to 2^{Act_e}$ for the environment. We can model the computation taking place in the system by means of a transition function $t : G \times Act \to G$, where $Act \subseteq Act_1 \times \ldots \times Act_n \times Act_e$ is the set of joint actions. Intuitively, given an initial state ι, the sets of protocols, and the transition function, we can build a (possibly infinite) structure that represents all the possible computations of the system. Many representations can be given to this structure; since in this paper we are only concerned with temporal epistemic properties, we shall find the following to be a useful one.

Definition 1 (Models). *Given a set of agents* $A = \{1, \ldots, n\}$, *a temporal epistemic* model *(or simply a* model*) is a pair* $M = (\mathcal{K}, \mathcal{V})$ *with* $\mathcal{K} = (G, W, T, \sim_1, \ldots, \sim_n, \iota)$, *where*

- *G is the set of the* global states *for the system (henceforth called simply* states*)*;
- $T \subseteq G \times G$ *is a total binary (successor) relation on G;*
- *W is a set of* reachable global states *from ι, i.e., $W = \{s \in G \mid (\iota, s) \in T^*\}$[3],*
- $\sim_i \subseteq G \times G$ *(i $\in$ A) is an* epistemic accessibility relation *for each agent $i \in A$ defined by $s \sim_i s'$ iff $l_i(s') = l_i(s)$, where the function $l_i : G \to L_i$ returns the local state of agent i from a global state s; obviously $\sim_i$ is an equivalence relation,*
- $\iota \in W$ *is the* initial state*;*
- $\mathcal{V} : G \longrightarrow 2^{\mathcal{PV}_K}$ *is a* valuation function *for a set of propositional variables $\mathcal{PV}_K$ such that* **true** $\in \mathcal{V}(s)$ *for all $s \in G$. $\mathcal{V}$ assigns to each state a set of propositional variables that are assumed to be true at that state.*

Note that in the definition above we include both all possible states and the subset of reachable states. The reason for this follows from having past modalities in the language (see the next section), which are defined over any possible global state so that a simple fixed point semantics for them can be given. Still, note that, if required, it is possible to restrict the range of the past modalities to reachable states only by insisting that the target state is itself reachable from the initial state.

By $|M|$ we denote the number of states of M, by $\mathbb{N} = \{0, 1, 2, \ldots\}$ the set of natural numbers and by $\mathbb{N}_+ = \{1, 2, \ldots\}$ the set of positive natural numbers.

Epistemic Relations. When we consider a group of agents, we are often interested in situations in which *everyone in the group* knows a fact α. In addition to this it is sometimes useful to consider other kinds of group knowledge. One of these is the one of *common knowledge*. A group of agents has common knowledge about α if everyone knows that α, and everyone knows that everyone knows α, and everyone knows that everyone knows that everyone knows that α, and so on. For example common knowledge is achieved following information broadcasting with no faults. A different notion is the one of *distributed knowledge* (sometimes referred to as "implicit knowledge", or "wise-man" knowledge). A fact α is distributed knowledge in a group of agents if it could be inferred by pooling together the information the agents have. We refer to [8] for an introduction to these concepts.

Let $\Gamma \subseteq A$. Given the epistemic relations for the agents in Γ, the union of Γ's accessibility relations defines the epistemic relation corresponding to the modality of everybody knows: $\sim_\Gamma^E = \bigcup_{i \in \Gamma} \sim_i$. $\sim_\Gamma^C$ denotes the transitive closure of $\sim_\Gamma^E$, and corresponds to the relation used to interpret the modality of common knowledge. Notice that from reflexivity of $\sim_\Gamma^E$ follows that $\sim_\Gamma^C$ is, in fact, the transitive and reflexive closure of $\sim_\Gamma^E$. The relation used to interpret the modality of distributed knowledge is given by taking the intersection of the relations corresponding to the agents in Γ.

---

[3] T^* denotes the reflexive and transitive closure of T.

Computations. A *computation* in M is a possibly infinite sequence of states $\pi = (s_0, s_1, \ldots)$ such that $(s_i, s_{i+1}) \in T$ for each $i \in \mathbb{N}$. Specifically, we assume that $(s_i, s_{i+1}) \in T$ iff $s_{i+1} = t(s_i, act_i)$, i.e., s_{i+1} is the result of applying the transition function t to the global state s_i, and a joint action act_i. All the components of act_i are prescribed by the corresponding protocols P_j for the agents at s_i. In the following we abstract from the transition function, the actions, and the protocols, and simply use T, but it should be clear that this is uniquely determined by the interpreted system under consideration. Indeed, these are given explicitly in the example in the last section of this paper. In interpreted systems terminology a computation is a *part* of a run; note that we do not require s_0 to be an initial state. For a computation $\pi = (s_0, s_1, \ldots)$, let $\pi(k) = s_k$, and $\pi_k = (s_0, \ldots, s_k)$, for each $k \in \mathbb{N}$. By $\Pi(s)$ we denote the set of all the infinite computations starting at s in M.

3 Computation Tree Logic of Knowledge with Past (CTL$_p$K)

Interpreted systems are traditionally used to give a semantics to an epistemic language enriched with temporal connectives based on linear time [8]. Here we use *Computation Tree Logic* (CTL) by Emerson and Clarke [7] as our basic temporal language and add an epistemic and past component to it. We call the resulting logic *Computation Tree Logic of Knowledge with Past* (CTL$_p$K).

Definition 2 (Syntax of CTL$_p$K**).** *Let $\mathcal{PV}_K$ be a set of propositional variables containing the symbol* **true**. *The set of* CTL$_p$K *formulas $\mathcal{FORM}$ is defined inductively by using the following rules only:*

- *every member p of $\mathcal{PV}_K$ is a formula,*
- *if α and β are formulas, then so are $\neg\alpha$, $\alpha \wedge \beta$ and $\alpha \vee \beta$,*
- *if α and β are formulas, then so are* $\mathrm{AX}\alpha$, $\mathrm{AG}\alpha$, *and* $\mathrm{A}(\alpha\mathrm{U}\beta)$,
- *if α is formula, then so are* $\mathrm{AY}\alpha$ *and* $\mathrm{AH}\alpha$,
- *if α is formula, then so is* $\mathrm{K}_i\alpha$, *for $i \in A$,*
- *if α is formula, then so are* $\mathrm{D}_\Gamma\alpha$, $\mathcal{C}_\Gamma\alpha$, *and* $\mathrm{E}_\Gamma\alpha$, *for $\Gamma \subseteq A$.*

The other modalities are defined by duality as follows:

- $\mathrm{EF}\alpha \stackrel{def}{=} \neg\mathrm{AG}\neg\alpha$, $\mathrm{EP}\alpha \stackrel{def}{=} \neg\mathrm{AH}\neg\alpha$, $\mathrm{EZ}\alpha \stackrel{def}{=} \neg\mathrm{AZ}\neg\alpha$, for $Z \in \{X, Y\}$,
- $\overline{\mathrm{K}}_i\alpha \stackrel{def}{=} \neg\mathrm{K}_i\neg\alpha$, $\overline{\mathrm{D}}_\Gamma\alpha \stackrel{def}{=} \neg\mathrm{D}_\Gamma\neg\alpha$, $\overline{\mathcal{C}}_\Gamma\alpha \stackrel{def}{=} \neg\mathcal{C}_\Gamma\neg\alpha$, $\overline{\mathrm{E}}_\Gamma\alpha \stackrel{def}{=} \neg\mathrm{E}_\Gamma\neg\alpha$.

Moreover, $\alpha \Rightarrow \beta \stackrel{def}{=} \neg\alpha \vee \beta$, $\alpha \Leftrightarrow \beta \stackrel{def}{=} (\alpha \Rightarrow \beta) \wedge (\beta \Rightarrow \alpha)$, and **false** $\stackrel{def}{=}$ $\neg$**true**. We omit the subscript Γ for the epistemic modalities if $\Gamma = A$, i.e., Γ is the set of all the agents. As customary X, G stand for respectively "at the next step", and "forever in the future". Y, H are their past counterparts "at the previous step", and "forever in the past". The *Until* operator U, precisely $\alpha\mathrm{U}\beta$, expresses that β occurs eventually and α holds continuously until then.

Definition 3 (Interpretation of CTL$_p$K). *Let* M $= (\mathcal{K}, \mathcal{V})$ *be a model with* $\mathcal{K} = (G, W, T, \sim_1, \ldots, \sim_n, \iota)$, $s \in G$ *a state,* π *a computation, and* α, β *formulas of* CTL$_p$K. M, $s \models \alpha$ *denotes that* α *is true at the state* s *in the model* M. M *is omitted, if it is implicitly understood. The relation* $\models$ *is defined inductively as follows:*

$$s \models p \qquad \textit{iff } p \in \mathcal{V}(s),$$
$$s \models \neg\alpha \qquad \textit{iff } s \not\models \alpha,$$
$$s \models \alpha \vee \beta \textit{ iff } s \models \alpha \textit{ or } s \models \beta,$$
$$s \models \alpha \wedge \beta \textit{ iff } s \models \alpha \textit{ and } s \models \beta,$$
$$s \models \text{AX}\alpha \qquad \textit{iff } \forall \pi \in \Pi(s) \; \pi(1) \models \alpha,$$
$$s \models \text{AG}\alpha \qquad \textit{iff } \forall \pi \in \Pi(s) \; \forall_{m \geq 0} \; \pi(m) \models \alpha,$$
$$s \models \text{A}(\alpha \text{U} \beta) \textit{ iff } \forall \pi \in \Pi(s) \; (\exists_{m \geq 0} \; [\pi(m) \models \beta \textit{ and } \forall_{j < m} \; \pi(j) \models \alpha]),$$
$$s \models \text{AY}\alpha \qquad \textit{iff } \forall s' \in G \; (\textit{if } (s', s) \in T, \textit{ then } s' \models \alpha),$$
$$s \models \text{AH}\alpha \qquad \textit{iff } \forall s' \in G \; (\textit{if } (s', s) \in T^*, \textit{ then } s' \models \alpha),$$
$$s \models \text{K}_i\alpha \qquad \textit{iff } \forall s' \in W \; (\textit{if } s \sim_i s', \textit{ then } s' \models \alpha),$$
$$s \models \text{D}_\Gamma\alpha \qquad \textit{iff } \forall s' \in W \; (\textit{if } s \sim_\Gamma^D s', \textit{ then } s' \models \alpha),$$
$$s \models \text{E}_\Gamma\alpha \qquad \textit{iff } \forall s' \in W \; (\textit{if } s \sim_\Gamma^E s', \textit{ then } s' \models \alpha),$$
$$s \models \mathcal{C}_\Gamma\alpha \qquad \textit{iff } \forall s' \in W \; (\textit{if } s \sim_\Gamma^C s', \textit{ then } s' \models \alpha).$$

Definition 4. (Validity) *A* CTL$_p$K *formula* φ *is valid in* M *(denoted* M $\models \varphi$*) iff* M, $\iota \models \varphi$, *i.e.,* φ *is true at the initial state of the model* M.

Notice that the past component of CTL$_p$K does not contain the modality *Since*, which is a past counterpart of the modality *Until* denoted by U. Extending the logic by *Since* is possible, but complicates the semantics, so this is not discussed in this paper.

4 Formulas in Conjunctive Normal Form and Quantified Boolean Formulas

In this section, we shortly describe Davis-Putnam-Logemann-Loveland approach [4] to checking satisfiability of formulas in conjunctive normal form (CNF), and show how to construct a CNF formula that is unsatisfiable exactly when a propositional formula α is valid. Having done so, we apply these two methods to compute a propositional formula equivalent to the quantified boolean formula $\forall v.\alpha$, where v is a vector of propositions. In order to do this we first give some basic definitions. The formalism in this section is from [11] and is reported here for completeness.

Let $\mathcal{PV}$ be a finite set of propositional variables. A *literal* is a propositional variable $p \in \mathcal{PV}$ or the negation of one: $\neg p, p \in \mathcal{PV}$. A *clause* is a disjunction of a set of zero or more literals $l[1] \vee \ldots \vee l[n]$. A disjunction of zero literals is taken to mean the constant **false**. A formula is in a *conjunctive normal form* (CNF) if it is a conjunction of a set of zero or more clauses $c[1] \wedge \ldots \wedge c[n]$. A conjunction of zero clauses is taken to mean the constant **true**. An *assignment* is a partial function from $\mathcal{PV}$ to {**true**, **false**}. An assignment is said to be *total* when its domain is $\mathcal{PV}$. A total assignment A is said to be *satisfying* for a formula α when $\alpha(A) = \mathbf{true}$, i.e., the value of α given by A is $\mathbf{true}$ (under the usual interpretation of the boolean connectives). We equate an assignment A with the conjunction of a set of literals, specifically the set containing $\neg p$ for all $p \in dom(A)$ such that $A(p) = \mathbf{false}$, and p for all $p \in dom(A)$ such that $A(p) = \mathbf{true}$.

For a given CNF formula α and an assignment A, an *implication graph* $\mathrm{IG}(A, \alpha)$ is a maximal directed acyclic graph (V, E), where V is a set of vertices, and E is a set of edges, such that:

- V is a set of literals,
- every literal in A is a root,
- for every vertex l not in A, the CNF formula α contains the clause
    $$cl(l, A, \alpha) \stackrel{def}{=} l \vee \bigvee_{m \in \{l' \in V : (l', l) \in E\}} \neg m,$$
- for all $p \in \mathcal{PV}$, V does not contain both p and $\neg p$.

Notice that the above conditions do not uniquely define the implication graph. We denote by A_α the assignment induced by the implication graph $\mathrm{IG}(A, \alpha)$, i.e., $A_\alpha = \bigwedge_{v \in V} v$, where V is a set of vertices of $\mathrm{IG}(A, \alpha)$. Observe that A_α is an extension of A. Furthermore, $\alpha \wedge A$ implies A_α.

Given two clauses of the form $c[1] = p \vee C_1$ and $c[2] = \neg p \vee C_2$, where C_1 and C_2 are disjunctions of literals, we say that the *resolvent* of $c[1]$ and $c[2]$ is $C_1 \vee C_2$, provided that $C_1 \vee C_2$ contains no contradictory literals, i.e., it does not contain a variable p and its negation $\neg p$. If this happens, the resolvent does not exist. Note that the resolvent of $c[1]$ and $c[2]$ is a clause that is implied by $c[1] \wedge c[2]$.

CNF formulas satisfy useful properties to check their satisfiability. Indeed, notice that a CNF formula is satisfied only when each of its clauses is satisfied individually. Thus, given a CNF formula α and an assignment A, if a clause in α has all its literals assigned value $\mathbf{false}$, then A cannot be extended to a satisfying assignment. A clause that has all its literals assigned to value $\mathbf{false}$ is called a *conflicting* clause. We also say that a clause is in *conflict* when all of its literals are assigned the value $\mathbf{false}$ under A_α. If there exists a clause in α such that the all but one of its literals have been assigned the value $\mathbf{false}$, then the remaining literal must be assigned the value $\mathbf{true}$ for this clause to be satisfied. In particular, in every satisfying assignment which is an extension of the assignment A, the unassigned literal must be $\mathbf{true}$. Such an unassigned literal is called *unit literal*, and the clause it belongs to is called a *unit clause*.

There are several algorithms for determining satisfiability of CNF formulas. Here, we use the algorithm proposed by Davis and Putnam and later modified by Davis, Logemann and Loveland [4]. The algorithm is based on the methods of *Boolean constraint propagation* (BCP) and *conflict-based learning* (CBL) and it is aimed at building a satisfying assignment for a given formula α in an incremental manner. The BCP technique is the most important part of the algorithm; it determines a logical consequence of the current assignment by building an implication graph and detecting unit clauses, and conflicting clauses. When a conflict is detected, as we mentioned above, the current assignment cannot be extended to a satisfying one. In this case, the technique of conflict-based learning is used to deduce a new clause that prevents similar conflicts from reoccurring. This new clause is called a *conflict clause* and is deduced by resolving the existing clauses using the implication graph as a guide.

The following is a generic conflict-based learning procedure that takes an assignment A, a CNF formula α, and a conflicting clause c and produces a conflict clause by repeatedly applying resolution steps until either a termination condition T is satisfied, or no further steps are possible. We elaborate on the condition T below when we discuss how the procedure *deduce* is used by the procedure *forall*.

procedure $deduce(c, A, \alpha)$,
while $\neg T$ and exists $l \in c$ such that $\neg l \notin A$
    let $c =$ resolvent of $cl(\neg l, A, \alpha)$ and c
return c

The resulting clause c is implied by α. Thus it can be added to α without changing its satisfiability.

In the following we show a polynomial-time algorithm that, given a propositional formula α, constructs a CNF formula which is unsatisfiable exactly when α is valid. The procedure works as follows. First, for every β subformula of the formula α (including α itself) we introduce a distinct variable l_β. If β is a propositional variable, then $l_\beta = \beta$. Next we assign a formula $\mathcal{CNF}(\beta)$ to every subformula β according to the following rules:

- if β is a variable then $\mathcal{CNF}(\beta) = \mathbf{true}$,
- if $\beta = \neg\phi$ then $\mathcal{CNF}(\beta) = \mathcal{CNF}(\phi) \wedge (l_\beta \vee l_\phi) \wedge (\neg l_\beta \vee \neg l_\phi)$,
- if $\beta = \phi \vee \varphi$ then $\mathcal{CNF}(\beta) = \mathcal{CNF}(\phi) \wedge \mathcal{CNF}(\varphi) \wedge (l_\beta \vee \neg l_\phi) \wedge (l_\beta \vee \neg l_\varphi) \wedge (\neg l_\beta \vee l_\phi \vee l_\varphi)$,
- if $\beta = \phi \wedge \varphi$ then $\mathcal{CNF}(\beta) = \mathcal{CNF}(\phi) \wedge \mathcal{CNF}(\varphi) \wedge (\neg l_\beta \vee l_\phi) \wedge (\neg l_\beta \vee l_\varphi) \wedge (l_\beta \vee \neg l_\phi \vee \neg l_\varphi)$,
- if $\beta = \phi \rightarrow \varphi$ then $\mathcal{CNF}(\beta) = \mathcal{CNF}(\phi) \wedge \mathcal{CNF}(\varphi) \wedge (l_\beta \vee l_\phi) \wedge (l_\beta \vee \neg l_\varphi) \wedge (\neg l_\beta \vee \neg l_\phi \vee l_\varphi)$.

It can be shown [11] that the formula α is valid when the CNF formula $\mathcal{CNF}(\alpha) \wedge \neg l_\alpha$ is unsatisfiable. This follows from the fact that there is a unique satisfying assignment A' of $\mathcal{CNF}(\alpha)$ consistent with A such that $A'(l_\alpha) = \alpha(A)$.

In our method, in order to have a more succinct notation for complex operations on boolean formulas, we also use *Quantified Boolean Formulas* (QBF), an extension of propositional logic by means of quantifiers ranging over propositions. In BNF: $\alpha ::= p \mid \neg\alpha \mid \alpha \wedge \alpha \mid \exists p.\alpha \mid \forall p.\alpha$. The semantics of the quantifiers is defined as follows:

- $\exists p.\alpha$ iff $\alpha(p \leftarrow \mathbf{true}) \vee \alpha(p \leftarrow \mathbf{false})$,
- $\forall p.\alpha$ iff $\alpha(p \leftarrow \mathbf{true}) \wedge \alpha(p \leftarrow \mathbf{false})$,

where $\alpha \in$ QBF, $p \in \mathcal{PV}$ and $\alpha(p \leftarrow \psi)$ denotes substitution with the formula ψ of every occurrence of the variable p in formula α.

We will use the notation $\forall v.\alpha$, where $v = (v[1],\dots,v[m])$ is a vector of propositional variables, to denote $\forall v[1].\forall v[2]\dots\forall v[m].\alpha$. Moreover, let $\alpha(w)$ be a QBF formula over the propositional variables of the vector $w = (w[1],\dots,w[m])$.

What is important here, is that for a given QBF formula $\forall v.\alpha$, we can construct a CNF formula equivalent to it by using the algorithm *forall* [11].

```
procedure forall(v, α), where v = (v[1], ..., v[m]) and α is a propositio-
nal formula
let φ = CNF(α) ∧ ¬lα, χ = true, and A = ∅
repeat
    if φ contains false, return χ
    else if some c in φ is in conflict
        add clause deduce(c, A, φ) to φ
        remove some literals from A
    else if Aφ is total
        choose a blocking clause c′
        remove literals of form v[i] and ¬v[i] from c′
        add c′ to φ and χ
    else
        choose a literal l such that l ∉ A and ¬l ∉ A
        add l to A
```

The procedure works as follows. Initially it assumes an empty assignment A, a formula χ to be **true** and ϕ to be a CNF formula $\mathcal{CNF}(\alpha) \wedge \neg l_\alpha$. The algorithm aims at building a satisfying assignment for the formula ϕ, i.e., an assignment that falsifies α. The search for an appropriate assignment is based on the Davis-Putnam-Logemann-Loveland approach. The following three cases may happen:

- A conflict is detected, i.e., there exists a clause in ϕ such that all of its literals are false in A_ϕ. So, the assignment A can not be extended to a satisfying one. Then, the procedure *deduce* is called to generate a conflict clause, which is added to ϕ, and the algorithm backtracks, i.e., it changes the assignment A by withdrawing one of the previously assigned literals.
- A conflict does not exist and A_ϕ is total, i.e., the satisfying assignment is obtained. In this case we generate a new clause which is false in the current assignment A_ϕ and whose complement characterizes a set of assignments falsifying the formula α. This clause is called a *blocking clause* and it must have the following properties:
  - it contains only input variables, i.e., the variables over which the input formula α is built,
  - it is false in the current assignment,
  - it is implied by $l_\alpha \wedge \mathcal{CNF}(\alpha)$.

A blocking clause could be generated using the conflict-based learning procedure, but we require the blocking clause to contain only input variables. To do this we use an implication graph, in which all the roots are input literals. Such a graph can be generated in the following way. Let A_ϕ be a satisfying assignment for ϕ, $A' = A_\phi \downarrow V$, i.e., A' is the projection of A_ϕ onto the input variables and let $\phi' = \mathcal{CNF}(\alpha) \wedge \chi$. It is not difficult to show that $A'_{\phi'} = A_\phi$, i.e., both the graphs $IG(A', \phi')$ and $IG(A, \phi)$ induce the same assignments. Furthermore, the variable l_α is in conflict in $IG(A', \phi')$, since ϕ contains the clause $\neg l_\alpha$. Thus, a clause $deduce(l_\alpha, A', \phi')$ is a blocking clause providing that it contains only input variables, what can be ensured by a termination condition T.

Next, in order to quantify universally over the variables $v[1], \ldots, v[m]$, the blocking clause is deprived of the variables either of the form $v[i]$ and the negation of these. This is sufficient as the blocking clause is a formula in CNF. Then, what remains is added to the formulas ϕ and χ and the algorithm continues, i.e., again finds a satisfying assignment for ϕ.
- The first two cases do not apply. Then, the procedure makes a new assignment A by giving a value to a selected variable.

On termination, when ϕ becomes unsatisfiable, χ is a conjunction of the blocking clauses and precisely characterizes $\forall v.\alpha$.

Theorem 1. *Let α be a propositional formula and $v = (v[1], \ldots, v[m])$ be a vector of propositions, then the QBF formula $\forall v.\alpha$ is logically equivalent to the CNF formula $forall(v, \alpha)$.*

The proof of the above theorem follows from the correctness of the algorithm *forall* (see [11]).

Example 1. We illustrate in a quite detailed way (as performed by a solver) some basic operations of the procedure *forall*. To make it simple, we explain these operations for a formula in CNF. So, let $\phi = (\neg v_1) \wedge (v_1 \vee v_4 \vee \neg v_5) \wedge (\neg v_2 \vee v_3) \wedge (v_4 \vee v_5)$ and assume that $\phi = \mathcal{CNF}(\alpha) \wedge \neg l_\alpha$ for some formula α. The aim of the procedure $forall(v_1, \alpha)$ is to find a formula in CNF equivalent to $\forall v_1.\alpha$. We will only show how one blocking clause is generated and added to ϕ and χ. Notice that at the start of the procedure the assignment of v_1 is implied as this variable is the only literal in a clause of ϕ and must be followed in order for the clause to be satisfied. Thus, we have $A = \{\neg v_1\}$. Now, the algorithm decides the assignment for another unassigned variable, say $A(v_2) = \mathbf{true}$. This implies the assignment of v_3, namely $A(v_3) = \mathbf{true}$, so that the clause $(\neg v_2 \vee v_3)$ is satisfied. Next, an assignment $A(v_4) = \mathbf{false}$ is decided, but notice that this implies both v_5 (because of the clause $(v_4 \vee v_5)$) and $\neg v_5$ (because of the clause $(v_1 \vee v_4 \vee \neg v_5)$) – a *conflict*. The implication graph is analysed (several algorithms can be applied [13]) and a *learned clause* $(v_1 \vee v_4)$ is generated and added to the working set of clauses (i.e., ϕ). Notice, that the variables v_2 and v_3 are not responsible for this conflict. The learned clause greatly reduces the number of assignments to be examined as the partial assignment $\{\neg v_1, \neg v_4\}$ is excluded from the future search irrespectively on valuations of the remaining variables. Next, the algorithm withdraws from the assignment of v_4. Notice that the learned clause implies $A(v_4) = \mathbf{true}$. Thus, a satisfying assignment that is found is $A_\varphi = \{\neg v_1, v_2, v_3, v_4, v_5\}$.

A *blocking clause* $(v_1 \lor \neg v_4)$ is generated and the literal v_1 is removed from this clause. We obtain the *blocking clause* $c' = (\neg v_4)$ and c' is added to ϕ and χ. The procedure keeps on going until ϕ does not contain *false*.

5   Fixed Point Characterization of CTL_pK

In this section we show how the set of states satisfying any CTL_pK formula can be characterized by a fixed point of an appropriate function. We follow and adapt, when necessary, the definitions given in [3].

Let $M = ((G, W, T, \sim_1, \ldots, \sim_n, \iota), \mathcal{V})$ be a model. Notice that the set 2^G of all subsets of G forms a lattice under the set inclusion ordering. Each element $G' \subseteq Q$ of the lattice can also be thought of as a *predicate* on G, where the predicate is viewed as being true for exactly the states in G'. The least element in the lattice is the empty set, which corresponds to the predicate **false**, and the greatest element in the lattice is the set G, which corresponds to **true**. A function τ mapping 2^G to 2^G is called a *predicate transformer*. A set $G' \subseteq G$ is a *fixed point* of a function $\tau : 2^G \to 2^G$ if $\tau(G') = G'$.

Whenever τ is monotonic (i.e., when $P \subseteq Q$ implies $\tau(P) \subseteq \tau(Q)$), τ has a least fixed point denoted by $\mu Z.\tau(Z)$, and a greatest fixed point, denoted by $\nu Z.\tau(Z)$. When τ is monotonic and $\bigcup$-continuous (i.e., when $P_1 \subseteq P_2 \subseteq \ldots$ implies $\tau(\bigcup_i P_i) = \bigcup_i \tau(P_i)$), then $\mu Z.\tau(Z) = \bigcup_{i \geq 0} \tau^i(\textbf{false})$. When τ is monotonic and $\bigcap$-continuous (i.e., when $P_1 \supseteq P_2 \supseteq \ldots$ implies $\tau(\bigcap_i P_i) = \bigcap_i \tau(P_i)$), then $\nu Z.\tau(Z) = \bigcap_{i \geq 0} \tau^i(\textbf{true})$ (see [18]).

In order to obtain fixed point characterizations of the modal operators, we identify each CTL_pK formula α with the set $\langle \alpha \rangle_M$ of states in M at which this formula is true, formally $\langle \alpha \rangle_M = \{s \in G \mid M, s \models \alpha\}$. If M is clear from the context we omit the subscript M. Furthermore, we define functions $AX, AY, K_i, E_\Gamma, D_\Gamma$ from 2^G to 2^G as follows:

- $AX(Z) = \{s \in G \mid \text{for every } s' \in G \text{ if } (s, s') \in T, \text{ then } s' \in Z\}$,
- $AY(Z) = \{s \in G \mid \text{for every } s' \in G \text{ if } (s', s) \in T, \text{ then } s' \in Z\}$,
- $K_i(Z) = \{s \in G \mid \text{for every } s' \in G \text{ if } (\iota, s') \in T^* \text{ and } s \sim s', \text{ then } s' \in Z\}$,
- $E_\Gamma(Z) = \{s \in G \mid \text{for every } s' \in G \text{ if } (\iota, s') \in T^* \text{ and } s \sim_\Gamma^E s', \text{ then } s' \in Z\}$,
- $D_\Gamma(Z) = \{s \in G \mid \text{for every } s' \in G \text{ if } (\iota, s') \in T^* \text{ and } s \sim_\Gamma^D s', \text{ then } s' \in Z\}$.

Observe that $\langle O\alpha \rangle = O(\langle \alpha \rangle)$, for $O \in \{AX, AY, K_i, E_\Gamma, D_\Gamma\}$. Then, the following temporal and epistemic operators may be characterized as the least or the greatest fixed point of an appropriate monotonic ($\bigcap$-continuous or $\bigcup$-continuous) predicate transformer.

- $\langle AG\alpha \rangle = \nu Z.\langle \alpha \rangle \cap AX(Z)$,
- $\langle A(\alpha U\beta) \rangle = \mu Z.\langle \beta \rangle \cup (\langle \alpha \rangle \cap AX(Z))$,
- $\langle AH\alpha \rangle = \nu Z.\langle \alpha \rangle \cap AY(Z)$,
- $\langle C_\Gamma \alpha \rangle = \nu Z.E_\Gamma(Z \cap \langle \alpha \rangle)$

The first three equations are standard (see [6], [3]), whereas the fourth one is defined analogously taking account that $\sim_\Gamma^C$ is the transitive, and reflexive closure of $\sim_\Gamma^E$.

6   Symbolic Model Checking on CTL_pK

Let $M = (\mathcal{K}, \mathcal{V})$ with $\mathcal{K} = (G, W, T, \sim_1, \ldots, \sim_n, \iota)$. Recall that the set of global states $G = \times_{i=1}^n L_i$ is the Cartesian product of the set of local states (without loss of generality we treat the environment as one of the agents).

We assume $L_i \subseteq \{0, 1\}^{n_i}$, where $n_i = \lceil \log_2(|L_i|) \rceil$ and let $n_1 + \ldots + n_n = m$, i.e., every local state is represented by a sequence consisting of 0's and 1's. Moreover, let D_i be a set of the indexes of the bits of the local states of each agent i of the global states, i.e., $D_1 = \{1, \ldots, n_1\}, \ldots, D_n = \{m - n_n + 1, \ldots, m\}$.

Let $\mathcal{PV}$ be a set of fresh propositional variables such that $\mathcal{PV} \cap \mathcal{PV}_K = \emptyset$, $F_{\mathcal{PV}}$ be a set of propositional formulas over $\mathcal{PV}$, and $lit : \{0, 1\} \times \mathcal{PV} \to F_{\mathcal{PV}}$ be a function defined as follows: $lit(0, p) = \neg p$ and $lit(1, p) = p$. Furthermore, let $w = (w[1], \ldots, w[m])$, where $w[i] \in \mathcal{PV}$ for each $i = 1, \ldots, m$, be a global state variable. We use elements of G as valuations[4] of global state variables in formulas of $F_{\mathcal{PV}}$. For example $w[1] \wedge w[2]$ evaluates to *true* for the valuation $q = (1, \ldots, 1)$, and it evaluates to *false* for the valuation $q = (0, \ldots, 0)$.

Now, the idea consists in using propositional formulas of $F_{\mathcal{PV}}$ to encode sets of states of G. For example, the formula $w[1] \wedge \ldots \wedge w[m]$ encodes the state represented by $(1, \ldots, 1)$, whereas the formula $w[1]$ encodes all the states, the first bit of which is equal to 1.

Next, the following propositional formulas are defined:

- $I_s(w) := \bigwedge_{i=1}^m lit(s_i, w[i])$.
  This formula encodes the state $s = (s_1, \ldots, s_m)$ of the model, i.e., $s_i = 1$ is encoded by $w[i]$, and $s_i = 0$ is encoded by $\neg w[i]$.
- $H(w, v) := \bigwedge_{i=1}^m w[i] \Leftrightarrow v[i]$.
  This formula represents logical equivalence between global state encodings, representing the fact that they represent the same state.
- $T(w, v)$ is a formula, which is true for a valuation $(s_1, \ldots, s_m)$ of $(w[1], \ldots, w[m])$ and a valuation $(s_1', \ldots, s_m')$ of $(v[1], \ldots, v[m])$ iff $((s_1, \ldots, s_m), (s_1', \ldots, s_m')) \in T$.

Our aim is to translate CTL_pK formulas into propositional formulas. Specifically, for a given CTL_pK formula β we compute a corresponding propositional formula $[\beta](w)$, which encodes those states of the system that satisfy the formula. Operationally, we work outwards from the most nested subformulas, i.e., the atoms. In other words, to compute $[O\alpha](w)$, where O is a modality, we work under the assumption of already having computed $[\alpha](w)$. To calculate the actual translations we use either the fixed point or the QBF characterization of CTL_pK formulas. For example, the formula $[AX\alpha](w)$ is equivalent to the QBF formula $\forall v.(T(w, v) \Rightarrow [\alpha](v))$. We can use similar equivalences for formulas $AY\alpha, K_i\alpha, D_\Gamma\alpha, E_\Gamma\alpha$. More specifically, we use the following three basic algorithms. The first one, implemented by the procedure *forall*, is used for formulas $O\alpha$ such that $O \in \{AX, AY, K_i, D_\Gamma, E_\Gamma\}$. This procedure eliminates the universal quantifier from a QBF formula representing a CTL_pK formula, and returns

---

[4] We identify 1 with *true* and 0 with *false*.

the result in a conjunctive normal form. The second algorithm, implemented by the procedure gfp_O, is applied to formulas $O\alpha$ such that $O \in \{AG, AH, \mathcal{C}_\Gamma\}$. This procedure computes the greatest fixed point. For the formulas of the form $A(\alpha U\beta)$ we use a third procedure, called lfp_{AU}, which computes the least fixed point. In so doing, given a formula β we obtain a propositional formula $[\beta](w)$ such that β is valid in the model M iff the conjunction $[\beta](w) \wedge I_\iota(w)$ is satisfiable, i.e., $\iota \in \langle \beta \rangle$. Below, we formalize the above discussion.

Definition 5 (Translation for UMC). *Given a* CTL_pK *formula* φ, *the propositional translation* $[\varphi](w)$ *is inductively defined as follows:*

- $[p](w) := \bigvee_{s \in \langle p \rangle} I_s(w),$ *for* $p \in \mathcal{PV}_\mathcal{K},$
- $[\neg\alpha](w) := \neg[\alpha](w),$
- $[\alpha \wedge \beta](w) := [\alpha](w) \wedge [\beta](w),$
- $[\alpha \vee \beta](w) := [\alpha](w) \vee [\beta](w),$
- $[AX\alpha](w) := forall(v, (T(w,v) \Rightarrow [\alpha](v))),$
  where $[\alpha](v)$ *denotes* $[\alpha](w)(w \leftarrow v)^5$,
- $[AY\alpha](w) := forall(v, (T(v,w) \Rightarrow [\alpha](v))),$
- $[K_i\alpha](w) := forall(v, ((H_i(w,v) \wedge \neg \, gfp_{AH}(\neg I_\iota(v))) \Rightarrow [\alpha](v))),$
- $[D_\Gamma\alpha](w) := forall(v, ((\bigwedge_{i \in \Gamma} H_i(w,v) \wedge \neg \, gfp_{AH}(\neg I_\iota(v))) \Rightarrow [\alpha](v))),$
- $[E_\Gamma\alpha](w) := forall(v, ((\bigvee_{i \in \Gamma} H_i(w,v) \wedge \neg \, gfp_{AH}(\neg I_\iota(v))) \Rightarrow [\alpha](v))),$
- $[AG\alpha](w) := gfp_{AG}([\alpha](w)),$
- $[A(\alpha U\beta)](w) := lfp_{AU}([\alpha](w), [\beta](w)),$
- $[AH\alpha](w) := gfp_{AH}([\alpha](w)),$
- $[\mathcal{C}_\Gamma\alpha](w) := gfp_{\mathcal{C}_\Gamma}([\alpha](w)).$

The algorithms gfp and lfp are based on the standard procedures computing fixed points.

procedure $gfp_{AG}([\alpha](w))$, where α is a CTL_pK formula
let $Q(w) = [\mathbf{true}](w),\ Z(w) = [\alpha](w)$
while $\neg(Q(w) \Rightarrow Z(w))$ is satisfiable
    let $Q(w) = Z(w),$
    let $Z(w) = forall(v, (T(w,v) \Rightarrow Z(v))) \wedge [\alpha](w)$
return $Q(w)$

The procedure gfp_{AH} is obtained by replacing in the above $forall(v, (T(w,v) \Rightarrow Z(v)))$ with $forall(v, (T(v,w) \Rightarrow Z(v)))$.

procedure $gfp_{\mathcal{C}_\Gamma}([\alpha](w))$, where α is a CTL_pK formula
let $Q(w) = [\mathbf{true}](w),$
    $Z(w) = forall(v, ((\bigvee_{i \in \Gamma} H_i(w,v) \wedge \neg gfp_{AH}(\neg I_\iota(v))) \Rightarrow [\alpha](v)))$
while $\neg(Q(w) \Rightarrow Z(w))$ is satisfiable
    let $Q(w) = Z(w),$

---

[5] Note that by $\alpha(w)(w \leftarrow v)$ we formally mean $[\alpha](w)(w[1] \leftarrow v[1]) \cdots (w[m] \leftarrow v[m])$, where $v = (v[1], \ldots, v[m])$ is a vector of propositional variables.

let $Z(w) = forall(v, (\bigvee_{i \in \Gamma} H_i(w,v) \land \neg gfp_{AH}(\neg I_\iota(v)) \Rightarrow (Z(v) \land [\alpha](v))))$
return $Q(w)$

procedure $lfp_{AU}([\alpha](w), [\beta](w))$, where α, β are CTL$_p$K formulas
let $Q(w) = [\textbf{false}](w)$, $Z(w) = [\beta](w)$
while $\neg(Z(w) \Rightarrow Q(w))$ is satisfiable
    let $Q(w) = Q(w) \lor Z(w)$,
    let $Z(w) = forall(v, (T(w,v) \Rightarrow Q(v))) \land [\alpha](w)$
return $Q(w)$

We now have all the ingredients in place to state the main result of this paper: modal satisfaction of a CTL$_p$K formula can be rephrased as propositional satisfaction of an appropriate conjunction. Note that the translation is sound and complete (details of the proof are not given here).

Theorem 2 (UMC for CTL$_p$K). *Let* M *be a model and* φ *be a* CTL$_p$K *formula. Then,* M $\models \varphi$ *iff* $[\varphi](w) \land I_\iota(w)$ *is satisfiable.*

Proof. Notice that $I_\iota(w)$ is satisfied only by the valuation $\iota = (\iota_1, \ldots, \iota_m)$ of $w = (w[1], \ldots, w[m])$. Thus $[\varphi](w) \land I_\iota(w)$ is satisfiable iff $[\varphi](w)$ is true for the valuation ι of w. On the other hand for a model M, M $\models \varphi$ iff M, $\iota \models \varphi$, i.e., $\iota \in \langle \varphi \rangle$. Hence, we have to prove that $\iota \in \langle \varphi \rangle$ iff $[\varphi](w)$ is true for the valuation ι of w. The proof is by induction on the complexity of φ. The theorem follows directly for the propositional variables. Next, assume that the hypothesis holds for all the proper sub-formulas of φ. If φ is equal to either $\neg \alpha$, $\alpha \land \beta$, or $\alpha \lor \beta$, then it is easy to check that the theorem holds.

For the modal formulas, let P be a set of states and $\alpha_P(w)$ a propositional formula such that $\alpha_P(w)$ is true for the valuation $s = (s_1, \ldots, s_m)$ of $w = (w[1], \ldots, w[m])$ iff $s \in P$. Note that given any P, α_P is well defined: since the set G of all states is finite, and one can take $\bigvee_{s \in P} I_s(w)$ as $\alpha_P(w)$. Consider φ to be of the following forms:

- $\varphi = AY\alpha$. We will prove that $\iota \in \langle AY\alpha \rangle$ iff the formula $[AY\alpha](w)$ is true for the valuation ι of w.
  First we prove that:
  (*) $s \in AY(P)$ iff the formula $\forall v.(T(v,w) \Rightarrow \alpha_P(v))$ is true for the valuation s of w.
  $s \in AY(P)$ iff $s \in \{s' \in G|$ for every $s'' \in G$ if $(s'', s') \in T$, then $s'' \in P\}$. On the one hand, $(s'', s') \in T$ iff $T(v,w)$ is true for the valuation s' of w and the valuation s'' of v. Moreover, $s'' \in P$ iff the formula $\alpha_P(v)$ is true for the valuation s'' of v. Thus $s \in AY(P)$ iff the formula $T(v,w) \Rightarrow \alpha_P(v)$ is true for the valuation s of w and every valuation s'' of v. Hence, $s \in AY(P)$ iff the QBF formula $\forall v.(T(v,w) \Rightarrow \alpha_P(v))$ is true for the valuation s of w.
  Therefore, $\iota \in \langle AY\alpha \rangle$ iff $\iota \in AY(\langle \alpha \rangle)$ iff (by the inductive assumption and (*)) the formula $(\forall v.(T(v,w) \Rightarrow [\alpha](v)))$ is true for the valuation ι of w iff (by Theorem 1) the propositional formula $forall(v, T(v,w) \Rightarrow [\alpha](v))$ is true for the valuation ι of w iff $[AY\alpha](w)$ is true for the valuation ι of w.

- $\varphi = \text{AX}\alpha$. The proof is analogous to the former case.
- $\varphi = \text{AH}\alpha$ We will show that $\iota \in \langle \text{AH}\alpha \rangle$ iff formula $[\text{AH}\alpha](w)$ is true for the valuation ι of w.

First we prove that:

(*) $s \in \nu Z.P \cap \text{AY}(Z)$ iff the formula $gfp_{AH}(\alpha_P(w))$ is true for the valuation s of w.

Let $\tau(Z) = P \cap \text{AY}(Z)$, then $s \in \nu Z.\tau(Z)$ iff $s \in \bigcap_{i \geq 0} \tau^i(G)$ (as $s \in \bigcap_{i \geq 0} \tau^i(\mathbf{true})$). Thus, $s \in \nu Z.\tau(Z)$ iff $s \in \tau^i(G)$ for the least i such that $\tau^i(G) \subseteq \tau^{i+1}(G)$ since for every $i \geq 0$ we have $\tau^{i+1}(G) \subseteq \tau^i(G)$. On the other hand, $s \in \tau(Z)$ iff formula $\alpha_P(w) \wedge \forall v.(T(v,w) \Rightarrow \alpha_Z(v))$ is true for the valuation s of w iff (by Theorem 1) formula $\alpha_P(w) \wedge forall(v, T(v,w) \Rightarrow \alpha_Z(v))$ is true for the valuation s of w.

Let $Z^0(w) = \alpha_P(w)$ and $Z^i(w) = \alpha_P(w) \wedge forall(v, (T(v,w) \Rightarrow Z^{i-1}(v)))$ for $i > 0$. Notice that $s \in \tau^i(G)$ iff $Z^i(w)$ is true for the valuation s of w. Moreover, $Q_i(w) = Z^{i-1}(w)$ and $Z_i(w) = Z^i(w)$ are invariants of the while-loop of the procedure $gfp_{AH}(\alpha_P(w))$. Hence on the termination, when $Q_{i_0}(w) \Rightarrow Z_{i_0}(w)$, where i_0 is the least i such that $Q_i(w) \Rightarrow Z_i(w)$, $gfp_{AH}(\alpha_P(w)) = Q_{i_0}(w)$ is a formula that is true for the valuation s of w iff $s \in \nu Z.\tau(Z)$.

Therefore, $\iota \in \langle \text{AH}\alpha \rangle$ iff $\iota \in \nu Z.\langle \alpha \rangle \cap \text{AY}(Z)$ iff (by the inductive assumption and (*)) the propositional formula $gfp_{AH}([\alpha](w))$ is true for the valuation ι of w iff propositional formula $[\text{AH}\alpha](w)$ is true for the valuation ι of w.

- $\varphi = \text{AG}\alpha \mid \mathcal{C}_\Gamma \alpha \mid \text{A}(\alpha \text{U}\beta)$. The proof is analogous to the former case.
- $\varphi = \text{K}_i \alpha$. In order to show that $\iota \in \langle \text{K}_i \alpha \rangle$ iff formula $[\text{K}_i \alpha](w)$ is true for the valuation ι of w, first we prove that:

(*) $s \in \text{K}_i(P)$ iff the formula $\forall v.(\neg gfp_{AH}(\neg I_\iota(v)) \wedge H_i(w,v) \Rightarrow \alpha_P(v))$ is true for the valuation s of w.

To this aim we prove the following two facts:

(**) $(\iota, s'') \in T^*$ iff $\neg gfp_{AH}(\neg I_\iota(v))$ is true for the valuation s'' of v.

Observe that $s'' \in G \backslash \{\iota\}$ iff $\neg I_\iota(v)$ is true for the valuation s'' of v. On the other hand $(\iota, s'') \notin T^*$ iff $s'' \in \nu Z.(G \backslash \{\iota\}) \cap \text{AY}(Z)$. Hence $(\iota, s'') \in T^*$ iff $s'' \notin \nu Z.(G \backslash \{\iota\}) \cap \text{AY}(Z)$ iff $gfp_{AH}(\neg I_\iota(v))$ is false for the valuation s'' of v iff $\neg gfp_{AH}(\neg I_\iota(v))$ is true for the valuation s'' of v.

(***) $s' \sim_i s''$ iff $H_i(w,v)$ is true for the valuation s' of w and the valuation s'' of v.

$s' \sim_i s''$ iff $l_i(s') = l_i(s'')$ iff $\bigwedge_{j \in D_i} s'_j = s''_j$ iff formula $\bigwedge_{j \in D_i} w[j] \Leftrightarrow v[j]$ is true for the valuation s' of w and the valuation s'' of v iff $H_i(w,v)$ is true for the valuation s' of w and the valuation s'' of v.

Thus by (**) and (***), $s \in \text{K}_i(P)$ iff for the valuation s of w and every valuation s'' of v formula $\neg gfp_{AH}(\neg I_\iota(v)) \wedge H_i(w,v) \Rightarrow \alpha_P(v)$ is true iff the QBF formula $\forall v.(\neg gfp_{AH}(\neg I_\iota(v)) \wedge H_i(w,v) \Rightarrow \alpha_P(v))$ is true for the valuation s of w.

Therefore, $\iota \in \langle \text{K}_i \alpha \rangle$ iff $\iota \in \text{K}_i(\langle \alpha \rangle)$ iff (by the inductive assumption and (*)) the formula $\forall v.(\neg gfp_{AH}(\neg I_\iota(v)) \wedge H_i(w,v) \Rightarrow [\alpha](v))$ is true for the valuation ι of w iff (by Theorem 1) the propositional formula

$forall(v, (\neg gfp_{AH}(\neg I_\iota(v)) \wedge H_i(w, v) \Rightarrow [\alpha](v)))$ is true for the valuation ι of w iff $[K_i\alpha](w)$ is true for the valuation ι of w.

- $\varphi = D_\Gamma\alpha \mid E_\Gamma\alpha$. The proof is analogous to the former case.

6.1 Optimizations of Algorithms

In our implementation we apply some optimizations to the fixed point computing algorithms described above. Precisely, we compute $[AG\alpha](w)$ and $[AH\alpha](w)$ by using the following *frontier set simplification method* [11]. Define the formula $(\forall v.\alpha) \downarrow \delta$, representing some propositional formula such that $\delta \wedge (\forall v.\alpha) \downarrow \delta$ is equivalent to $\delta \wedge \forall v.\alpha$. The formula $(\forall v.\alpha) \downarrow \delta$ is computed using the procedure *forall* with a slight modification. Next, we compute $[AG\alpha](w)$ as the conjunction of the following sequence: $Z_1(w) = [\alpha](w)$, $Z_{i+1}(w) = (\forall v.(T(w, v) \Rightarrow Z_i(v))) \downarrow \bigwedge_{j=1}^{i} Z_j(w)$. The sequence converges when $\bigwedge_{j=1}^{i} Z_j(w) \Rightarrow forall(v, (T(w, v) \Rightarrow Z_i(v)))$, in which case $Z_{i+1}(w)$ is the constant **true**. The procedure $fssm_{AG}$ for computing $[AG\alpha](w)$ is as follows.

procedure $fssm_{AG}([\alpha](w))$, where α is a CTL$_p$K formula
let $Z(w) = Q(w) = [\alpha](w)$
while $Z(w) \neq$ **true**
    let $Z(w) = (\forall v.(T(w, v) \Rightarrow Z(v))) \downarrow Q(w)$
    let $Q(w) = Q(w) \wedge Z(w)$
return $Q(w)$

The procedure $fssm_{AH}$ for computing $[AH\alpha](w)$ is obtained by replacing in the above $(\forall v.(T(w, v) \Rightarrow Z(v))) \downarrow Q(w)$ with $(\forall v.(T(v, w) \Rightarrow Z(v))) \downarrow Q(w)$. Similar procedure can be obtained for computing formulas $[C_\Gamma\alpha](w)$.

7 Example of Train, Gate and Controller

In this section we exemplify the procedure above by discussing the scenario of the train controller system (adapted from [20]). The system consists of three agents: two trains (agents 1 and 3), and a controller (agent 2). The trains, one Eastbound, the other Westbound, occupy a circular track. At one point, both tracks pass through a narrow tunnel. There is no room for both trains to be in the tunnel at the same time. Therefore the trains must avoid this to happen. There are traffic lights on both sides of the tunnel, which can be either red or green. Both trains are equipped with a signaller, that they use to send a signal when they approach the tunnel. The controller can receive signals from both trains, and controls the colour of the traffic lights. The task of the controller is to ensure that the trains are never both in the tunnel at the same time. The trains follow the traffic lights signals diligently, i.e., they stop on red.

We can model the example above with an interpreted system as follows. The local states for the agents are:

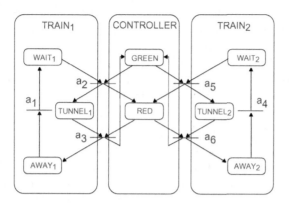

Fig. 1. The local transition structures for the two trains and the controller

- $L_{train_1} = \{away_1, wait_1, tunnel_1\}$,
- $L_{controller} = \{red, green\}$,
- $L_{train_2} = \{away_2, wait_2, tunnel_2\}$.

The set of global states is defined as $G = L_{train_1} \times L_{controller} \times L_{train_2}$. Let $\iota = (away_1, green, away_2)$ be the initial state. We assume that the local states are numbered in the following way: $away_1 := 1$, $wait_1 := 2$, $tunnel_1 := 3$, $red; = 4$, $green := 5$, $away_2 := 6$, $wait_2 := 7$, $tunnel_2 := 8$ and the agents are numbered as follows: $train_1 := 1$, $controller := 2$, $train_2 := 3$. Thus we assume a set of agents A to be the set $\{1, 2, 3\}$.

Let $Act = \{a_1, ..., a_6\}$ be a set of joint actions. For $a \in Act$ we define the preconditions $pre(a)$, postconditions $post(a)$, and the set $agent(a)$ containing the numbers of the agents that may change local states by executing a.

- $pre(a_1) = \{1\}, post(a_1) = \{2\}, agent(a_1) = \{1\}$,
- $pre(a_2) = \{2, 5\}, post(a_2) = \{3, 4\}, agent(a_2) = \{1, 2\}$,
- $pre(a_3) = \{3, 4\}, post(a_3) = \{1, 5\}, agent(a_3) = \{1, 2\}$,
- $pre(a_4) = \{6\}, post(a_4) = \{7\}, agent(a_4) = \{3\}$,
- $pre(a_5) = \{5, 7\}, post(a_5) = \{4, 8\}, agent(a_5) = \{2, 3\}$,
- $pre(a_6) = \{4, 8\}, post(a_6) = \{5, 6\}, agent(a_6) = \{2, 3\}$.

In our formulas we use the following two propositional variables in_tunnel_1 and in_tunnel_2 such that $in\_tunnel_1 \in \mathcal{V}(s)$ iff $l_{train_1}(s) = tunnel_1$, $in\_tunnel_2 \in \mathcal{V}(s)$ iff $l_{train_2}(s) = tunnel_2$, for $s \in G$.

We now encode the local states in binary form in order to use them in the model checking technique. Given that agent $train_1$ can be in 3 different local states we shall need 2 bits to encode its state; in particular we shall take: $(0, 0) = away_1$, $(1, 0) = wait_1$, $(0, 1) = tunnel_1$. Similarly for the agent $train_2$: $(0, 0) = away_2$, $(1, 0) = wait_2$, $(0, 1) = tunnel_2$. The modelling of the local states of the controller requires only one bit: $(0) = green$, $(1) = red$. In view of this a global state is modelled by 5 bits. For instance the initial state

$\iota = (away_1, green, away_2)$ is represented as a tuple of 5 0's. Notice that the first two bits of a global state encode the local state of agent 1, the third bit encodes the local state of agent 2, and two remaining bits encode the local state of agent 3. We represent this by taking: $D_1 = \{1, 2\}$, $D_2 = \{3\}$, $D_3 = \{4, 5\}$.

Let $w = (w[1], ..., w[5])$, $v = (v[1], ..., v[5])$ be two global state variables. We define the following propositional formulas over w and v:

- $I_\iota(w) := \bigwedge_{j \in D_1 \cup D_2 \cup D_3} \neg w[j]$,
  this formula encodes the initial state,
- $H_i(w, v) := \bigwedge_{j \in D_i} w[j] \Leftrightarrow v[j]$,
  the formula $H_i(w, v)$, where $i \in A$, represents logical equivalence between local states of agent i at two global states represented by variables w and v,
- $p_1(w) := \neg w[1] \wedge \neg w[2]$, $p_2(w) := w[1] \wedge \neg w[2]$, $p_3(w) := \neg w[1] \wedge w[2]$, $p_4(w) := w[3]$, $p_5(w) := \neg w[3]$, $p_6(w) := \neg w[4] \wedge \neg w[5]$, $p_7(w) := w[4] \wedge \neg w[5]$, $p_8(w) := \neg w[4] \wedge w[5]$,
  the formula $p_j(w)$, for $j = 1, \ldots, 8$, encodes a particular local state of an agent.

For $a \in Act$, let $B_a := \bigcup_{i \in A \setminus agent(a)} D_i$ be the set of the labels of the bits that are not changed by the action a, then

- $T(w, v) := \bigvee_{a \in Act} \left(\bigwedge_{j \in pre(a)} p_j(w) \wedge \bigwedge_{j \in post(a)} p_j(v) \wedge \bigwedge_{j \in B_a} (w[j] \Leftrightarrow v[j]) \right) \vee$
  $(\bigwedge_{a \in Act} \bigvee_{j \in pre(a)} (\neg p_j(w)) \wedge \bigwedge_{j \in D_1 \cup D_2 \cup D_3} (w[j] \Leftrightarrow v[j]))$.
  Intuitively, $T(w, v)$ encodes the set of all couples of global states s and s' represented by variables w and v respectively, such that s' is reachable from s, i.e., either there exists a joint action which is available at s and s' is the result of execution a at s or there is not such an action and s' equals s. Notice that the above formula is composed of two parts. The first one encodes the transition relation of the system whereas the second one adds self-loops to all the states without successors. This is necessary in order to satisfy the assumption that T is total.

Consider now the following formulas:

- $\alpha_0 = \neg AX(\neg in\_tunnel_1)$,
- $\alpha_1 = AG(in\_tunnel_1 \Rightarrow K_{train_1}(\neg in\_tunnel_2))$,
- $\alpha_2 = AG(\neg in\_tunnel_1 \Rightarrow (\neg K_{train_1} in\_tunnel_2 \wedge \neg K_{train_1}(\neg in\_tunnel_2)))$,

where in_tunnel_1 (respectively in_tunnel_2) is a proposition true whenever the local state of $train_1$ is equal to $tunnel_1$ (respectively the local state of $train_2$ is equal to $tunnel_2$).

The first formula states that agent $train_1$ may at the next step be in the tunnel. The second formula expresses that when the agent $train_1$ is in the tunnel, it knows that agent $train_2$ is not in the tunnel. The third formula expresses that when agent $train_1$ is away from the tunnel, it does not know whether or not agent $train_2$ is in the tunnel.

As discussed above, the translation of propositions in_tunnel_1 and in_tunnel_2 is as follows:

- $[in\_tunnel_1](w) = \neg w[1] \wedge w[2]$,
- $[in\_tunnel_2](w) = \neg w[4] \wedge w[5]$.

Next, we show how to translate the formula α_0:

$$[\alpha_0](w) = [\neg AX(\neg in\_tunnel_1)](w) = \neg [AX(\neg in\_tunnel_1)](w).$$

The formula $[AX(\neg in\_tunnel_1)](w)$ is computed as follows:
$[AX(\neg in\_tunnel_1)](w) = forall(v, T(w, v) \Rightarrow [\neg in\_tunnel_1](v)) = forall(v, T(w, v) \Rightarrow (\neg(\neg v[1] \wedge v[2]))) = forall(v, T(w, v) \Rightarrow (v[1] \vee \neg v[2]))$.

Consequently $[\alpha_0](w) = \neg forall(v, T(w, v) \Rightarrow (v[1] \vee \neg v[2]))$ and $[\alpha_0](w) \wedge I_\iota(w) = \neg forall(v, T(w, v) \Rightarrow (v[1] \vee \neg v[2])) \wedge I_\iota(w) = ((w[1] \wedge \neg w[2] \wedge \neg w[3]) \vee (\neg w[1] \wedge w[2] \wedge \neg w[3] \wedge \neg w[5]) \vee (\neg w[1] \wedge w[2] \wedge w[3] \wedge \neg w[4]) \vee (\neg w[1] \wedge w[2] \wedge \neg w[3] \wedge \neg w[4] \vee w[5])) \wedge I_\iota(w) = $ **false**. Therefore α_0 is not valid in the model.

But, both the formulas α_1 and α_2 are valid in the model since
$[\alpha_1](w) \wedge I_\iota(w) = true \wedge I_\iota(w) = \neg w[1] \wedge \neg w[2] \wedge \neg w[3] \wedge \neg w[4] \wedge \neg w[5]$ and
$[\alpha_2](w) \wedge I_\iota(w) = (\neg w[1] \vee \neg w[2]) \wedge I_\iota(w) = \neg w[1] \wedge \neg w[2] \wedge \neg w[3] \wedge \neg w[4] \wedge \neg w[5]$.

This corresponds to our intuition.

8   Preliminary Experimental Results

In this section we describe an implementation of the UMC algorithm and present some preliminary experimental results for selected benchmark examples.

Our tool, unbounded model checking for interpreted systems, is a new module of the verification environment VerICS [5]. The tool takes as input an interpreted system and a CTL$_p$K formula φ and produces a set of states (encoded symbolically), in which the formula holds. The implementation consists of two main parts: the translation module and the *forall* module. According to the detailed description in former sections, each subformula ψ of φ is encoded (by the translation module) by a QBF formula which characterizes all the states at which ψ holds. In case of checking a modal formula, the corresponding QBF formula is then evaluated by the *forall* module, which is implemented on the top of the SAT solver Zchaff [13]. The whole tool is written in C++ making intensive use of STL libraries.

The tests presented below have been performed on a workstation equipped with the AMD Athlon XP+ 2400 MHz processor and 2 GB RAM running under Linux Redhat. For each of the results we present the time (in seconds) used by VerICS and Zchaff, and give RAM (in kB) consumed during the computation.

8.1   Train, Gate and Controller - Example Parameterized

The first example we have tested is the train, gate and controller system presented in Section 7. In order to show how the algorithm copes with the combinatorial explosion, this example is parameterized with the number of trains N. For a given $N \in \{2, 4, 6\}$, we have generalized the property α_2 of Section

Table 1. Experimental results for Train-Gate-Controller

		$\alpha_2(N)$		
N	CNF clauses	UMC-mem	UMC-time	SAT-time
2	557	2260 kB	0.12 s	0.01 s
4	5214	8376 Mb	1.51 s	0.01 s
6	58489	64 MB	46.55 s	0.01 s

7 to N trains: $\alpha_2(N) = \mathrm{AG}(\neg in\_tunnel_1 \Rightarrow (\neg K_{train_1} \bigwedge_{i=2..N} \neg in\_tunnel_i \wedge \neg K_{train_1} \bigvee_{i=2..N} in\_tunnel_i))$.

The results (time and memory consumption) are presented in the Table 1. *SAT-time* denotes the amount of time necessary to determine by means of unmodified `Zchaff` whether the obtained set of states contains an initial state (this is a SAT problem).

8.2 Attacking Generals

The second analyzed example is a scenario of the coordinated attack problem, often discussed in the area of MAS, distributed computing as well as epistemic logic. It concerns coordination of agents in the presence of unreliable communication. It is also known as the coordinated attack problem [8].

For the purpose of this paper, we choose a particular joint protocol for the scenario and verify the truth and falsehood of particular formulas that capture its key characteristics. The variant we analyse is the following (for more detailed protocol description we refer to [10]) :

> After having studied the opportunity of doing so, general A may issue a request-to-attack order to general B. A will then wait to receive an acknowledgment from B, and will attack immediately after having received it. General B will not issue request-to-attack orders himself, but if his assistance is requested, he will acknowledge the request, and will attack after a suitable time for his messenger to reach A (*assuming no delays*) has elapsed. A joint attack guarantees success, and any non-coordinated attack causes defeat of the army involved (Fig. 2).

Figure 2 presents three scenarios for the agents involved in the coordinated attack problem. The rounded boxes represent locations (local states), while the arrows denote transitions between locations. The beginning location for each agent is in bold. The transitions sharing labels are executed simultaneously (i.e., synchronize). The local states for the agents are listed below:

- $L_{General_A} = \{wait_A, order_A, ack_A, win_A\}$,
- $L_{General_B} = \{wait_B, order_B, ready_B, win_B, fail_B\}$,
- $L_{Environment} = \{wait_E, order_E, ack_E, ack\_lost_E\}$.

In our formulas we use the following propositional variables: $attack_A$ and $attack_B$ meaning that corresponding General has made the decision of attacking the enemy, $success_A$ and $success_B$ meaning the victory of each General

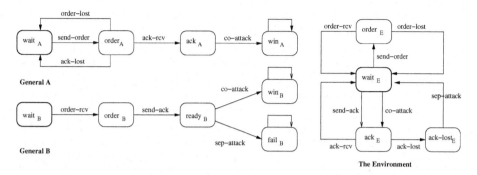

Fig. 2. The attacking generals scenarios

and finally $fail_B$ which denotes the defeat of General B (and both Generals). For $s \in G$:

- $attack_A \in \mathcal{V}(s)$ iff $l_{General_A}(s) \in \{win_A, ack_A\}$
- $success_A \in \mathcal{V}(s)$ iff $l_{General_A}(s) \in \{win_A\}$
- $attack_B \in \mathcal{V}(s)$ iff $l_{General_B}(s) \in \{order_B, win_B, ready_B, fail_B\}$
- $success_B \in \mathcal{V}(s)$ iff $l_{General_B}(s) \in \{win_B\}$
- $fail_B \in \mathcal{V}(s)$ iff $l_{General_B}(s) \in \{fail_B\}$

Below we present some properties we test for the coordinated model problem. Results of the tests are listed for each property in the same way as in the previous example.

- $\beta_1 = \text{AG}(attack_B \Rightarrow K_A K_B attack_A)$
- $\beta_2 = \text{EF}(\mathcal{C}_{\{AB\}}(attack_A \wedge attack_B))$

The property β_1 states that if the general B decides to attack, then the general A knows that B knows that A will attack the enemy. The property β_2 expresses that there is a possibility of achieving common knowledge about the decision of attacking the enemy. The experimental results for this example are given in the Table 2.

9   Conclusions

Verification of multi-agent systems is quickly becoming an active area of research. In the case of model checking, plain temporal verification is not sufficient because

Table 2. Experimental results for the coordinated attack problem

Property	CNF clauses	UMC-memory	UMC-time	SAT-time
β_1	917	1488 kB	1.08 s	0.02 s
β_2	971	2300 kB	1.54 s	0.01 s

of the variety of modalities that are commonly used to specify multi-agent systems. In this paper we have extended the state-of-the-art of the area by providing a model checking theory to perform unbounded model checking on a temporal epistemic language interpreted on interpreted systems. This surpasses the possibilities available already with other SAT-based approaches, namely bounded model checking, in that it is possible to check the full CTLK language, not just its existential fragment.

It should be noted that our tool provides only a preliminary implementation of UMC. The major problem we found was that blocking clauses are defined only over input variables V. This often seemed to be a too finer description and lead to generating exponentially many clauses (as can be seen in Table 1). We have found that the Alternative Implication Graph $IG(A', \phi')$ usually gives shorter blocking clauses only for simple formulas, while formulas encoding "real" UMC problems produce clauses over all literals of V. In future work we shall investigate the conjecture of K. McMillan stating that by allowing in blocking clauses literals corresponding not only to state vectors, but also to subformulas, one could obtain a dramatic improvement in performance.

References

1. A. Biere, A. Cimatti, E. Clarke, and Y. Zhu. Symbolic model checking without BDDs. In *Proc. of TACAS'99*, volume 1579 of *LNCS*, pages 193–207. Springer-Verlag, 1999.
2. P. Blackburn, M. de Rijke, and Y. Venema. *Modal Logic*, volume 53 of *Cambridge Tracts in Theoretical Computer Science*. Cambridge University Press, 2001.
3. E. M. Clarke, O. Grumberg, and D. A. Peled. *Model Checking*. The MIT Press, Cambridge, Massachusetts, 1999.
4. M. Davis, G. Logemann, and D. Loveland. A machine program for theorem proving. *Journal of the ACM*, 5(7):394–397, 1962.
5. P. Dembiński, A. Janowska, P. Janowski, W. Penczek, A. Półrola, M. Szreter, B. Woźna, and A. Zbrzezny. VerICS: A tool for verifying Timed Automata and Estelle specifications. In *Proc. of the 9th Int. Conf. on Tools and Algorithms for the Construction and Analysis of Systems (TACAS'03)*, volume 2619 of *LNCS*, pages 278–283. Springer-Verlag, 2003.
6. E. A. Emerson and E. M. Clarke. Characterizing correctness properties of parallel programs using fixpoints. In *Proc. of the 7th Int. Colloquium on Automata, Languages and Programming (ICALP'80)*, volume 85 of *LNCS*, pages 169–181. Springer-Verlag, 1980.
7. E. A. Emerson and E. M. Clarke. Using branching-time temporal logic to synthesize synchronization skeletons. *Science of Computer Programming*, 2(3):241–266, 1982.
8. R. Fagin, J. Y. Halpern, Y. Moses, and M. Y. Vardi. *Reasoning about Knowledge*. MIT Press, Cambridge, 1995.
9. J. Halpern and M. Vardi. *Model checking vs. theorem proving: a manifesto*, pages 151–176. Artificial Intelligence and Mathematical Theory of Computation. Academic Press, Inc, 1991.

10. A. Lomuscio, T. Łasica, and W. Penczek. Bounded model checking for interpreted systems: Preliminary experimental results. In *Proc. of the 2nd NASA Workshop on Formal Approaches to Agent-Based Systems (FAABS'02)*, volume 2699 of *LNAI*, pages 115–125. Springer-Verlag, 2003.

11. K. L. McMillan. Applying SAT methods in unbounded symbolic model checking. In *Proc. of the 14th Int. Conf. on Computer Aided Verification (CAV'02)*, volume 2404 of *LNCS*, pages 250–264. Springer-Verlag, 2002.

12. R. van der Meyden and H. Shilov. Model checking knowledge and time in systems with perfect recall. In *Proceedings of Proc. of FST&TCS*, volume 1738 of *Lecture Notes in Computer Science*, pages 432–445, Hyderabad, India, 1999.

13. M. Moskewicz, C. Madigan, Y. Zhao, L. Zhang, and S. Malik. Chaff: Engineering an efficient SAT solver. In *Proc. of the 38th Design Automation Conference (DAC'01)*, pages 530–535, June 2001.

14. W. Penczek and A. Lomuscio. Verifying epistemic properties of multi-agent systems via bounded model checking. *Fundamenta Informaticae*, 55(2):167–185, 2003.

15. F. Raimondi and A. Lomuscio. Symbolic model checking of deontic interpreted systems via OBDD's. In *Proceedings of DEON04, Seventh International Workshop on Deontic Logic in Computer Science*, volume 3065 of *Lecture Notes in Computer Science*. Springer Verlag, May 2004.

16. F. Raimondi and A. Lomuscio. Towards model checking for multiagent systems via OBDD's. In *Proceedings of the Third NASA Workshop on Formal Approaches to Agent-Based Systems (FAABS III)*, Lecture Notes in Computer Science. Springer Verlag, April 2004. To appear.

17. F. Raimondi and A. Lomuscio. Verification of multiagent systems via ordered binary decision diagrams: an algorithm and its implementation. In *Proceedings of the Third International Joint Conference on Autonomous Agents and Multiagent Systems (AAMAS'04)*, July 2004.

18. A. Tarski. A lattice-theoretical fixpoint theorem and its applications. *Pacific Journal of Mathematics*, 5:285–309, 1955.

19. W. van der Hoek and M. Wooldridge. Model checking knowledge and time. In *Proc. of the 9th Int. SPIN Workshop (SPIN'02)*, volume 2318 of *LNCS*, pages 95–111. Springer-Verlag, 2002.

20. W. van der Hoek and M. Wooldridge. Tractable multiagent planning for epistemic goals. In *Proc. of the 1st Int. Conf. on Autonomous Agents and Multi-Agent Systems (AAMAS'02)*, volume III, pages 1167–1174. ACM, July 2002.

21. M. Wooldridge. Computationally grounded theories of agency. In E. Durfee, editor, *Proceedings of ICMAS, International Conference of Multi-Agent Systems*. IEEE Press, 2000.

22. M. Wooldridge. *An introduction to multi-agent systems*. John Wiley, England, 2002.

23. M. Wooldridge, M. Fisher, M. Huget, and S. Parsons. Model checking multiagent systems with mable. In *Proceedings of the First International Conference on Autonomous Agents and Multiagent Systems (AAMAS-02)*, Bologna, Italy, July 2002.

Towards Symbolic Model Checking for Multi-agent Systems via OBDD's

Franco Raimondi and Alessio Lomuscio

Department of Computer Science,
King's College London,
London, UK
{franco, alessio}@dcs.kcl.ac.uk

Abstract. We present an algorithm for model checking temporal-epistemic properties of multi-agent systems, expressed in the formalism of interpreted systems. We first introduce a technique for the translation of interpreted systems into boolean formulae, and then present a model-checking algorithm based on this translation. The algorithm is based on OBDD's, as they offer a compact and efficient representation for boolean formulae.

1   Introduction

Theoretical investigations in the area of multi-agent systems (MAS) have traditionally focused on *specifications*. Various logics have been explored to give formal foundations to MAS, particularly for *mental attitudes* [1] of agents, such as knowledge, belief, desire, etc. To consider the temporal evolution of these attitudes, temporal logics such as CTL and LTL [2] have been included in MAS formalisms, thereby producing combinations of temporal logic with, for example, epistemic, doxastic, and deontic logics.

Although it is important to investigate formal tools for specifying MAS, the problem of *verification* of MAS must also be taken into account to ensure that systems behave as they are supposed to. *Model checking* is a well-established verification technique for distributed systems specified by means of temporal logics [3, 2]. The problem of model checking is to verify whether a logical formula φ expressing a certain required property is true in a model M representing the system, that is establishing whether or not $M \models \varphi$. This approach can also be applied to MAS, where in this case M is a semantical model representing the evolutions of the MAS, and φ is a formula expressing temporal-intentional properties of the agents. Recent work along these lines includes [4], in which Wooldridge et al. present the MABLE language for the specification of MAS. In this work, modalities are translated as nested data structures (in the spirit of [5]). Bordini et al. [6] use a modified version of the AgentSpeak(L) language [7] to specify agents and to exploit existing model checkers. For verification purposes, both the works of Wooldridge et al. and of Bordini et al. translate the MAS specification into a SPIN specification [8] to perform the verification. The works of van der Meyden and Shilov [9], and van der Meyden and Su [10], are concerned

M.G. Hinchey et al. (Eds.): FAABS 2004, LNAI 3228, pp. 213–221, 2005.
© Springer-Verlag Berlin Heidelberg 2005

with verification of interpreted systems. They consider the verification of a particular class of interpreted systems, namely the class of synchronous distributed systems with perfect recall. An algorithm for model checking is introduced in the first paper using automata, and [10] suggests the use of OBDD's for this approach.

The aim of this paper is to present an algorithm for model checking epistemic and temporal properties of interpreted systems [11]. This differs from previous work by treating all the modalities explicitly in the verification process. We focus on temporal-epistemic model checking because the verification of epistemic properties (and their temporal evolution) is crucial in many scenarios, including communication protocols and security protocols.

Interpreted systems are a formalism for representing epistemic properties of MAS and their evolution with time. The algorithm that we present does not involve the translation into existing model checkers, it is fully *symbolic*, and it is based on boolean functions. Boolean functions can be represented and manipulated efficiently by means of OBDD's, as it has been shown for CTL model checking [12].

The rest of the paper is organised as follows: in Section 2 we briefly review OBDD's-based model checking and the formalism of interpreted systems. In Section 3.1 we present the translation of interpreted systems into boolean formulae, while in Section 3.2 we introduce an algorithm based on this translation. We provide a proof of the correctness of the algorithm in Section 3.3. We conclude in Section 4.

2   Preliminaries

2.1   CTL Model Checking and OBDD's

Given a model M and a formula φ in some logic, the problem of *model checking* involves establishing whether or not $M \models \varphi$ holds. Tools have been built to perform this task automatically, where M is a model of some temporal logic [3, 2, 8]. SMV [12] and SPIN [8] are two well-known model checkers; in these tools the model is given indirectly by means of a program P. It is not efficient to build explicitly the model M represented by P, because M has a size which is exponential in the number of variables of P (this fact is known as the *state explosion problem*). Instead, various techniques have been developed to perform *symbolic model checking*, which is the problem of model checking where the model M is not described or computed in extension. Techniques for symbolic model checking mostly use either automata [8], or OBDD's [13] for the representation of all the parameters needed by the algorithms. For the purpose of this paper, we will only consider symbolic model checking of the temporal logic CTL using OBDD's [14].

CTL is a logic used to reason about the evolution of a system represented as a *branching* path. Given a countable set of propositional variables $\mathcal{P} = \{p, q, \ldots\}$, CTL formulae are defined as follows:

$$\varphi ::= p \mid \neg\varphi \mid \varphi \vee \varphi \mid EX\varphi \mid EG\varphi \mid E(\varphi U \varphi)$$

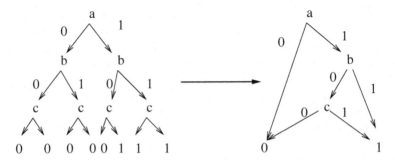

Fig. 1. OBDD representation for $a \land (b \lor c)$

where the temporal operator X means in the next state, G means globally and U means until. Each temporal operator is pre-fixed by the existential quantifier E. Thus, for example, $EG(\varphi)$ means that "there exists a path in which φ is globally true". Traditionally, other operators are added to the syntax of CTL, namely AX, EF, AF, AG, AU (notice the "universal" quantifier A over paths, dual of E). These operators can be derived from the operators introduced here [2]. The semantics of CTL is given via a model $M = (S, R, \mathcal{V}, I)$ where $S = \{s_0, s_1, \ldots\}$ is a set of states, $R \subseteq S \times S$ is a binary relation, $\mathcal{V} : \mathcal{P} \to 2^S$ is an evaluation function, and $I \subseteq S$ is a set of initial states. A *path* π is a sequence of states $\pi = \{s_0, s_1, \ldots\}$ such that $s_0 \in I$ and $\forall i, (s_i, s_{i+1}) \in R$. A state s_i in a path π is denoted with π_i. Satisfaction in a state is defined inductively as follows:

$$s \models p \qquad \text{iff } s \in \mathcal{V}(p),$$
$$s \models EX\varphi \quad \text{iff there exists a path } \pi \text{ such that } \pi_i = s \text{ and } \pi_{i+1} \models \varphi,$$
$$s \models EG\varphi \quad \text{iff there exists a path } \pi \text{ such that } \pi_i = s \text{ and } \pi_{i+j} \models \varphi$$
$$\text{for all } j \geq 0.$$
$$s \models E(\varphi U \psi) \text{ iff there exists a path } \pi \text{ such that } \pi_i = s \text{ and a } k \geq 0 \text{ such}$$
$$\text{that } \pi_{i+k} \models \psi \text{ and } \pi_{i+j} \models \varphi \text{ for all } 0 \leq j < k.$$

OBDD's (Ordered Binary Decision Diagrams) are an efficient representation for the manipulation of boolean functions. As an example, consider the boolean function $a \land (b \lor c)$. The truth table of this function would be 8 lines long. Equivalently, one can evaluate the truth value of this function by representing the function as a directed graph, as exemplified on the left-hand side of Figure 1. As it is clear from the picture, under certain assumptions, this graph can be simplified into the graph pictured on the right-hand side of Figure 1. This "reduced" representation is called the OBDD of the boolean function.

Besides offering a compact representation of boolean functions, OBDD's of different functions can be composed efficiently: in [13] algorithms are provided for the manipulation and composition of OBDD's.

The idea of CTL model checking using OBDD's is to represent states of the model and relations by means of boolean formulae. A CTL formula is identified with a set of states, i.e. the states of the model satisfying the formula. As set of states can be represented as a boolean formula, each CTL formula can be characterised by a boolean formula. Thus, the problem of model checking for CTL is reduced to the construction of boolean formulae. This is achieved by composing OBDD's, or by computing fix-points of operators on OBDD's; we refer to [2] for the details. By means of this approach large systems have been checked, including hardware and software components.

2.2   Interpreted Systems

An interpreted system is a semantic structure representing the temporal evolution of a system of agents. Each agent i $(i = \{1, \ldots, n\})$ is characterised by a set of *local states* L_i and by a set of actions Act_i that may be performed. Actions are performed in compliance with a protocol $P_i : L_i \rightarrow 2^{Act_i}$; notice that this definition allows for non-determinism. A tuple $g = (l_1, \ldots, l_n) \in L_1 \times \ldots, L_n$, where $l_i \in L_i$ for each i, is called a *global state* and gives a snapshot of the system. Given a set I of *initial global states*, the evolution of the system is described by n evolution functions[1]: $t_i : L_1 \times \ldots \times L_n \times Act_1 \times \ldots \times Act_n \rightarrow L_i$ In this formalism the environment in which agents "live" is usually modeled by means of a special agent E; we refer to [11] for more details.

The set I, t_i and the protocols P_i generate a set of *runs*. Formally, a run π is a sequence of global states $\pi = (g_0, g_1, \ldots)$ such that $g_0 \in I$ and, for each pair $(g_j, g_{j+1}) \in \pi$, there exists a set of actions a enabled by the protocols such that $t(g_j, a) = g_{j+1}$. $G \subseteq (L_1 \times \ldots \times L_n)$ denotes the set of *reachable* global states.

Given a set of agents $A = \{1, \ldots, n\}$ with corresponding local states, protocols, and transition functions, a countable set of propositional variables $\mathcal{P} = \{p, q, \ldots\}$, and a valuation function for the atoms $\mathcal{V} : \mathcal{P} \rightarrow 2^G$, an *interpreted system* is a tuple $IS = (G, I, \Pi, \sim_1, \ldots, \sim_n, \mathcal{V})$. In the above G is the finite set of reachable global states for the system, $I \subseteq G$ is the set of initial states, and Π is the set of possible runs in the system. The binary relation $\sim_i, i \in A$, is defined by $g \sim_i g'$ iff $l_i(g) = l_i(g')$, i.e. if the local state of agent i is the same in g and in g'. Some issues arise with respect to the generation of the reachable states in the system given a set of protocols and transition relations; since they do not influence this paper we do not report them here.

Interpreted systems semantics can be used to interpret formulae of a temporal language enriched with epistemic operators [11]. Here we assume a temporal tree structure to interpret CTLK formulae [15]. The syntax of CTLK is defined in terms of a countable set of propositional variables $\mathcal{P} = \{p, q, \ldots\}$ and using the following modalities:

$$\varphi ::= p \mid \neg\varphi \mid \varphi \vee \varphi \mid EX\varphi \mid EG\varphi \mid E(\varphi U\varphi) \mid K_i\varphi$$

The modalities AX, EF, AF, AG, AU are derived in the standard way. Further, given a set of agents Γ, two group modalities can be introduced: $E_\Gamma\varphi$ and $C_\Gamma\varphi$ denote, respectively, that every agent in the group knows φ, and that φ is *common knowledge* in the group (see [11] for details).

---

[1] This definition is equivalent to the definition of a single evolution function t as in [11].

Given an interpreted system IS, a global state g, and a formula φ, the semantics of CTLK is defined as follows:

$$
\begin{array}{lll}
IS, g \models p & \text{iff } g \in \mathcal{V}(p), \\
IS, g \models \neg\varphi & \text{iff } g \not\models \varphi, \\
IS, g \models \varphi_1 \vee \varphi_2 & \text{iff } g \models \varphi_1 \text{ or } g \models \varphi_2, \\
IS, g \models EX\varphi & \text{iff there exists a run } \pi \text{ such that} \\
& \quad \pi_i = g \text{ for some } i, \text{ and } \pi_{i+1} \models \varphi, \\
IS, g \models EG\varphi & \text{iff there exists a run } \pi \text{ such that} \\
& \quad \pi_i = g \text{ for some } i, \text{ and } \pi_j \models \varphi \text{ for all } j \geq i. \\
IS, g \models E(\varphi U \psi) & \text{iff there exists a run } \pi \text{ such that} \\
& \quad \pi_i = g \text{ for some } i, \text{ and a } k \geq 0 \text{ such that } \pi_{i+k} \models \psi \\
& \quad \text{and } \pi_j \models \varphi \text{ for all } i \leq j < i + k, \\
IS, g \models K_i\varphi & \text{iff } \forall g' \in G, \ g \sim_i g' \text{ implies } g' \models \varphi \\
IS, g \models E_\Gamma\varphi & \text{iff } \forall g' \in G, \ g \sim_\Gamma^E g' \text{ implies } g' \models \varphi \\
IS, g \models C_\Gamma\varphi & \text{iff } \forall g' \in G, \ g \sim_\Gamma^G g' \text{ implies } g' \models \varphi
\end{array}
$$

In the definition above, π_j denotes the global state at place j in run π. Other temporal modalities can be derived, namely AX, EF, AF, AG, AU. We write $IS \models \varphi$ if, for every global state $g \in G$, $IS, g \models \varphi$. We refer to [11, 15] for more details.

3   A Model Checking Algorithm for CTLK

The main idea of this paper is to use algorithms based on OBDD's to verify temporal and epistemic properties of multi-agent systems, in the spirit of traditional model checking for temporal logics. To this end, it is necessary to encode all the parameters needed by the algorithms by means of boolean functions, and then to represent boolean functions by means of OBDD's. As this last step can be performed automatically using software libraries that are widely available, in this paper we introduce only the translation of interpreted systems into boolean formulae (Section 3.1). In Section 3.2 we present an algorithm based on this translation for the verification of CTLK formulae.

3.1   Translating an Interpreted System into Boolean Formulae

The local states of an agent can be encoded by means of boolean variables (a boolean variable is a variable that can assume just one of the two values 0 or 1). The number of boolean variables needed for each agent is $nv(i) = \lceil log_2 |L_i| \rceil$. Thus, a global state can be identified by means of $N = \sum_i nv(i)$ boolean variables: $g = (v_1, \ldots, v_N)$. The evaluation function $\mathcal{V}$ associates a set of global states to each propositional atom, and so it can be seen as a boolean function. The protocols, too, can be expressed as boolean functions (actions being represented with boolean variables $(a_1, \ldots, a_M)$ similarly to global states).

The definition of t_i in Section 2.2 can be seen as specifying a list of *conditions* $c_{i,1}, \ldots, c_{i,k}$ under which agent i changes the value of its local state. Each $c_{i,j}$

relates conditions on global state and actions with the value of "next" local state for i.

$$t_i = c_{i,1} \vee \ldots \vee c_{i,k}$$

We assume that the last condition $c_{i,k}$ of t_i prescribes that, if none of the conditions $c_{i,j}(j < k)$ is true, then the local state for i does not change. This assumption is key to keep compact the description of an interpreted system, as in this way only the conditions that are actually causing a change need to be listed.

The algorithm presented in Section 3.2 requires the definition of a boolean function $R_t(g, g')$ representing a temporal relation between g and g'. $R_t(g, g')$ can be obtained from the evolution function t_i as follows. First, we introduce a *global* evolution function t:

$$t = \bigwedge_{i \in \{1,\ldots,n\}} t_i = \bigwedge_{i \in \{1,\ldots,n\}} (c_{i,1} \vee \ldots \vee c_{i,k_i})$$

Notice that t is a boolean function involving two global states and a joint action $a = (a_1, \ldots, a_M)$. To abstract from the joint action and obtain a boolean function relating two global states only, we can define R_t as follows:

$R_t(g, g')$ iff $\exists a \in Act : t(g, a, g')$ is true and each local action $a_i \in a$ is enabled by the protocol of agent i in the local state $l_i(g)$.

The quantification over actions above can be translated into a propositional formula using a disjunction (see [12, 3] for a similar approach to boolean quantification):

$$R_t(g, g') = \bigvee_{a \in Act} [(t(g, a, g') \wedge P(g, a)]$$

where $P(g, a)$ is a boolean formula imposing that the joint action a must be consistent with the agents' protocols in global state g. R_t gives the desired boolean relation between global states.

3.2   The Algorithm

In this section we present the algorithm SAT_{CTLK} to compute the set of global states in which a CTLK formula φ holds, denoted with $[[\varphi]]$. The following are the parameters needed by the algorithm:

- the boolean variables $(v_1, \ldots, v_N)$ and $(a_1, \ldots, a_M)$ to encode global states and joint actions;
- the boolean functions $P_i(v_1, \ldots, v_N, a_1, \ldots, a_M)$ to encode the protocols of the agents;
- the function $\mathcal{V}(p)$ returning the set of global states in which the atomic proposition p holds. We assume that the global states are returned encoded as a boolean function of $(v_1, \ldots, v_N)$;
- the set of initial states I, encoded as a boolean function;

- the set of reachable states G. This can be computed as the fix-point of the operator $\tau = (I(g) \vee \exists g'(R_t(g', g) \wedge Q(g')))$ where $I(g)$ is true if g is an initial state and Q denotes a set of global states. The fix-point of τ can be computed by iterating $\tau(\emptyset)$ by standard procedure (see [12]);
- the boolean function R_t to encode the temporal transitions;
- n boolean functions R_i to encode the accessibility relations $\sim_i$ (these functions are easily defined using equivalence on local states of G).
- the boolean function R_E^Γ to encode $\sim_E^\Gamma$, defined by $R_E^\Gamma = \bigwedge_{i \in \Gamma} R_i$.

The algorithm is as follows:

$$SAT_{CTLK}(\varphi) \{$$
φ is an atomic formula: return $\mathcal{V}(\varphi)$;
φ is $\neg\varphi_1$: return $G \setminus SAT_{CTLK}(\varphi_1)$;
φ is $\varphi_1 \wedge \varphi_2$: return $SAT_{CTLK}(\varphi_1) \cap$
$\quad SAT_{CTLK}(\varphi_2)$;
φ is $EX\varphi_1$: return $EX_{CTLK}(\varphi_1)$;
φ is $E(\varphi_1 U \varphi_2)$: return $EU_{CTLK}(\varphi_1, \varphi_2)$;
φ is $EG\varphi_1$: return $EG_{CTLK}(\varphi_1)$;
φ is $K_i\varphi_1$: return $K_{CTLK}(\varphi_1, i)$;
φ is $E_\Gamma\varphi_1$: return $E_{CTLK}(\varphi_1, \Gamma)$;
φ is $C_\Gamma\varphi_1$: return $C_{CTLK}(\varphi_1, \Gamma)$;
$$\}$$

In the algorithm above, EX_{CTLK}, EG_{CTLK}, EU_{CTLK} are the standard procedures for CTL model checking [2] in which the temporal relation is R_t and, instead of temporal states, global states are considered. The procedures $K_{CTLK}(\varphi, i)$ and $E_{CTLK}(\varphi, \Gamma)$ and $C_{CTLK}(\varphi, \Gamma)$ are presented below.

$$K_{CTLK}(\varphi, i) \{$$
$\quad X = SAT_{CTLK}(\neg\varphi)$;
$\quad Y = \{g \in G | K_i(g, g') \text{ and } g' \in X\}$
$\quad$ return $\neg Y$;
$$\}$$

$$E_{CTLK}(\varphi, \Gamma) \{$$
$\quad X = SAT_{CTLK}(\neg\varphi)$;
$\quad Y = \{g \in G | R_\Gamma^E(g, g') \text{ and } g' \in X\}$
$\quad$ return $\neg Y$;
$$\}$$

```
C_CTLK(φ,Γ) {
    X = SAT_CTLK(φ);
    Y = G;
    while ( X != Y ) {
        X = Y;
        Y = {g ∈ G|R_Γ^E(g,g') and g' ∈ Y and g' ∈ SAT_CTLK(φ)}
    }  return Y;
}
```

The procedure $C_{CTLK}(\varphi, \Gamma)$ is based on the equivalence [11]

$$C_\Gamma \varphi = E_\Gamma(\varphi \wedge C_\Gamma \varphi)$$

which implies that $[[C_\Gamma\varphi]]$ is the fix-point of the (monotonic) operator $\tau(Q) = [[E_\Gamma(\varphi \wedge (Q))]]$. Hence, $[[C_\Gamma\varphi]]$ can be obtained by iterating $\tau(G)$.

Notice that all the parameters can be encoded as OBDD's. Moreover, all the operations inside the algorithms can be performed on OBDD's as presented in [13].

To check that a formula holds in a model, it is enough to check whether or not the result of SAT_{CTLK} is equivalent to the set of reachable states.

3.3   Correctness of the Algorithm

The algorithm presented in Section 3.2 is sound and complete.

Theorem 1. *For every CTLK formula φ, $IS \models \varphi$ iff $SAT_{CTLK}(\varphi) \equiv G$. (i.e. iff the set of states computed by the algorithm is the set of reachable states G).*

Proof. (=>): by induction on the structure of φ. We consider here the epistemic operators (a proof for the temporal operators can be found in [2]). Let $\varphi = K_i(\psi)$ and let $IS, g \models K_i(\psi)$. This means that $IS, g' \models \psi$ for all $g' \in G$ s.t. $g \sim_i g'$. By the induction step, $g' \in [[\psi]]$; also we have $R_i(g, g')$ by definition of R_i. This implies that $g \in [[K_i(\psi)]]$, i.e. $g \in [[\varphi]]$. The proof for E_Γ is similar. The proof of correctness for common knowledge follows from the correctness of the fix-point characterisation of C_Γ[11].

(<=): straightforward, as the induction steps above are symmetrical.   □

4   Conclusion

Temporal logic model checking using OBDD's [12] is one of the most successful techniques for the verification of distributed systems. In the last decade, this methodology has been used for the verification of both software and hardware components.

In this paper we have presented an algorithm for the verification of temporal-epistemic properties based on the manipulation of boolean functions. The methodology presented here encodes directly a MAS (specified in the formalism of interpreted systems) by means of boolean formulae; then, the algorithm allows for the (fully symbolic) verification of temporal-epistemic properties. Moreover, the algorithm allows for the verification of two group modalities (E_Γ and C_Γ) and is not restricted to a particular class of interpreted systems, nor to a particular class of formulae. We are currently implementing the algorithm and in the future we aim at testing epistemic and temporal properties of various scenarios from the MAS literature. This will help in evaluating the efficiency of the algorithm.

References

1. McCarthy, J.: Ascribing mental qualities to machines. In Ringle, M., ed.: Philosophical Perspectives in Artificial Intelligence. Humanities Press, Atlantic Highlands, New Jersey (1979) 161–195
2. Huth, M.R.A., Ryan, M.D.: Logic in Computer Science: Modelling and Reasoning about Systems. Cambridge University Press, Cambridge, England (2000)
3. Clarke, E.M., Grumberg, O., Peled, D.A.: Model Checking. The MIT Press, Cambridge, Massachusetts (1999)
4. Wooldridge, M., Fisher, M., Huget, M.P., Parsons, S.: Model checking multi-agent systems with MABLE. In Gini, M., Ishida, T., Castelfranchi, C., Johnson, W.L., eds.: Proceedings of the First International Joint Conference on Autonomous Agents and Multiagent Systems (AAMAS'02), ACM Press (2002) 952–959
5. Benerecetti, M., Giunchiglia, F., Serafini, L.: Model checking multiagent systems. Journal of Logic and Computation **8** (1998) 401–423
6. Bordini, R.H., Fisher, M., Pardavila, C., Wooldridge, M.: Model checking AgentSpeak. In: Proceedings of the Second International Joint Conference on Autonomous Agents and Multiagent Systems (AAMAS'03). (2003)
7. Rao, A.S.: AgentSpeak(L): BDI agents speak out in a logical computable language. Lecture Notes in Computer Science **1038** (1996) 42–58
8. Holzmann, G.J.: The model checker spin. IEEE transaction on software engineering **23** (1997)
9. van der Meyden, R., Shilov, N.V.: Model checking knowledge and time in systems with perfect recall. FSTTCS: Foundations of Software Technology and Theoretical Computer Science **19** (1999)
10. van der Meyden, R., Su, K.: Symbolic model checking the knowledge of the dining cryptographers. Submitted (2002)
11. Fagin, R., Halpern, J.Y., Moses, Y., Vardi, M.Y.: Reasoning about Knowledge. The MIT Press, Cambridge, Massachusetts (1995)
12. McMillan, K.: Symbolic model checking: An approach to the state explosion problem. Kluwer Academic Publishers (1993)
13. Bryant, R.E.: Graph-based algorithms for boolean function manipulation. IEEE Transaction on Computers (1986) 677–691
14. Burch, J.R., Clarke, E.M., McMillan, K.L., Dill, D.L., Hwang, L.J.: Symbolic model checking: 10^{20} states and beyond. Information and Computation **98** (1992) 142–170
15. Penczek, W., Lomuscio, A.: Verifying epistemic properties of multi-agent systems via model checking. Fundamenta Informaticae **55** (2003) 167–185

Formal Consistency Verification of Deliberative Agents with Respect to Communication Protocols

Jaime Ramírez and Angélica de Antonio

Technical University of Madrid, Madrid, Spain
{jramirez, angelica}@fi.upm.es
http://decoroso.ls.fi.upm.es

Abstract. The aim of this paper is to show a method that is able to detect inconsistencies in the reasoning carried out by a deliberative agent. The agent is supposed to be provided with a hybrid Knowledge Base expressed in a language called CCR-2, based on production rules and hierarchies of frames, which permits the representation of non-monotonic reasoning, uncertain reasoning and arithmetic constraints in the rules. The method can give a specification of the scenarios in which the agent would deduce an inconsistency. We define a scenario to be a description of the initial agent's state (in the agent life cycle), a deductive tree of rule firings, and a partially ordered set of messages and/or stimuli that the agent must receive from other agents and/or the environment. Moreover, the method will make sure that the scenarios will be valid w.r.t. the communication protocols in which the agent is involved.

1 Introduction

The purpose of this paper is to show a method to verify the consistency of the reasoning that a deliberative agent can perform. We assume the agent to comprise a knowledge base (KB) expressed in a knowledge representation formalism called CCR-2.

The CCR-2 formalism is valid to represent hybrid KBs that combine production rules with hierarchies of frames. This formalism allows us to represent non-monotonic reasoning, uncertain reasoning, and arithmetic constraints in the rules.

We assume that the agent whose reasoning is checked needs to carry out a reasoning process for deciding its next action according to its goals. The agent's knowledge can fall into three different categories: *acquired knowledge, innate knowledge* or *deduced knowledge*. The acquired knowledge is made up of acquired facts, that is, information coming from its perception or requested to other agents; the innate knowledge is made up of knowledge that the agent knows since the beginning of its life; and the deduced knowledge is formed by the facts deduced by firing rules. It is clear that, as the reasoning process evolves, the agent may obtain contradictory acquired facts from different sources w.r.t.

M.G. Hinchey et al. (Eds.): FAABS 2004, LNAI 3228, pp. 222–237, 2005.

previously acquired facts. In this case, the new knowledge would replace the obsolete knowledge. However, the agent should not be allowed to deduce a set of contradictory facts from the acquired facts and the innate facts.

The proposed method finds scenarios in which the agent would deduce an inconsistency. A scenario consists of a description of the initial agent's state (in the agent life cycle), a deductive tree of rule firings, and a partially ordered set of messages and/or stimuli (expressed as schemas) that the agent must receive from other agents and/or the environment to achieve the execution of the deductive tree. A scenario permits the execution of a deductive tree of rule firings that will deduce a set of semantically contradictory facts. We assume the agent's state to be a set of innate facts, acquired facts (from the sources mentioned above) and/or deduced facts, that is, it is a Fact Base (FB). Basically, the partially ordered set of messages and/or stimuli schemas, included as part of a scenario, will represent precedence dependencies between the messages/stimuli required in the reasoning. This set will be checked w.r.t. the communication protocols in which the verified agent is involved, so as to warrant the precedence dependencies can be satisfied by the specification of the communication protocols.

Some methods or tools designed to detect inconsistencies in a Knowledge Base System (KBS) (mostly rule-based systems) build a model of the KBS (Graph, Petri Net, etc.), and execute the model for each valid input, in order to identify possible inconsistencies during the reasoning process. This approach in many cases turns to be computationally very costly. Thus, we decided to adopt another approach in which the starting point is one of the inconsistencies that might be possibly deduced by the verified KBS, and the goal is to compute a description of the scenarios in which the KBS included in the agent would deduce that inconsistency. This approach takes some ideas from the ATMS designed by de Kleer [1], since it uses the concept of label as a way to represent a description of a set of FBs. Other methods for verifying rule-based systems that follow a similar approach were proposed in [2] [3] [4] [5] [6] [7] [8].

Section 2 explains some points related to the agent's KB and inconsistencies that are verified by this method, and the hypotheses that will be assumed in the operation of the method. In section 3 it is described how this method specifies the way in which an agent deduces an inconsistency, if possible. In section 4, the procedure for detecting an inconsistency is explained, and in section 5, a small example of application is shown. We end with some conclusions about our work, and some future works that will be derived from this work.

2   Scope

Our method receives as inputs a CCR-2 KB (the agent's KB), a classification of the possible facts that the agent can manage, an Integrity Constraint (IC) to be checked, and a set of communication protocol specifications.

CCR-2 (also called GKR) [9] supports the representation of production rules and a high number of object types in the FB: frame classes and instances, relationships, propositions, attribute values and attribute identifiers. A rule's antecedent in CCR-2 is a Disjunctive Normal Form (DNF) formula made up of literals. A *literal* is an atom, a negated atom or a linear arithmetic inequation over attribute values and/or certainty factors. An *atom* states something about some object in the FB. In CCR-2 a rule's consequent contains a list of actions that can modify the state of an object, create or destroy objects while executing the KB system included in the agent. This last characteristic allows us to represent some types of non monotonic reasoning. As it is possible to declare variables as relationships and propositions in the rules, the antecedent of a rule is a second order logic formula. Nevertheless, the actions of the rules can not change the type of a relationship or a proposition, therefore CCR-2 supports a limited representation of the second order logic. Moreover, uncertain reasoning can be represented in CCR-2 by associating certainty factors to attribute values, to tuples in a relationship or to propositions.

The CCR-2 KBs can use two kinds of management of the negation: closed world assumption (CWA) or 3-valued logic. The kind of negation management determines: when a fact can be considered true or false; what is the effect of the actions; how the facts and actions can be chained during the KBS execution; and which pairs of actions are contradictory. For instance, in the 3-valued logic there are three truth values: true, false and unknown; while a fact will be false if its negation appears in the FB, a fact will be unknown if neither it nor its negation appear in the FB; moreover, the action $Add(\neg p)$ deduces the fact $\neg p$, and the pair of actions $Add(p)$ and $Add(\neg p)$ are contradictory. It must be highlighted that the action $Add(\neg p)$ cannot be employed under CWA.

The rules are assumed to execute with forward chaining or backward chaining under conflict set resolution. The rules are structured in groups whose activation or inhibition is controlled by metarules. When a rule is fired, we assume the sequential execution of all the actions belonging to the consequent of the rule.

We assume that two kinds of facts can appear during the agent's execution: *static facts* and *dynamic facts*. A static fact is a fact whose truth value changes neither from true to false nor from false to true during the reasoning process, whereas the truth value of a dynamic fact actually may change those ways. In this sense, some acquired facts will be dynamic facts. Moreover, facts representing innate knowledge are assumed to be static. The method needs to know both whether a literal is static or dynamic, and whether a literal is acquired, innate or deduced, so a classification must be provided.

2.1   Defining Inconsistencies: Integrity Constraints

An IC defines a consistency criterion over input data, output data or input and output data. The IC form is:

$$\exists x_1 \in T_1 \exists x_2 \in T_2 ... \exists x_n \in T_n \exists()x_{n+1} \in T_{n+1} \exists()x_{n+2} \in T_{n+2} ... \exists x_{n+m} \in T_{n+m}$$
$$A \Rightarrow \bot$$

where A is a second order logic formula in DNF that includes conditions over whatever types of CCR-2 objects. Each literal in A has an associated scope, which specifies whether the literal is related to input data (acquired literal or innate literal), or output data (deducible literal). For the variables in A, two kinds of quantifiers can be employed: the existential quantifier (with the classical meaning) and the restricted existential quantifier (denoted as $\exists()x$).

> An IC $\exists x \in T(A(x) \Rightarrow \perp)$ is violated if at least one object in the class T that is included in the FB satisfies the conditions imposed over the variable x in the formula A.
>
> An IC $\exists()x \in T(A(x) \Rightarrow \perp)$ is violated if every object in the class T that is included in the FB satisfies the conditions imposed over the variable x in the formula A and <u>only</u> those conditions.

This semantics for the restricted existential quantifier permits the detection of knowledge gaps. Lets see an example of an IC with a restricted existential quantifier:

$$\exists()x \in PATIENT$$
$$Is\_Ill(x, FLU), (x.Fever = high) \Rightarrow \perp$$

Clearly, having a high fever is not enough to deduce that a patient has flu. So, if a KBS can violate this IC, it is likely that there is a knowledge gap in the KB, that is, the KBS needs more rules.

2.2 Specifying Interaction with the Environment and Other Agents

Nowadays, different notations can be employed to specify communication protocols: AUML interaction diagrams[1] or state machines as in [10]. For the purpose of the proposed method, state machines are more suitable as the checking of the scenarios w.r.t. the protocols must be automated. Hence, a state machine view for the verified agent must also be supplied as an input to our method. Each state transition of the state machine owns a label that describes how the messages/stimuli that fire the transition are. This label is expressed in terms of message/stimulus schemas.

In addition to the state machine, a correspondence between message/stimulus schemas and acquired literals must be supplied. If a message/stimulus schema corresponds to a set of acquired literals $\{l_i\}_{i=1,..,n}$, any message/stimulus that matches that schema contains a model for the formula $\exists x_1 \exists x_2 ... \exists x_n (\bigwedge_{i=1,...,n} l_i)$ where $x_1, x_2, ..., x_n$ are all the free variables in $\bigwedge_{i=1,...,n} l_i$. This latter formula can be also viewed as a query.

2.3 Assumed Non-monotonic Reasoning

CCR-2 rules can introduce new facts in the agent's state, but they can also delete already existing facts. This provides the agent's designer with the capability of building agents with non-monotonic reasoning. So, we could find production

---

[1] http://www.auml.org/

rules of the form $p \rightarrow Del(p)$ under CWA. This kind of rules (when p is assumed to be provided) are not admissible in a RB from the point of view of classical logic or default logic [11], since they are logical inconsistencies. However, if we examine these rules from the point of view of temporal logic [12], and we rewrite them as $\neg p$ *atnext* p (where the intended meaning for the operator *atnext* is: $\neg p$ holds at the next time point that p holds), then these rules should be perfectly admissible in a RB. From our perspective, production rules should be interpreted as rules of the form $\neg p$ *atnext* p. If we admit rules of the form $\neg p$ *atnext* p, we situate ourselves quite far from the concept of inconsistency as defined in other works, so we are going to clarify the meaning of inconsistency in this work:

A deductive tree T that deduces a pair of facts F and F' is *consistent* iff:

(a) T does not contain a set of contradictory static facts, or
(b) the deductive subtree of T that deduces F does not deduce F' in the end, and vice versa.

This definition implies that the deductive subtree that deduces a fact F must not deny the other fact F' that must hold at the same time than F, and vice versa.

When the agent executes a reasoning process, a deductive tree is evaluated and a sequence of rules is fired. A deductive tree defines a partial order for rule firings, so many sequences correspond to a certain deductive tree. The definition showed above is not more than a structural property to be fulfilled by the deductive trees built by the agent that we want to verify using our method. We will call this property $Tree\_Consistency(dt)$ where dt is a deductive tree that is a *tree of rule firings* defined recursively by means of the constructor *tree* and the constant NIL_TREE (empty tree). As our method will simulate the agent's reasoning, it will discard any deductive process that implies the creation of an invalid deductive tree. Next, we will define this property formally:

$$Tree\_Consistency(dt) \equiv Tree\_Consistency\_Aux1(Boundary(dt))$$
$$\wedge \, Tree\_Consistency\_Aux2(dt, \emptyset)$$

$$Tree\_Consistency\_Aux1(B) \equiv$$
$$\neg(\exists is \in INCONSISTENT\_SETS \; is \subset \bigcup_{r \in B} Assumed\_Facts(r))$$
$$Tree\_Consistency\_Aux2(dt, scope) \equiv (dt = NIL\_TREE) \vee$$
$$\exists r \exists a_1, \exists a_2 ... \exists a_n (dt = tree(r, (a_1, a_2, ..., a_n))),$$
$$scope\_in\_rule = scope \setminus Deduced\_Facts(r),$$
$$\neg((\exists f \in scope\_in\_rule, \exists f' \in Assumed\_Facts(r), (f = \neg f')) \vee$$
$$(\exists f \in Deduced\_Facts(r), \exists f' \in scope, (f = \neg f'))),$$
$$Tree\_Consistency\_Aux2(a_1, scope\_in\_rule \cup Assumed\_Facts(r)),$$
$$Tree\_Consistency\_Aux2(a_2, scope\_in\_rule \cup Assumed\_Facts(r)),$$
$$\cdots\cdots\cdots\cdots\cdots\cdots\cdots$$
$$Tree\_Consistency\_Aux2(a_n, scope\_in\_rule \cup Assumed\_Facts(r)))$$

where $INCONSISTENT\_SETS$ is the set of the different inconsistencies to be considered, the function $Boundary(dt)$ returns the set of

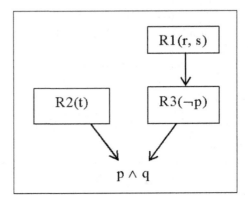

Fig. 1. Example of an invalid deductive tree

rule firings that are leaves of the tree dt, the function $Deduced\_Facts(r)$ returns the facts deduced by the rule firing r and the function $Assumed\_Facts(r)$ returns the static facts that must hold to permit the rule firing r.

In the definition above, the property $Tree\_Consistency\_Aux1$ specifies the condition (1) in the definition of consistent deductive tree above, and the property $Tree\_Consistency\_Aux2$ specifies the condition (2).

Lets see an example of an inconsistent RB. Lets take the production rules $R1: r, s \to Del(p); R2: t \to Add(p); R3: \neg p \to Add(q)$ under CWA. In the figure 1 we can see the deductive tree for the conjunction $p \wedge q$ that is supposed to be the antecedent of another rule. The facts p and q are deducible and all the other facts are non-deducible. Obviously (see rule R3), in order to deduce q, $\neg p$ must be deduced beforehand, and after having deduced $\neg p$ it is not possible to deduce p. This example deserves an additional comment. If we assume that the rules are executed with forward chaining and we fire them in the sequence [R1, R3, R2] then the facts p and q will be both true in the final FB. However, if the rules are fired in the following sequence [R2, R1, R3] then the facts $\neg p$ and q will be present in the final FB. With the first sequence, the fact q was deduced first, and then the fact p; with the second sequence the facts were deduced the other way round. Our definition of inconsistency includes situations like this one, when the truth values of the goal facts depend on the order in which they are deduced.

Lets see an example of a RB that is consistent according to our definition, but inconsistent according to other definitions. Lets take the production rules $R1: n, u \to Add(q); R2: s, \neg q \to Add(q); R3: q, m, t \to Del(p); R4: v \to Del(q)$ under CWA. In the figure 2 we can see the deductive tree for the conjunction $\neg p \wedge q$ that is supposed to be the antecedent of another rule. We want to deduce the $\neg p$ and q, and all the other facts are non-deducible. We can see that there are six different sequences of rules that correspond to the deductive tree of the figure 2. However, among them, only three sequences are feasible ([R4, R2, R1, R3], [R1, R3, R4, R2] and [R1, R4, R2, R3]), and all of these three sequences deduce the same truth values for p and q.

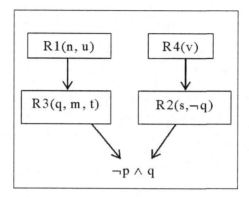

Fig. 2. Example of a valid deductive tree

According to the above definition of inconsistency, it is clear that the proposed method will not be able to verify some non-monotonic KBS. In particular, all the KBS whose deductive trees do not follow the consistency definition exposed above, for instance, the planners of STRIPS type.

3   Requirements for Getting an Inconsistency: Scenario

The aim of the proposed method, as it was explained in the section 1, will be to compute scenarios for an inconsistency described by an IC. Each scenario is formed by a description of the initial agent's state, a deductive tree of rule firings, and a partially ordered set of messages and/or stimuli. The proposed method will construct an object called *subcontext* to specify how the initial agent's state must be and which deductive tree must be executed in order to yield an inconsistency. There may be different initial agent's states and different deductive trees that lead to the same inconsistency. All the different ways to violate a certain IC will be specified by means of an object called *context*. Thus, a context will be composed of n subcontexts. In turn, a subcontext is defined as a pair *(environment, deductive tree)* where an environment is made up of a set of *metaobjects*, and a deductive tree is a tree of rule firings.

A metaobject describes the characteristics that one object which can be present in the agent's state should have. For each type of CCR-2 object there will be a different type of metaobject: metaproposition, metaframe, metarelationship, metaattribute and metaid-attribute. In order to describe a CCR-2 object, a metaobject must include a set of constraints on the characteristics of the CCR-2 object. Some CCR-2 objects may include references to other CCR-2 objects (for example, a frame instance can have references to attributes and a relationship can include tuples of references to frame instances), so the counterpart metaobjects will contain references to other metaobjects. In the table below, the attributes of each type of metaobject are shown. The value of these attributes will represent the constraints described by each metaobject.

CCR-2 Object	Metaobject	Attributes of the Metaobject
Frame	Metaframe	(identifier, is_restricted_exist, instance_of, subclass_of, metaattributes, metarelationships)
Attribute	Metaattribute	(identifier, is_restricted_exist, metaframe, metaid-attributes, value_conditions, cf_conditions)
Id-Attribute	Metaid-attribute	(identifier, is_restricted_exist)
Relationship	Metarelationship	(identifier, is_restricted_exist, type, tuples, conditions_for_each_tuple)
Proposition	Metaproposition	(identifier, is_restricted_exist, type, truth_value, conditions)

Given that certain constraints expressed as arithmetic inequations can affect the attribute values and the certainty factors associated with CCR-2 objects, a different kind of metaobject called *condition* will represent them. Conditions will also appear in environments, together with metaobjects, and they will be referenced from and contain references to the metaobjects that participate in them. Considering the references among metaobjects and conditions, there can be one or more networks of metaobjects and conditions in one environment. Figure 3 illustrates an example of an environment describing a FB in which the formula $\neg Has(X, Water) \wedge X.temperature \geqslant 80$ is true, where the variable X is declared as an instance of the frame *Car*. If there exists a CCR-2 object in the FB, for each metaobject in the environment, that satisfies all the requirements imposed on it, then the given formula will hold in the FB.

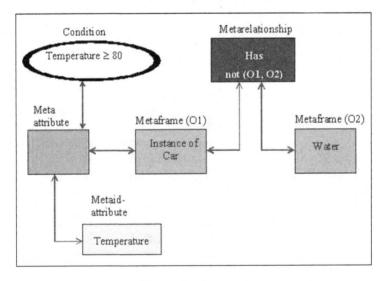

Fig. 3. Environment

3.1   Temporal Labels and Constraints

A *goal* h is a pair (l, A) where l is a literal and A is a set of metaobjects associated with the object names and variables in l, that specifies the FBs in which the literal l is satisfied. Moreover, a goal (l, A) is static/dynamic/deducible/acquired/innate iff the literal l is static/dynamic/deducible/acquired/innate.

For the purpose of executing a deductive tree, it may be required that a dynamic acquired fact f holds in a rule, and later on, that the fact $\neg f$ holds in another rule. This situation may yield an apparently contradictory environment. To determine if it is a real contradiction, *temporal labels* will be associated with some constraints included in the goals (l, A) and (l', A') that entail f and $\neg f$ respectively, to represent that these constraints must be satisfied in different rule firings (or moments). Each temporal label associated with a constraint identifies the rule firing where the constraint must be satisfied, and specifies that the constraint comes from a dynamic acquired fact. From these labels, the method will specify, as part of the resulting scenario, that a message/stimulus that matches schema M and allows literal l to hold must be received <u>before</u> a message/stimulus that matches schema M' and allows literal l' to hold is received, formally $M < M'$. *Temporal constraints*, like the one stated in the previous sentence, will define a partially ordered set of messages and/or stimuli schemas, in which the relationship $<$ expresses temporal precedence.

For each static acquired literal included in the KB, it will be required to produce a temporal constraint to establish that the message/stimulus (according to a schema) allowing the static acquired literal to hold must be received before the end of the deductive process. Consequently, to permit the proposed method to obtain the proper temporal constraints later, some temporal labels must also be associated with the constraints derived from static acquired literals. Besides, these labels must specify that the constraints have been obtained from a static acquired fact.

Moreover, the method has to generate temporal constraints to establish that some messages/stimuli allowing static acquired literals to hold must be received before the message/stimulus that allows a certain dynamic acquired literal to hold. Lets see the conditions in which these temporal constraints must be generated. Let $(R1, R2, ..., RN)$ be the sequence of rules that are fired as a result of evaluating a deductive tree according to the control mechanisms. Let Ri *s.t.* $1 \leqslant i < N$ be a rule whose antecedent requires the dynamic acquired literal Ld to hold, and let M be a message/stimulus schema that entails Ld; let Rj *s.t.* $i < j \leqslant N$ be a rule whose antecedent requires the dynamic acquired literal $\neg Ld$ to hold, and let M' be a message/stimulus schema that entails $\neg Ld$. Then, it is clear that any message/stimulus schema $M1$ that entails a static acquired literal Ls belonging to the antecedent of a rule Rk *s.t.* $i \leqslant k < j$ must satisfy $M1 < M'$. The rationale for generating these temporal constraints will become clearer in the section 5 when an example is shown.

4 Description of the Method

Computing the scenarios associated with an IC requires three steps:

1. Computing the context associated with the IC without taking into account the control mechanisms, and considering all the rules to form a unique group.
2. Computing the scenarios from the context associated with the IC and the control mechanisms.
3. Discarding invalid scenarios w.r.t. the communication protocols.

4.1 Computing the Context Associated with the IC

Basically, the first step can be divided into two phases. In the first phase, the AND/OR decision tree associated with the IC is expanded following a backward chaining simulation of the real rule firings. The leaves of this tree are rules that only contain acquired facts in their antecedents. At this point, the difference between a deductive tree and a AND/OR decision tree should be explained. While a deductive tree can be viewed as one way and only one way for achieving a certain goal (that is, for deducing a bound formula or for firing a rule), an AND/OR decision tree comprises one or more deductive trees, therefore it specifies one or more ways to achieve a certain goal. During the first phase, metaobjects are built and propagated from a rule to another one. In this propagation, some constraints are added to the metaobjects due to the rule literals and the declaration part of the rules/IC, and some constraints are removed from the metaobjects due to the rule actions. In addition to the metaobjects, a set of assumed propositions and tuples (SAPT) are propagated and updated.

In the second phase, the AND/OR decision tree is contracted by means of context operations, and metaobjects associated with non-deducible facts and conditions associated with inequations are inserted in the subcontexts. Lets define the following contexts operations: creation of a context, concatenation of a pair of contexts and combination of a list of contexts.

Contexts Operation

a) *Creation*: a context with an unique subcontext is created from a non-deducible goal $g = (l, A)$ and a rule r: $C(g, r) = \{(E, NIL\_TREE)\}$ where the environment E comprises all the metaobjects included in g. The rule r must be a rule that comprises the literal l in its antecedent. If the literal l is not innate (so it is related to a message/stimulus), some constraints of the metaobjects must be labelled with a temporal label indicating that these constraints must be satisfied at least in the firing of the rule r; in particular, constraints that state the truth value of a metaproposition, and constraints that state the truth value of a tuple in a metarelationship. The literal l will hold in any agent's state that satisfies all the constraints specified in E.

b) *Concatenation of a pair of contexts*: let C_1 and C_2 be a pair of contexts and $Conc(C_1, C_2)$ be the context resulting from the concatenation, then: $Conc(C_1, C_2) = C_1 \cup C_2$.

c) *Combination of a list of contexts*: Let $C_1, C_2, ..., C_n$ be the list of contexts, and $Comb(C_1, C_2, ..., C_n)$ be the context resulting from the combination. The form of this resulting context is: $Comb(C_1, C_2, ..., C_n) = \{(E_{k1} \cup E_{k2}... \cup E_{kn}, DT_{k1} * DT_{k2}... * DT_{kn})$ s.t. $(E_i, DT_i) \in C_i\}$

    c.1) *Union of environments* $(E_i \cup E_j)$: this operation consists of the union of the sets of metaobjects E_i and E_j. After the union of two sets, it is necessary to check whether any pair of metaobjects can be merged. A pair of metaobjects will be merged if they contain a pair of constraints c_1 and c_2 respectively such that c_1 and c_2 specify the same name. As a result of this fusion, the new metaobject could be invalid if it contains contradictory constraints not coming from dynamic acquired facts. In this case, the resulting environment will be invalid, and it will be discarded. Finally, if the resulting environment represents an invalid initial agent state, then this environment will also be discarded. Moreover, after the union of two environments, it is also necessary to check whether the resulting set of conditions can be satisfied or, in others words, whether the resulting set of conditions is feasible.

    c.2) *Combination of deductive trees* $(DT_i * DT_j)$: let DT_i and DT_j be deductive trees, then $DT_i * DT_j$ is the deductive tree that results from constructing a new tree whose root node represents an empty rule firing, and whose two subtrees are DT_i and DT_j.

Basically, the creation operation is employed to work out the context associated with a non-deducible goal; the combination operation is employed to work out the context associated with a conjunction of literals from the contexts associated with the literals; and the concatenation operation is employed to work out the context associated with a disjunction from the contexts associated with the formulas involved in the disjunction.

These two phases are explained in detail in [13]. However, there are some differences between the current step and the process explained in [13]. These differences are related mainly to the context operations and the treatment of acquired facts and deductive trees. In [13] is explained a method for verifying an isolated KB System, so acquired facts are not considered, and the KB System is assumed to deal only with innate knowledge (external facts in [13]), and deduced knowledge.

4.2   Computing the Scenarios

In the second step of the method, a different scenario is derived from each subcontext in the context associated with the IC by adding a partially ordered set of messages and/or stimuli to the subcontext. In this step, some subcontexts may be discarded if they are impossible w.r.t. the control mechanisms. The partial order on the message/stimulus schemas reflects the temporal constraints derived from the control mechanisms and the deductive tree. These temporal constraints are generated as it was explained in the section 3. It may happen that more than one message/stimulus schema entails the same literal, so this aspect must be taken into account in building the temporal constraints to be added to the partially ordered set.

4.3 Discarding Invalid Scenarios w.r.t. the Communication Protocols

In the previous steps, some scenarios have been computed for an IC. However, it may happen that some scenario obtained in the previous step describes impossible sequences of messages or stimuli w.r.t. the communication protocols. In order to check this, at least one path that satisfies all the temporal constraints must be found in the state machine. The first state of this path must be the state in which the agent begins its reasoning process.

5 Example of Application

In this section we will show how the method can be applied to a small example. We will assume a deliberative agent that executes the sequence of rules that appears in the figure 4. For the sake of clarity and conciseness, the rules and the IC of this example are not represented in the CCR-2 format, and all the facts are propositional. In this example, the facts q and $\neg q$ are dynamic acquired facts entailed by the messages M and M' respectively, whereas the fact $\neg r$ is a static acquired fact entailed by the stimulus S. Moreover, the fact s belongs to the agent's innate knowledge, and the facts t and p are deducible.

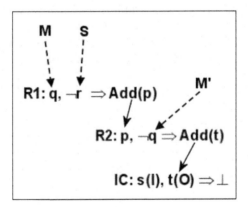

Fig. 4. Example with an IC and two rules

The method begins expanding the AND/OR decision tree. First of all, it is necessary to bind each variable of the IC and each referenced object to a metaobject. Some constraints are derived from each IC literal, and they are added to the metaobjects (in this case, metapropositions). The resulting metapropositions are:

$$PROP1 = (id \rightarrow s, truth\_value \rightarrow true)$$
$$PROP2 = (id \rightarrow t, truth\_value \rightarrow true)$$

In addition to the metaobjects, the SAPT is created. This set contains the names of the propositions included in the IC and the tuples of relationships whose names appear in the IC that are not associated with dynamic acquired facts. So, initially, SAPT = $\{s, t\}$, because neither the fact s nor the fact t are dynamic acquired facts. The aim of the SAPT is to warrant the consistency of the non-monotonic reasoning in the sense explained in the section 2, concretely the second point of the consistency definition in section 2. The SAPT plays the role of the *scope* parameter in the definition of the *Tree_Consistency* property. Unluckily, the metaobjects alone cannot warrant the consistency in all the cases. For example, if the SAPT is not used in the example of the section 2 (see figure 1), the inconsistency would not be detected in the simulation of the agent's execution, and that deductive tree would not be discarded.

Obtaining the context of an IC implies obtaining the context associated with each literal included in the IC. If it corresponds to a non-deducible goal, its context is created (see Creation Operation in section 4.1). In order to compute the context of a deducible goal, the method has to generate the contexts associated with all the rules that deduce the goal (conflict set), and then it has to concatenate them (in the contraction phase). To decide whether a rule deduces a goal, it is needed to check whether there exists any action in the rule that is unifiable with the goal. In the example of the figure 4, the IC comprises an innate literal (input literal) and a deducible literal (output literal). So, the method finds a rule (R2) to deduce the deducible literal.

In general, a CCR-2 rule premise contains a list of conjunctions joined by disjunction operators. Hence, to compute the context of a rule it is needed to calculate the context of each conjunction, and then they have to be concatenated (in the contraction phase). In order to compute the context of a conjunction, it is required to compute the context of each literal included in the conjunction. A pre-processing similar to that of an IC is performed over each conjunction before computing the contexts of the included literals. As a result of this, new metaobjects and conditions appear and some constraints are added to the metaobjects. In the rule R2, the metapropositions *PROP3(p)* and *PROP4(¬q)* are created.

The rule R2 contains only one conjunction with two literals p and $\neg q$. While p is a deducible fact, $\neg q$ is a dynamic acquired fact. In this example, the rule R1 can be employed to deduce the fact p. In the rule R1, the metapropositions *PROP5(q)* and *PROP6(¬r)* are created.

The SAPT propagated from the IC is updated while processing the rule R2, so now SAPT = $\{s, p\}$, since t is deleted by the action of the rule R1, and $\neg q$ is a dynamic acquired fact. If the antecedent of the rule R2 had comprised the fact $\neg s$, a conflict would have been detected when updating the SAPT, and the rule R2 would have been discarded. Finally, the SAPT in the rule R1 is SAPT = $\{s\}$.

Once the AND/OR decision tree has been expanded completely, the tree is contracted by using the context operations, and the constraints generated for the non-deducible goals (inside the metaobjects) are propagated forward from the leaves of the AND/OR decision tree to the IC. Thus, all these constraints are collected in the context associated with the IC. In the example, the contexts associated with the non-deducible facts s, q, $\neg r$ and $\neg q$ are created, and next, the necessary combination operations are carried out until the context associated

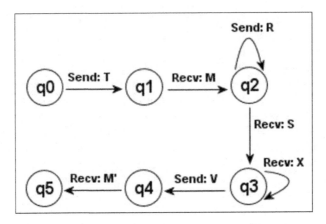

Fig. 5. Fragment of the Agent's State Machine

with the IC is computed. Every time a context is obtained from a combination operation in a rule R, this rule R is added to each deductive tree of the context as the new root node.

It is worth mentioning that while computing $Comb(C(p), C(\neg q))$ in the rule R2, an apparent conflict is detected between the metapropositions *PROP4* and *PROP5*, as they require different truth values for the same proposition q. However, there is no contradiction, since the facts q and $\neg q$ are dynamic acquired facts, that is, the contradictory facts may hold in different moments. Hence, these metapropositions are merged, and the new metaproposition *PROP7* is yielded:

$$PROP7 = (id \rightarrow q, truth\_value \rightarrow \{true(R1, dynamic), false(R2, dynamic)\})$$

After applying the first step of the method, the resulting context associated with the IC is: $C(IC) = \{(\{PROP1, PROP6, PROP7\}, tree(R1, [tree(R2, nil)])$ $)\}$, where these metapropositions are defined as:

$$PROP7 = (id \rightarrow q, truth\_value \rightarrow \{true(R1, dynamic), false(R2, dynamic)\})$$
$$PROP6 = (id \rightarrow r, truth\_value \rightarrow false(R1, static))$$
$$PROP1 = (id \rightarrow s, truth\_value \rightarrow true)$$

Next, in the second step, according to the control mechanism, it is determined that this deductive tree is evaluated by firing the sequence of rules $[R1, R2]$. Taking this into account, the following temporal constraints are derived from the metapropositions: $M < M'$, because the message M must be received before the message M', in order to allow the fact q to hold first, and then to allow the fact $\neg q$ to hold later; and $S < M'$, because the stimulus S must be received before the message M', since, otherwise, the rule $R1$ will not be able to be fired before the rule $R2$. Thus, the partially ordered set is $\{M < M', S < M'\}$, and the scenario is $(C(IC), \{M < M', S < M'\})$

Finally, in the third step, the scenario is checked w.r.t. the agent's state machine, which describes the agent behaviour. We can see a fragment of this state machine in the figure 5. The reasoning process is supposed to begin in the state $q0$. It is clear that there is a path that satisfies all the temporal constraints imposed in the scenario, so the scenario is consistent with the state machine.

6   Conclusion

In this paper, a formal method to verify the consistency of the reasoning process of a deliberative agent w.r.t. communication protocols has been presented. To the best of our knowledge, there is no other method or tool that also addresses this kind of verification. It is also noteworthy that the agent to be verified encompasses a hybrid KB that permits the representation of non-monotonic reasoning and arithmetic constraints.

7   Future Work

Mainly, there are two aspects of the proposed method that we want to improve: first, the validation of the deductive tree w.r.t. control mechanisms, more concretely, w.r.t. metarules; and second, the deletion of redundancy in the sets of temporal constrains by taking into account transitive dependencies and other aspects.

Moreover, we are working on the adaptation of the proposed method so that it can be applied to verify agents whose knowledge domain is expressed in a wide known ontology like OWL[2].

References

1. de Kleer, J.: An assumption based TMS. Artificial Intelligence **28** (1986) 127–162
2. Rousset, M.: On the consistency of knowledge bases: The COVADIS system, Proceedings ECAI-88, Munich, Alemania (1988) pp. 79-84.
3. Ginsberg, A.: Knowledge-base reduction: A new approach to checking knowledge bases for inconsistency and redundancy, Proceedings of the AAAI-88 (1988) pp. 585-589.
4. de Antonio, A.: Sistema para la verificación estructural y detección de inconsistencias en bases de conocimientos. Final year project, Facultad de Informática, UPM (1990)
5. Meseguer, P.: Incremental verification of rule-based expert systems, Proceedings of the 10th. European Conference on AI (ECAI'92) (1992) pp. 840-844.
6. Dahl, M., Williamson, K.: A verification strategy for long-term maintenance of large rule-based systems, Workshop Notes of the AAAI92 WorkShop on Verification and Validation of expert Systems (1992) pp. 66-71.
7. Ayel, M.: Protocols for consistency checking in expert system knowledge bases, Proceedings of the European Conference on Artificial Intelligence (ECAI'88) (1988) pp. 220-225.

---

[2] http://www.w3.org/TR/owl-features/

8. Ayel, M., Laurent, J.P.: Validation, Verification and Test of Knowledge-Based Systems: SACCO-SYCOJET: Two Different Ways of Verifying Knowledged-Based Systems. John Wiley publishers (1991)

9. de Antonio, A., Cardeñosa, J., Martínez, L.: GKR: A generic model of knowledge representation. Volume II, Student Abstracts., Proceedings of the 12th National Conference on Artificial Intelligence (AAAI94) (1994) pp. 1438.

10. d'Inverno, M., Kinny, D., Luck, M.: Interaction protocols in agentis, Third International Conference on Multi-Agent Systems (ICMAS98) (1998) 261–268

11. Antoniou, G.: Verification and correctness issues for nonmonotonic knowledge bases. International Journal of Intelligent Systems **12** (1997) 725–738

12. Kröger, F.: Temporal Logic of Programs. Springer-Verlag (1987)

13. Ramírez, J., de Antonio, A.: Knowledge base semantic verification based on contexts propagation, Notes of the AAAI-01 Symposium on Model-based Validation of Intelligence (2001)

F-OWL: An Inference Engine for the Semantic Web[1]

Youyong Zou, Tim Finin, and Harry Chen

Computer Science and Electrical Engineering,
University of Maryland, Baltimore County,
1000 Hilltop Circle, Baltimore MD 21250
{yzou1, finin, hchen4}@cs.umbc.edu

Abstract. Understanding and using the data and knowledge encoded in semantic web documents requires an inference engine. F-OWL is an inference engine for the semantic web language OWL language based on F-logic, an approach to defining frame-based systems in logic. F-OWL is implemented using XSB and Flora-2 and takes full advantage of their features. We describe how F-OWL computes ontology entailment and compare it with other description logic based approaches. We also describe TAGA, a trading agent environment that we have used as a test bed for F-OWL and to explore how multiagent systems can use semantic web concepts and technology.

1 Introduction

The central idea of the Semantic Web [Berners-Lee 2001] is to publish documents on the World Wide Web defined and linked in a way that make them both human readable and machine understandable. Human readable means documents in the traditional sense which are intended for machine display and human consumption. Machine understandable means that the data has explicitly been prepared for machine reasoning and reuse across various applications. Realizing the semantic web vision requires well defined languages that can model the meaning of information on the Web as well as applications and services to publish, discover, process and annotate information encoded in them. This involves aspects from many areas, including knowledge representation and reasoning, databases, information retrieval, digital libraries, multi-agent systems, natural language processing and machine learning. The Web Ontology Language OWL [Patel-Schneider, 2003] is part of the growing stack of W3C recommendations related to the Semantic Web. OWL has its origins in DAML+OIL [Hendler 2000] and includes a set of three increasingly complex sublanguages: OWL-Lite, OWL-DL and OWL-Full.

OWL has a model-theoretic semantics that provides a formal meaning for OWL ontologies and instance data expressed in them. In addition, to support OWL-Full, a second model-theoretic semantics has been developed as an extension to the RDF's semantics, grounding the meaning of OWL ontologies as RDF graphs. An OWL inference engine's core responsibilities are to adhere to the formal semantics in

---

[1] This work was partially supported by the Defense Advanced Research Projects Agency under contract F30602-97-1-0215 and by the   National Science Foundation under award IIS-0242403.

processing information encoded in OWL, to discover possible inconsistencies in OWL data, and to derive new information from known information. A simple example demonstrates the power of inference: Joe is visiting San Francisco and wants to find an Italian restaurant in his vicinity. His wireless PDA tries to satisfy his desire by searching for a thing of type *restaurant* with a *cuisineType* property with the value *Italian*. The goodPizza restaurant advertises its cuisine type as *Pizza*. These cannot be matched as keywords or even using a thesaurus, since *Italian* and *Pizza* are not equivalent in all contexts. The restaurant ontology makes things clearer: *Pizza rdfs: SubClassOf ItalianCuisine*. By using an inference engine, Joe's PDA can successfully determine that the restaurant goodPizza is what he is looking for. F-OWL, an inference engine for OWL language, is designed to accomplish this task.

In the next section, we outline the functional requirement of the OWL inference engine. Section three describes F-OWL, the OWL inference engine in Frame Logic that we have developed. Section four explained how F-OWL is used in a multi-agent test bed for trading agents. Chapters five and six conclude this paper with a discussion of the work and results and an outline of some potential future research.

2   OWL Engine

An inference engine is needed for the processing of the knowledge encoded in the semantic web language OWL. An OWL inference engine should have following features:

- **Checking ontology consistency.** An OWL concept ontology (e.g., terms defined in the "Tbox") imposes a set of restrictions on the model graph. The OWL inference Engine should check the syntax and usage of the OWL terms and ensure that the OWL instances (e.g., assertions in the "Abox") meet all of the restrictions.
- **Computing entailments.** Entailment, including satisfiability and subsumption, are essential inference tasks for an OWL inference engine.
- **Processing queries.** OWL inference engines need powerful, yet easy-to-use, language to support queries, both from human users (e.g., for debugging) and software components (e.g., for software agents).
- **Reasoning with rules.** Rules can be used to control the inference capability, to describe business contracts, or to express complex constrictions and relations not directly supported by OWL. An OWL inference engine should provide a convenient interface to process rules that involve OWL classes, properties and instance data.
- **Handling XML data types.** XML data types can be used directly in OWL to represent primitive kinds of data types, such as integers, floating point numbers, strings and dates. New complex types can be defined using base types and other complex types. An OWL inference Engine must be able to test the satisfiability of conjunctions of such constructed data types.

The OWL language is rooted in description logic (DL), a family of knowledge representation languages designed for encoding knowledge about concepts and concept hierarchies. Description Logics are generally given a semantics that make them subsets of first-order logic. Therefore, several different approaches based on those logics have been used to design OWL inference engines:

- **Using a specialized description logic reasoner.** Since OWL is rooted in description logic, it is not surprising that DL reasoners are the most widely used tools for OWL reasoning. DL reasoners are used to specify the terminological hierarchy and support subsumption. It has the advantage of being decidable. Three well-known systems are FaCT [Horrocks, 1999], Racer [Haarslev 2001] and Pellet. They implement different types of description logic. Racer system implements SHIQ(D) using a Tableaux algorithm. It is a complete reasoner for OWL-DL and supports both Tbox and Abox reasoning. The FaCT system implements SHIQ, but only support Tbox reasoning. Pellet implements SHIN(D) and includes a complete OWL-lite consistency checker supporting both Abox and Tbox queries.
- **Using full first order logic (FOL) theorem prover.** OWL statements can be easily translated into FOL, enabling one to use existing FOL automated theorem provers to do the inference. Examples of this approach include Hoolet (using the Vampire [Riazanov, 2003] theorem prover) and Surnia (using Otter theorem prover). In Hoolet, for example, OWL statements are translated into a collection of axioms which is then given to the Vampire theorem prover for reasoning.
- **Using a reasoner designed for a FOL subset.** A fragment of FOL and general logic based inference engine can also be used to design the OWL inference engine. Horn Logic is most-widely used because of its simplicity and availability of tools, including Jena, Jess, Triple and F-OWL (using XSB). Other logics, like higher-order logic in F-OWL (using Flora), can also be used.

As the following sections describe, F-OWL has taken the third approach. An obvious advantage is that many systems have been developed that efficiently reason over expressive subsets of FOL and are easy to understand and use.

3   F-OWL

F-OWL is a reasoning system for RDF and OWL that is implemented using the XSB logic programming system [Sagonas, 1994] and the Flora-2 [Kifer, 1995] [Yang 2000] extension that provides an F-logic frame-based representation layer. We have found that XSB and Flora-2 not only provide a good foundation in which to implement an OWL reasoner but also facilitate the integration of other reasoning mechanisms and applications, such as default reasoning and planners.

XSB is a logic programming system developed at Stony Brook University. In addition to providing all the functionality of Prolog, XSB contains several features not usually found in Logic Programming systems, including tabling, non-stratified negation, higher order constructs, and a flexible preprocessing system. Tabling is useful for recursive query computation, allowing programs to terminate correctly in many cases where Prolog does not. This allows, for example, one to include "if and only if" type rules directly. XSB supports for extensions of normal logic programs through preprocessing libraries including a sophisticated object-oriented interface called Flora-2. Flora-2 is itself a compiler that compiles from a dialect of Frame logic into XSB, taking advantage of the tabling, HiLog [Chen 1995] and well-founded semantics for negation features found in XSB. Flora-2 is implemented as a set of run-time libraries and a compiler that translates a united language of F-logic and HiLog into tabled Prolog code. HiLog is the default syntax that Flora-2 uses to represent function terms and predicates. Flora-2 is a sophisticated object-oriented knowledge base language and application development platform. The programming language supported by Flora-2 is a dialect of F-logic with numerous extensions, which include a natural way to do meta-programming in the style of HiLog and logical updates in the style of Transaction Logic. Flora-2 was designed with extensibility and flexibility in mind, and it provides strong support for modular software design through its unique feature of dynamic modules.

F-OWL is the OWL inference engine that uses a Frame-based System to reason with OWL ontologies. F-OWL is accompanied by a simple OWL importer that reads an OWL ontology from a URI and extracts RDF triples out of the ontology. The extracted RDF triples are converted to format appropriate for F-OWL's frame style and fed into the F-OWL engine. It then uses flora rules defined in flora-2 language to check the consistency of the ontology and extract hidden knowledge via resolution.

A model theory is a formal theory that relates expressions to interpretation. The RDF model theory [Hayes 2003] formalizes the notion of inference in RDF and provides a basis for computing deductive closure of RDF graphs. The semantics of OWL, an extension of RDF semantics, defines bindings, extensions of OWL interpretations that map variables to elements of the domain:

- The vocabulary V of the model is composed of a set of **URI**'s.
- LV is the set of *literal values* and XL is the mapping from the literals to LV.
- A *simple interpretation I* of a vocabulary V is defined by:
  - A non-empty set IR of resources, called the domain or universe of I.
  - A mapping IS from V into IR
  - A mapping $IEXT$ from IR into the power set of IR X (IR union LV) i.e. the set of sets of pairs $<x,y>$ with x in IR and y in IR or LV. This mapping defines the properties of the triples. $IEXT(x)$ is a set of pairs which identify the arguments for which the property is true, i.e. a binary relational extension, called the *extension* of x.

Informally this means that every **URI**[2] represents a resource that might be a page on the Internet but not necessarily; it might also be a physical object. A property is a relation; this relation is defined by an extension mapping from the property into a set. This set contains pairs where the first element of a pair represents the subject of a triple and the second element represents the object of a triple. With this system of extension mapping the property can be part of its own extension without causing paradoxes.

Take the triple:*goodPizza :cuisineType :Pizza* from the pizza restaurant in the introduction as example. In the set of **URI**'s there will be terms (i.e., classes and properties) like: *#goodPizza, #cuisineType, #pizza, #Restanrant, #italianCuisine*, etc.

---

[2] The W3C says of URIs: "Uniform Resource Identifiers (URIs, aka URLs) are short strings that identify resources in the web: documents, images, downloadable files, services, electronic mailboxes, and other resources." By convention, people understand many URIs as denoting objects in the physical world.

These are part of the vocabulary *V*. The set **IR** of resources include instances that represent resources on the internet or elsewhere, like *#goodPizza*, , etc. For example the class *#Restanrant* might represent the set of all restaurants. The **URI** refers to a page on the Internet where the domain **IR** is defined. Then there is the mapping *IEXT* from the property *#cuisineType* to the set {(*#goodPizza*, *#Pizza*),(*#goodPizza*, *#ItalianCuisine*)} and the mapping *IS* from *V* to *IR*: *:goodPizza* → *#goodPizza*, *:cuisineTYpe* → *#cuisineType*.

A rule *A →B* is satisfied by an interpretation *I* if and only if every binding that satisfies the antecedent *A* also satisfies the consequent *B*. An ontology *O* is satisfied by an interpretation *I* if and only if the interpretation satisfies every rules and facts in the ontology. A model is satisfied if none of the statements within contradict each other. An ontology O is consistent if and only if it is satisfied by at least one interpretation. An ontology O_2 is entailed by an ontology O_1 if and only if every interpretation that satisfies O_1 also satisfies O_2.

One of the main problems in OWL reasoning is ontology entailment. Many OWL reasoning engines, such as Pellet and SHOQ, follow an approach suggested by Ian Horrocks [Horrocks 2003]. By taking advantage of the close similarly between OWL and description logic, the OWL entailment can be reduced to knowledge base satisfiability in the SHOIN(D) and SHIF(D). Consequently, existing mature DL reasoning engines such as Racer [Haarslev 2001] can provide reasoning services to OWL. Ora Lassila suggested a *"True RDF processor"* [Lassila 2002] in his implementation of Wilbur system [Lassila 2001] in which entailment is defined via the generation of a deductive closure from an RDF graph composed of triples. The proving of entailment becomes the building and searching of closure graph.

With the support of forward/backward reasoning from XSB and frame logic from Flora, F-OWL takes the second approach to compute the deductive closure of a set of RDF or OWL statements. The closure is a graph consisting of every triples *<subject, predicate, object>* that satisfies *{subject, object }* ⇒ *IEXT(I(predicate))*. This is defined as:

$$<subject, predicate, object> \Rightarrow KB \Leftrightarrow \{subject, object\} \Rightarrow IEXT(I(predicate))$$

Where *KB* is the knowledge base, *I(x)* is the interpretation of a particular graph, and *IEXT(x)* is the binary relational extension of property as defined in [Hayes 2002].

F-OWL is written in the Flora-2 extension to XSB and consists of the following major sets of rules:

- A set of rules that reasons over the data model of RDF/RDF-S and OWL;
- A set of rules that maps XML DataTypes into XSB terms;
- A set of rules that performs ontology consistency checks; and
- A set of rules that provides an interface between the upper Java API calls to the lower layer Flora-2/XSB rules.

F-OWL provides command line interface, a simple graphical user interface and a Java API to satisfy different requirements. Using F-OWL to reason over the ontology typically consists of the following four steps:

- Loading additional application-related rules into the engine;
- Adding new RDF and OWL statements (e.g., ontologies or assertions) to the engine. The triples (subject, predicate, object) on the OWL statements are translated into 2-ply frame style: subject(predicate, object)@model;
- Querying the engine. The RDF and OWL rules are recursively applied to generate all legal triples. If a query has no variables, a True answer is returned when an interpretation of the question is found. If the question includes variable, the variables is replaced with values from the interpretation and returned;
- The ontology and triples can be removed if desired. Else, the XSB system saves the computed triples in indexed tables, making subsequent queries faster.

4 F-OWL in TAGA

Travel Agent Game in Agentcities (TAGA) [Zou 2003] is a travel market game developed on the foundation of FIPA technology and the Agentcities infrastructure. One of its goals is to explore and demonstrate how agent and semantic web technology can support one another and work together.

TAGA extends and enhances the Trading Agent Competition scenario to work in Agentcities, an open multiagent systems environment of FIPA compliant systems. TAGA makes several contributions: auction services are added to enrich the Agentcities environment, the use of the semantic web languages RDF and OWL improve the interoperability among agents, and the OWL-S ontology is employed to support service registration, discovery and invocation. The FIPA and Agentcities standards for agent communication, infrastructure and services provide an important foundation in building this distributed and open market framework. TAGA is intended as a platform for research in multiagent systems, the semantic web and/or automated trading in dynamic markets as well as a self contained application for teaching and experimentation with these technologies. It is running as a continuous open game at http://taga.umbc.edu/ and source code is available on Sourceforge for research and teaching purposes.

The agents in TAGA use OWL in various ways in communication using the FIPA agent content language (ACL) and also use OWL-S as the service description language in FIPA's directory facilitators. Many of the agents in the TAGA system use F-OWL directly to represent and reason about content presented in OWL. On receiving an ACL message with content encoded in OWL, a TAGA agent parses the content into triples, which are then loaded into the F-OWL engine for processing.

When an agent receives an incoming ACL message, it computes the meaning of the message from the ACL semantics, the protocols in effect, the content language and the conversational context. The agent's subsequent behavior, both internal (e.g., updating its knowledge base) and external (e.g., generating a response) depends on the correct interpretation of the message's meaning. Thus, a sound and, if possible, complete understanding the *semantics* of the key communication components (i.e., ACL, protocol, ontologies, content language, context) is extremely important.   In TAGA, the service providers are independent and autonomous entities, which making it difficult to enforce a design decision that all use exactly the same ontology or protocol. For example, the Delta Airline service agent may have its own view of travel business and uses class and property terms that extend an ontology used in the industry. This situation parallels that for the semantic web as a whole – some amount of diversity is inevitable and must be panned for lest our systems become impossibly brittle.

Many of the agents implemented in TAGA system use F-OWL to represent and reason about the message content presented in RDF or OWL. Upon receiving an ACL message with content in RDF or OWL, a TAGA agent parses the content into triples, which are then loaded into the FOWL engine for processing.

The message's meaning (communicative act, protocol, content language, ontologies and context) all play a part in the interpretation. For example, when an agent receives a query message that uses the query protocol, the agent searches its knowledge base for matching answers and returns an appropriate inform message. TAGA uses multiple models to reflect the multiple namespaces and ontologies used in the system. The agent treats each ontology as an independent model in the F-OWL engine.

F-OWL has many usages in TAGA, including the following.

- **As knowledge base.** Upon receiving an ACL message with content encoded in OWL, agents in TAGA parse the content into triples and feeds them into their F-OWL engine. The information can be easily retrieved by submitting queries in various query languages.
- **As reasoning engine.** The agent can answer more questions with the help of F-OWL engine, for example, the restaurant can answer the question "what is the average price of a starter" after it understands that "starter" is *sameAs* "appetizer".
- **As a service matchmaker.** FIPA platforms provide a *directory facilitator* service which matches service requests against descriptions of registered services. We have extended this model by using OWL-S as a service description language. F-OWL manages the service profiles and tries to find the best match based on description in the service request.
- **As an agent interaction coordinator.** The interaction protocol can be encoded into an ontology file using OWL language. F-OWL will advise the agents what to respond based on received messages and context.

5 Discussion

This section describes the design and implementation of F-OWL, an inference engine for OWL language. F-OWL uses a Frame-based System to reason with OWL ontologies. F-OWL supports consistency checking of the knowledge base, extracts hidden knowledge via resolution and supports further complex reasoning by importing rules. Based on our experience in using F-OWL in several projects, we found it to be a fully functional inference engine that was relatively easy to use and able to integrate with multiple query languages and rule languages.

There have been lots of works on the OWL inference engine, from semantic web research community and description logic community. The following table compares F-OWL with some of them:

Table 1. Comparison of F-OWL and other OWL Inference Engine

	F-OWL	**Racer**	**FaCT**	**Pellet**	**Hoolet**	**Sur-nia**	**Tri-ple**
Logic	Horn, Frame, Higher Order	Description Logic	DL	DL	Full FOL	Full FOL	Horn Logic
Support	OWL-Full	OWL-DL	OWL-DL	OWL-DL	OWL-DL	OWL-Full	RDF
Based on	XSB/Flora	Lisp	Lisp	Java	Vampire	Otter	XSB
XML Datatype	Yes	Yes	No	Yes	No	No	No
Decidable	No	Yes	Yes	Yes	No	No	Yes
Complete consistency checker	No	Yes (OWL-Lite)	Yes	Yes(OWL-Lite)	No	No	No
Interface	Java, GUI, Command Line	DIG, Java, GUI	DIG, Command Line	DIG, Java	Java	Python	Java
Query	Frame style, RDQL	Racer query language		RDQL			Horn logic style
Known Limitation	Poor scaling		No Abox support		Poor scaling	Poor scaling	Only support RDF

The first thing to notice in Table 1 is that the description logic based system can only support reasoning over OWL-Lite and OWL-DL statements but not OWL-Full. OWL-Full is a full extension of RDF, which needs the supporting of terminological cycle. For example, a class in OWL-Full can also be an individual or property. The cyclic terminological definitions can be recognized and understood in horn logic or frame logic system.

Table 1 shows that only three DL-based owl inference engines, which are all use a Tableau based algorithms [Baader 2000], are decidable and support complete consistency checking (at least in OWL-Lite). However, [Balaban 1993] argues that DL only forms a subset of F-Logic. The three kinds of formulae in the description logic can be transformed into first class objects and n-ary relationships. F-Logic is able to provide a full account for DL without losing any semantics and descriptive nature. We understand that our current F-OWL approach is neither decidable nor complete. However, a complete F-Logic based OWL-DL reasoner is feasible.

The table also shows that F-OWL system doesn't scale well when dealing with large datasets, because of the incompleteness of the reasoner. Actually, none of the OWL inference engines listed here scales well when dealing with the OWL test case wine ontology[3] which defines thousands of classes and properties and a relatively modest number of individuals. Further research is needed to improve the performance and desirability.

Comparing with other OWL inference engines, F-OWL has several unique features: tabling, support for multiple logical models or reasoning, and a pragmatic orientation.

Tabling. XSB's tabling mechanism gives F-OWL the benefits of a forward chaining system in a backward chaining environment. The triples in a model are computed only when the system needs to know whether or not they are in the model. Once it is established that a triple is in the current model, it is added to the appropriate table, obviating the need to prove that it is in the model again. This mechanism can have a significant impact on the system's performance. While the first few queries may take a long time, subsequent queries tend to be very fast. This is an interesting compromise between a typical forward-only reasoning system and backward-only reasoning systems.

Multiple Logics. F-OWL supports Horn logic, frame logic and a kind of higher-order logic; all inherited from the underlying XSB and Flora substrates. Working together, these logic frameworks improve F-OWL's performance and capabilities. For example, the F-logic supports non-monotonic (default) reasoning. Another example is higher-order logic. The semantics of higher-order logics, in general, are difficult and in many cases not suitable for practical applications. XSB's Hilog, however, is a simple syntactic extension of first-order logic in which variables can appear in the position of a predicate. In many cases, this simplifies the expression of the statements, rules and constraints, improving the writability and readability of F-OWL and associated programs.

Pragmatic Approach. The aim of F-OWL system is to be a practical OWL reasoner, not necessary a complete OWL reasoner. So F-OWL system provides various interface to access the engine and supports multiple query and rule languages.

In the open web environment, it is generally assumed that the data are not complete and not all facts are known. We will research how this fact affects the implementation of inference engine. In the semantic web an inference engine may not necessarily serve to generate proofs but should be able to check proofs. We will work on using F-OWL to resolve trust and proof in semantic web.

In a stand-alone system inconsistencies are dangerous but can be controlled to a certain degree. However, controlling the inconsistencies in the Semantic Web is a lot more difficult. During the communication, ontology definition origin from other agents, who is unknown beforehand, may be asserted. Therefore special mechanisms are needed to deal with inconsistent and contradictory information in the Semantic Web. There are two steps: detecting the inconsistency and resolving the inconsistency.

---

[3] The wine ontology is used as a running example in the W3C's OWL Web Ontology Language Guide and is available at http://www.w3.org/TR/owl-guide/wine.owl.

The detection of the inconsistency is based on the declaration of inconsistency in the inference engine. The restriction, which imposes the possible values and relation that the ontology elements can have, leads to the inconsistency. For example, *owl:equivalentClass*: imposes a restriction on the resource which the subject is same class as. *owl:disjointWith* imposes a restriction on the resource which the subject is different from. The triples *(a owl:equivalentClass b)* and *(a owl:disjointWith b)* is not directly lead to an inconsistency until applying the detection rule: *(A owl:equivalentClass B) & (A owl:disjointWith B)*➜*inconsistency*.

When inconsistencies are detected, Namespaces can help tracing the origin of the inconsistencies. John posted "all dogs are human" at his web site, while "all dogs are animal" appears in daml.org's ontology library. It is clear that the second is more trustable. Every web site are identified and treated unequivocally in the semantic web. The inference engine contacts trust system to evaluate the creditability of the namespaces. [Klyne 2002] and [Golbeck 2003] enlist lots of works and brilliant ideas about how to maintain the trust system in the semantic web. Once having the trust evaluation result, the agent could take three different actions: (a) accept the one suggested by the inference engine; (b) reject both as none of them is trustable; (c) ask the human user to select.

6   Conclusion

This paper describes the design and implementation of F-OWL, an inference engine for OWL language. F-OWL uses a Frame-based System to reason with OWL ontologies. F-OWL supports consistency checking, extracts hidden knowledge via resolution and supports further complex reasoning by importing rules. While using it in TAGA user case, we find that F-OWL is a full functional inference engine and easy to use with the support of multiple query languages and rule languages.

In the open web environment, it is generally assumed that the data are not complete and not all facts are known. We will research how this fact affects the implementation of inference engine. In the semantic web an inference engine may not necessarily serve to generate proofs but should be able to check proofs. We will work on using F-OWL to resolve trust and proof in semantic web in the future.

References

[Baader 2000] Franz Baader and Ulrike Sattler: "An Overview of Tableau Algorithms for Description Logics", Proceeding of Tableau 2000, RWTH Achen.

[Balaban 1993] Mira Balaban: "The F-Logic Approach for Description Languages", Ben-Gurion University of Negev Technical Report FC-93-02, 1993.

[Berners-Lee 2001] Tim Berners-Lee, James Hendler and Ora Lassila: "The Semantic Web", Scientific America , May 2001.

[Chen 1995] Weidong Chen, Michael Kifer and David Warren: "Logical Foundations of Object-Oriented and Frame-Based Languages", Journal of ACM, May 1995.

[Golbeck 2003] Jennifer Golbeck, Bijan Parsia, and James Hendler: "Trust networks on the semantic web", Proceedings of Cooperative Intelligent Agents 2003, Helsinki, Finland, August 2003.

[Haarslev 2001] Volker Haarslev and Ralf Moller: "Racer system description", Proceeding of International Joint Conference on Automated Reasoning, Volume 2083, page 701-705, Springer 2001

[Hayes 2003] Pat Hayes, RDF Semantics, W3C working Draft, 2003.

[Hendler 2000] James Hendler and Deborah L. McGuinness, "The DARPA Agent Markup Language." IEEE Intelligent Systems, Trends and Controversies, page. 6-7, November/December 2000.

[Horrocks, 1999] I. Horrocks. FaCT and iFaCT. In P. Lambrix, A. Borgida, M. Lenzerini, R. Möller, and P. Patel-Schneider, editors, *Proceedings of the International Workshop on Description Logics (DL'99)*, pages 133-135, 1999.

[Horrocks, 2003] Ian Horrocks and Peter F. Patel-Schneider. Reducing OWL entailment to description logic satisfiability. In Dieter Fensel, Katia Sycara, and John Mylopoulos, editors, *Proc. of the 2003 International Semantic Web Conference (ISWC 2003)*, number 2870 in Lecture Notes in Computer Science, pages 17-29. Springer, 2003.

[Kifer, 1995] M. Kifer, G. Lausen, and J. Wu. Logical foundations of object-oriented and frame-based languages. JACM, 42(4):741--843, Jul. 1995.

[Klyne 2002] G.Klyne: "Framework for Security and Trust Standards", SWAD-Europe, December 2002.

[Lassila 2001] Ora Lassila: Enabling Semantic Web Programming by Integrating RDF and Common lisp", Proceeding of first Semantic Web Working Symposium, Stanford University, 2001.

[Lassila 2002] Ora Lassila: "Taking the RDF Model Theory Out for a Spin", First International Semantic Web Conference (ISWC 2002), Sardinia (Italy), June 2002.

[Patel-Schneider 2003] Peter F. Patel-Schneider, Pat Hayes and Ian Horrocks, OWL Web Ontology Language Semantics and Abstract Syntax, W3C working Draft, 2003.

[Riazanov, 2003] A.Riazanov, Implementing an Efficient Theorem Prover, PhD thesis, University of Manchester, 2003.

[Sagonas, 1994] Kostantinos Sagonas, Terrance Swift, and David S. Warren: XSB as an efficient deductive database engine, In ACM Conference on Management of Data (SIGMOD), 1994.

[Yang, 2000] Guizhen Yang and Michael Kifer. FLORA: Implementing an efficient DOOD system using a tabling logic engine. Proceedings of Computational Logic --- CL-2000, number 1861 in LNAI, pp 1078--1093. Springer, July 2000.

[Zou, 2003] Youyong Zou, Tim Finin, Li Ding, Harry Chen, and Rong Pan, TAGA: Trading Agent Competition in Agentcities, Workshop on Trading Agent Design and Analysis, held in conjunction with the Eighteenth International Joint Conference on Artificial Intelligence, Monday, 11 August, 2003, Acapulco MX.

[Zou, 2004] Youyong Zou, Agent-Based Services for the Semantic Web, Ph.D. Dissertation, University of Maryland, Baltimore County, August, 2004.

Model-Driven Architecture for Agent-Based Systems

Denis Gračanin[1], H. Lally Singh[1], Shawn A. Bohner[1], and Michael G. Hinchey[2]

[1] Department of Computer Science,
Virginia Polytechnic Institute and State University, Blacksburg, VA 24061, USA
{gracanin, lally, sbohner}@vt.edu
[2] NASA Goddard Space Flight Center, Greenbelt, MD 20771, USA
Michael.G.Hinchey@nasa.gov

Abstract. The Model Driven Architecture (MDA) approach uses a platform-independent model to define system functionality, or requirements, using some specification language. The requirements are then translated to a platform-specific model for implementation. An agent architecture based on the human cognitive model of planning, the Cognitive Agent Architecture (Cougaar) is selected for the implementation platform. The resulting Cougaar MDA prescribes certain kinds of models to be used, how those models may be prepared and the relationships of the different kinds of models. Using the existing Cougaar architecture, the level of application composition is elevated from individual components to domain level model specifications in order to generate software artifacts. The software artifacts generation is based on a metamodel. Each component maps to a UML structured component which is then converted into multiple artifacts: Cougaar/Java code, documentation, and test cases.

1 Introduction

Agent-based systems provide a foundation for development of large scale applications like logistics management, battlefield management, supply-chain management, to mention some. An example of agent-based systems is Cougaar (Cognitive Agent Architecture). Cougaar provides a software architecture for distributed agent-based applications in domains characterized by hierarchical decomposition, tracking of complex tasks, generation and maintenance of dynamic plans [1, 2].

The ability to develop very complex applications comes with a price. It takes a lot of effort and learning in order to have complete understanding and ability to effectively use such agent-based systems. A domain expert must closely collaborate with the developer in order to fully utilize an agent-based system for a particular domain. It is very unlikely that a domain expert will have sufficient understanding of the underlying agent-based system.

A Model Driven Architecture (MDA) based approach can be used to automatically generate software artifacts and to significantly simplify application

M.G. Hinchey et al. (Eds.): FAABS 2004, LNAI 3228, pp. 249–261, 2005.

development [3, 4, 5]. The domain expert can specify requirements in a familiar, platform-independent format that hides platform-specific details.

The MDA approach can be used for developing applications using the Cougaar agent-based architecture. Cougaar components can be composed into a General Cougaar Application Model (GCAM) and develop a General Domain Application Model (GDAM) for specifying and automatically generating software applications. This approach is discussed in the paper.

The remainder of the paper is organized as follows. Section 2 briefly describes Cougaar and its capabilities. Section 3 describes the use of the MDA approach for Cougaar-based applications. Section 4 discusses the Cougaar-based MDA model while Section 5 describes the implementation. Section 6 concludes the paper.

2   Cougaar Agent-Based System

Cougaar is a "large–scale workflow engine built on a component-based distributed agent architecture" [1]. It is deployed as a *society* of *agents*, which communicate and work together to solve a problem. A Cougaar society is a set of agents running on one or more interconnected computers, all working together to solve a common class of problems. The problem may be partitioned into subproblems, in which case the responsible subset of agents is called a *community*. A society may have one or more communities within.

The relationship between societies, communities, and agents is not a strict one, a society may directly contain both agents and communities. While a society has a real-world representation, a set of computers running a Cougaar system, a community is only notational in nature.

A Cougaar agent is a first-class member of a Cougaar Society [1] and it contains a Blackboard and one or more Plugins. While the specific purpose of any agent is chosen by the system developer; the objective is for a single agent to represent a single organizational entity or a part thereof.

At the most basic level, an agent consists of two parts: a Blackboard and a set of Plugins (Figure 1). The former is a container of objects, with a subscription-based change notification mechanism; the latter is a set of responders to these notifications, with the ability to change the contents of the Blackboard.

The Blackboard serves as the communications backbone connecting the Plugins together. More importantly, it serves as the entry point for any incoming messages to the agent as a whole, which are then picked up by the Plugins for handling. All instance-specific behavior of the agent is implemented within the Plugin. A Plugin listens to add, remove, and change events on the Blackboard. Evaluating the objects involved in the event, the Plugin may respond by performing some computation, changes to the Blackboard, or some external work.

A Cougaar Node conceptually encapsulates a set of agents. Agents can collaborate with other agents in the same Node or with agents in other Nodes. However, it is not a direct collaboration. Instead, Cougaar Tasks are allocated to Cougaar Organizations, which are representations of agents in the local Blackboard. The subscription mechanism allows agents to use Tasks to exchange messages

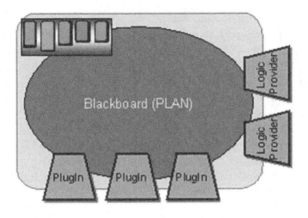

Fig. 1. Cougaar Agent Structure [1]

(objects). The Cougaar communication infrastructure then ensures that the Task is sent to the destination Organization's (i.e. agent's) Blackboard.

3   Cougaar Model Driven Architecture (CMDA)

The MDA approach advocates converting a Platform Independent Model (PIM) into a Platform Specific Model (PSM) through a series of transformations, where the PIM is iteratively made more platform specific, ending in the PSM. The PIM is used to represent a system's business functionality without including any technical aspects. The PIM allows Subject Matter Experts (SMEs) to work at the domain layer. However the current technologies may not offer the required richness to implement the complex transformation rules. For example, the Unified Modeling Language (UML), the foundation for MDA, lacks in the required precision and formalization.

While the development of PIM and PSM UML models might be easy, a blind adaptation of the MDA approach might create problems during the development of mapping rules and transformations. It should be noted that the MDA approach advocates for a Computational Independent Model (CIM) that needs to be transformed into a PIM. Since UML uses different representations for each of the models, the translation between models is more like translation between natural languages, the mappings are not necessarily exact. Further, while the learning curve associated with UML is fairly low, the SMEs nevertheless need to learn a new technical language and need to "move" out of their work environment.

The productivity of Cougaar system developers can be improved by using the MDA approach. The Cougaar MDA (CMDA) attempts to provide fully automated generation of software artifacts and simplifies Cougaar-based application development by providing two important abstraction layers. The first layer is the Generic Domain Application Model (GDAM) layer. The GDAM represents the

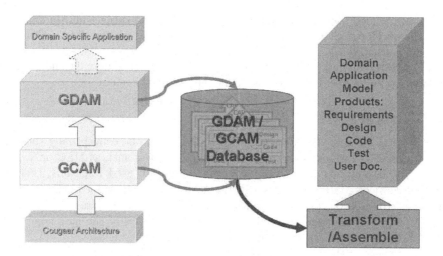

Fig. 2. Basic CMDA Approach

PIM and encompasses the representation of generic agent and domain specific components found in the domain workflow. The second layer, Generic Cougaar Application Model (GCAM) reflects the PSM or Cougaar architecture. The user specifies the intended Cougaar system using workflow paradigm and the system is then refined using GDAM and GCAM models.

The GDAM layer implements the PIM based on a representation that SMEs are comfortable with and results in a proper mapping to the PSM. The goal is to make this mapping as automated as possible, while having human-in-the-loop as a fallback mechanism to correct any mapping imperfections. The initial versions of the tool might force the developers to fine-tune the generated PSM to a certain extent, but it is hoped that as the tools and algorithms advance, such fine-tuning would be less and less necessary.

The GDAM layer specifies the structure and semantic information that the tool uses to ensure that the developer has annotated the GDAM model properly. Furthermore, the layer provides all information required by the tool to produce a more specific but still platform-independent PIM that includes details of desired semantics, and guides choices that the approach/tool will have to make (Figure 2).

In order to develop a tool based on the proposed approach, the following assumptions and constraints were formulated after detailed research.

- Fully automated software artifacts (requirements, design document, code, and test cases) generation is a desirable goal.
- The generated requirements are partial in nature.
- The validation of generated code and the generation of test cases are of lower priority.
- The development of tools and implementation mechanisms are of lower priority than formulating the "recipes" for transformations.

- The intended users of the system are developers and subject matter experts.
- The developer should be fully aware of the Cougaar system, its capabilities, and constraints.
- The SMEs should have sufficient knowledge about the domain and a basic understanding of the requirements of the intended system.

3.1   GDAM Layer

The General Domain Application Model (GDAM) can be conceptually thought to be similar to various programming language libraries such as MFC or Swing. The libraries abstract and modularize the commonly used functions, thereby helping Subject Matter Experts (SMEs) to focus on encoding business logic. However, the abstractions achieved by class libraries, which are written in an implementation language, are limited by the capabilities of the language. Further, SMEs have to work at the implementation language level.

The genesis of the GDAM layer can be traced to the need to allow SMEs to develop systems at the domain layer using current technologies and simple transformation rules. In short, GDAM allows SMEs to represent the specifications of the system in a platform-independent, domain-specific language that can be transformed, without losing information, into specifications of how applications will be implemented in the Cougaar platform. Further, GDAM provides a set of components and patterns representing the different kinds of generic domain elements that can be assembled to specify the application.

There are two, potentially conflicting, implications of the GDAM functionality. First, SMEs should be allowed to capture their domain knowledge and application requirements in a manner that is computationally independent. Second, there should be a well defined structure and relationships among requirements to allow for an automatic and mechanic transformation of the requirements/constraints into an internal, platform independent, GDAM representation that can be later transformed into a platform specific GCAM representation. To reduce this potential conflict, the following decisions were made:

- The transformation between the computational independent and platform independent representations should be a lightweight one. In other words, the platform independent transformation should subsume computational independent representation thus requiring only a simple transformation between the two.
- The business logic, i.e. the "semantic" of the application must be embedded within the computational independent representation enforce constraints. The constraint language must be simple and easily transformed into code that can be integrated within the platform.
- The configuration and deployment of the application is treated separately from the application requirements because that is inherently platform specific. While every effort will be made to make it as generic as possible, some platform specific information may be necessary.
- User interactions and user interface represent a separate challenge. Automatic or semi-automatic user interface generation based on the application

requirements is not a unique one, i.e. there can be many different user interface designs. Such designs can be customized based on the SMEs preferences. While this effort is outside of the scope of the project, some considerations will be provided for possible future research.

The development of GDAM is an iterative and evolutionary process. In addition to the general system wide assumptions, the following assumptions are specific to the GDAM layer.

- The current scope is restricted to the development of some of the indispensable generic domain components that pertain to the logistics domain.
- The GDAM components development is an evolutionary process and it is not expected or possible to develop each and every GDAM component.
- The developer and the SME will work together, sitting side-by-side if required, while developing the GDAM model of the intended system.
- Developers will collaborate with subject matter experts to develop and update the system with GDAM components that are required and not available.

3.2 GCAM Layer

The GCAM is an abstraction layer above the Cougaar code that represents an application's design. Therefore, the GCAM hides the Cougaar code implementation while providing a platform specific "environment." One of the important issues is a separation between the GDAM and the GCAM levels. The GDAM level represents requirements and the GCAM level represents design. Each level performs one mapping. The GDAM level maps from requirements to design which then serves as input to the GCAM level. The GCAM level then maps from the provided design to code. Therefore, the GCAM level is taking as an input the design (GCAM representation) that contains constraints, references to the GCAM components, etc. A repository of components contains detailed descriptions of individual GCAM components in a form of "beans." The GCAM engine is assembling the code segments of the GCAM components from the repository and augments them with code generated from constraints and other design information. The resulting code, combined with the configuration information, provides a developed application.

In addition to the general system-wide assumptions, the following assumptions are specific to the GCAM layer. The GCAM is essentially a design level representation of the Cougaar system.

- As the Cougaar system is revised, the revisions will be reflected in the GCAM layer.
- The developers will write Cougaar code to encode details that cannot be represented using GCAM components.
- The code generated by the system is not intended to be modified by developers. The code generator is optimized for runtime performance and simplicity.
- The GCAM engine does not have optimization capabilities and hence the generated code might not be as efficient as manually written code. The GCAM engine does not support model debugging capabilities.

4 CMDA Model

The GDAM requirements necessitated the development of a model representation that is both versatile (to represent domain information) and familiar (to the SMEs and developers). Based on studies conducted, there is enough confidence to choose workflow as the medium to represent the generic domain model. Workflow is familiar to both SMEs and developers and charts out the working mechanism of the intended system. Further, the structure of the workflow (essentially boxes and arrows) is both generic (to represent most domain information) and extensible (to support addition and modifications of GDAM components). However, it should be noted that workflow does not capture all the requisite information. The information that is not captured includes:

– Deployment and configuration information,
– Information pertaining to GUI such as screen layout and user interactions,
– Domain and system level constraints, and
– Business rules.

It is necessary to develop and refine the software artifact generation mechanism based on the information that is captured using workflow. Information that cannot be represented using workflow can be captured either by extending the workflow model (to record domain and system wide constraints) or by creating "threads" that will "run" in parallel to the workflow thread.

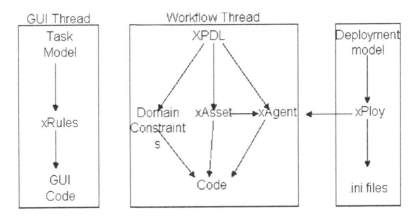

Fig. 3. Threads

Figure 3 shows the different threads that exist in the developed tool. The threads are designed to capture information pertaining to (1) workflow, (2) GUI layout, and (3) deployment. While the structure and semantics of the workflow thread are known, the details about the GUI and deployment threads are being worked out. The GDAM model representation consists of the Task model for GUI, Workflow model for Agent code and deployment model for deployment

code. The models are transformed into corresponding XML representations by
the GDAM engine. The GCAM engine reads in the XML, aggregates and corre-
lates the information to produce the code artifacts. Higher priority is assigned
to developing and refining the workflow thread.

4.1   Domain Components Presented in Cougaar

The Cougaar system provides mechanisms to encode domain knowledge directly
in the code. The domain that Cougaar implement is the planning domain for
which the generic-domain components present in the Cougaar were identified.
As the project moves forwards, more detailed study will be performed. The two
important domain components found in the Cougaar were the task component
and the asset component.

Cougaar defines a task as "A requirement or request from one agent to another to perform or plan a particular operation." The tasks are implemented
in the planning domain library and are used by agents to let another agents
perform a job or plan the execution of a job.

Cougaar defines an asset as "Resources assigned to the task." Any asset
instance will have two key attributes: (1) a reference to its prototype and (2) a
reference to the item identification property group. Assets are also implemented
in the planning domain library.

4.2   GDAM Representation

Current Cougaar application development practices were analyzed and used to
define the GDAM representation. The workflow model is the computational
model used by Cougaar developers. Some of its functionality has been already
incorporated in the Cougaar based code. As a consequence, the workflow model
and its underlying XML Processing Description Language (XPDL) format have
been selected as for specifying application requirements [6]. The underlying platform independent GDAM model subsumes the workflow model by using the
basic components of the workflow model as templates for the part of GDAM
components.

XPDL, defined by Workflow Management Coalition (WfMC), provides a
framework for implementing business process management and workflow engines, and for designing, analyzing, and exchanging business processes. Further,
XPDL is extensible and versatile to handle information used by a variety of
different tools.

While XPDL provides excellent mechanisms to define and record workflow
processes, certain customization was needed. The customizations include:

- **Type Declarations:** The type declarations were used to define the assets
  at the domain level. The SMEs will define and declare the primitive types,
  PropertyGroups (PGs) and assets using type declarations. The primitive
  types or elements within a PG were recorded as basic type in XPDL. The
  basic types were then grouped into a record type, which will represent the

PG. The PGs are then grouped into a record to form the asset. The type declarations in XPDL provide all the capabilities required to define an asset.

- **Abstractions:** The generic notations of XPDL were abstracted to represent Cougaar concepts. The agents were represented using participants and the behavior of the agents was described using activities. The transitions represented the tasks generated by agents.
- **Constraint enforcement:** The condition tags present in the XPDL was extended to support constraint representations. While XPDL has many useful features, it lacks some of the required structure and constraint capabilities. As a consequence, the Object Constraint Language (OCL) is selected to capture this information [7]. The OCL constraints are includes in the XPDL as pre- and post- conditions thus eliminating free-text constraints from the original XDPL format.
- **Extended attributes:** The extended attributes section was used to describe Cougaar specific semantics such as tasks, assets, and allocations.

It should be noted that care was taken to extend the XPDL without breaking the XML structure defined by the WfMC. This was done to allow the XPDL file to be loaded in any standard workflow editor that supports XPDL.

4.3   GDAM Components

The current structure and semantics of GDAM components have provision to specify constraints (pre and post conditions), documentation section, need revisions to incorporate fragments of design diagrams, mapping criteria. The workflow component describes the participant, activity and transition elements.

The participant component which is used to represent Agent is defined in XPDL under the participants tag. Each participant has two attributes: ID (unique Id used to reference the participant within the workflow model) and Name (user specified name, which need not be unique). The participant component also contains the tag ParticipantType which is used by XPDL to identify the type of participant.

The activity component is used to describe the behavior. The activity component, described inside Activities tag, consist of two attributes: ID (unique Id used to reference the activity within the workflow model) and Name (user specified name, which need not be unique). The activity component provides details about a particular behavior, which are mapped into Plugins during transformations. The Activity component has a performer tag to identify which Agent's behavior is being defined, a transition restrictions tag to reference the constraints of a particular Task, and an extended attribute namely asset, to identify the asset used by the activity. An activity component can occur more than once in the workflow model. The first occurrence of activity component is mapped into a new Plugin and subsequent occurrences result in appending the Plugin's behavior. The Plugin's behavior is appended by appending the subscription and action subsets of the Plugin.

The asset component is used to describe the resources attached to tasks. The asset component is described using XPDL's type declarations. The primitive types or elements within a PG were recorded as a basic type in XPDL. The basic types were then grouped into a record type, which will represent the PG. The PGs are then grouped into a record to form the asset. The TypeDeclaration tag, consist of two attributes: ID (unique Id used to reference the type within the workflow model) and Name (user specified name, which need not be unique). The TypeDeclaration also lists whether the type is basic type or record type. If the type is a record, the members of the record are listed.

5   CMDA Implementation

The graphical user interface (GUI) for the developed CMDA tool has been implemented as an editor using the Eclipse IDE [8, 9]. The editor allows editing and validation of XPDL data in both text and graphical formats. The XPDL is loaded into the editor, with the workflow displayed. The editor connects to the repository of components. The user drags components from a palette (representing what's available in the repository) assigning the activity to a new instance of the component, which can have all its properties set in a GUI. The editor shows any validation errors detected by the validating compiler. The instantiation data (component name and property values) are stored in the XPDL as extended attributes. The editor also shows a set of available resources, which can be assigned to each activity. As these resources are assigned, they are stored in the XPDL as extended attributes. Completed requirements include a fully defined components with parameters, roles, and deployment data.

Since the entire system is a component itself, with deployment information added, the editor is used to edit any inner component as well. The components are defined in a UML-like XML-based language [10] where an XML schema is defined for specifying components that can be automatically converted to an EMF [11] model. Eclipse's EMF is a modeling system similar to the Meta-Object Facility (MOF) [11]. Those similarities enable the use of EMF and the related tools for easy conversion to a UML representation. The UML representation, in addition to documentation generation, provides a better understanding of the application under development.

The characteristics of the metamodel are determined from the parameterization of Cougaar components, related constraints and properties. Components must define properties that can be queried and derived. Interconnected components work together as agents and societies of agents. Composition of components is specified using graphs describing interconnected and configured components. The graphs can be saved and reused.

Generation of Cougaar/Java code, documentation, requirements, and test cases depends on components that must provide information for artifact generation Deployment of components requires assignment of hosts and other computational, storage, or other type of resources while maintaining Java compatibility.

Each component maps to a UML structured component with template parameters (Figure 4). The compiler validates components and generates artifacts. The validation insures that the component is valid and is suitable for artifact

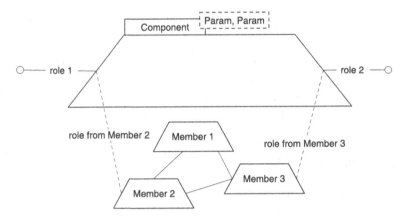

Fig. 4. Component

generation. Artifact generation creates Java code, documentation, test cases, and requirements data.

Components have named parameters that are defined like a very small subset of an XML schema [12]. Parameters specify a **name** attribute, which is matched when given a value. Parameters may also define a parent parameter, thus allowing sets of parameters and a cardinality. This allows variable numbers of sets of parameters, giving a reasonable configuration language for components.

The metamodel directly provides constraint data through constraints given in the component definition, and implicitly through the typed connections and defined restrictions on the various metamodel elements. The compiler verifies constraints to assure that a valid system can be generated.

The compiler considers the entire system as one top level component. The components are grouped into a tree of instantiations (component names coupled with values for all of their parameters) that is traversed by the compiler. The compiler calls the relevant profile mapping at each node to generate corresponding artifacts. The profile mappings use either an XML tree for the component definition or a set of EMF objects representing them in memory. The former is the serialized form of the latter. Extensible Stylesheet Language (XSL) Transformations (XSLT) [13] or Eclipse's Java Emitter Template (JET) [14, 15] are then used to generate the artifacts.

Components can specify *roles*, named interconnections with other components, that specify data types sent and received over them. Roles are special types of parameters which are fully initialized only with references to other component instances. They also cannot have inner roles or any such hierarchy.

Deployment data is considered a special type of non-hierarchical parameter. Deployment data are not fixed values. They are expression usable for deriving the value when the system is deployed.

Components can specify inner *member* components to define the inner structure. These member components are initialized and connected together. Their parameters, connections, and deployment information have static values or OCL expressions [7] based on the component's parameter data. OCL expressions provide additional information to object-oriented models, including constraints, queries, referencing values, stating conditions and business rules. Each value is expressed using OCL constants or using OCL expressions that allow their derivation. The component can define its properties as the values of properties in its member components, possibly with some modification and renaming.

While the definitions immediately provide useful descriptions of the system, they do not directly provide code, test cases, etc. The compiler, in some cases, needs "help" from the component definitions to create code, test cases, and related artifacts. Each component specifies the name of a *Profile Mapping* that links the component to a set of definitions for how the artifacts are generated. Each profile mapping handles different categories of components, such as Cougaar Plugins, Agents, or Societies.

6   Conclusions

Cougaar is complex requiring considerable mappings and transforms. MDA provides a systematic way of capturing requirements and mapping them from PIM to PSM and ultimately to the code level. The developed CMDA framework is an MDA based approach for the Cougaar agent-based architecture. It enables automatic transformation of the application requirements, expressed in the XPDL format, into a platform-independent, GDAM representation. The artifacts are generated from models assembled using components that contain information related to requirement, design, code, test and documentation details for that component, along with transformation information. Platform-specific GCAM components are derived from the metamodel and then converted into Cougaar/Java code. The CMDA combines assembly approach with transformations to generate the artifacts. While the CMDA-based approach uses the Cougaar architecture, it is applicable to other agent-based architectures.

Acknowledgements

This work has been supported, in part, by the DARPA STTR grant "AMIIE Phase II — Cougaar Model Driven Architecture Project," (Cougaar Software, Inc.) subcontract number CSI-2003-01. We would like to acknowledge the efforts, ideas, and support that we received from our research team including Todd Carrico, Sandy Ronston, Tim Tschampel, and Boby George.

References

1. —: Cougaar architecture document. Technical report, BBN Technologies (2004) Version for Cougaar 11.2.
2. —: Cougaar developers' guide. Technical report, BBN Technologies (2004) Version for Cougaar 11.2.

3. Arlow, J., Neustadt, I.: Enterprise Patterns and MDA: Building Better Software with Archetype Patterns and UML. Addison-Wesley, Boston (2003)

4. Kleppe, A., Warmer, J., Bast, W.: MDA Explained: The Model Driven Architecture: Practice and Promise. Addison-Wesley, Boston (2003)

5. Weis, T., Ulbrich, A., Geihs, K.: Model metamorphosis. IEEE Software **20** (2003) 46–51

6. Workflow Management Coalition: (Workflow process definition interface – XML process definition language (XPDL)) http://www.wfmc.org/standards/docs/TC-1025_10_xpdl_102502.pdf.

7. Warmer, J., Kleppe, A.: Object Constraint Language, The: Getting Your Models Ready for MDA. Second edn. Addison Wesley Professional (2004)

8. Clayberg, E., Rubel, D.: Eclipse: Building Commercial-Quality Plug-ins. The Eclipse Series. Addison-Wesley, Boston (2004)

9. Gamma, E., Beck, K.: Contributing to Eclipse: Principles, Patterns, and Plug-Ins. The Eclipse Series. Addison-Welsey, Boston (2004)

10. Rumbaugh, J., Jacobson, I., Booch, G.: The Unified Modeling Language Reference Manual. Addison-Wesley Publishing Co. (2004)

11. Budinsky, F., Steinberg, D., Merks, E., Ellersick, R., Grose, T.J.: Eclipse Modeling Framework. Addison-Wesley Publishing Co. (2003)

12. McLaughlin, B.: Java^{TM} & XML Data Binding. O'Reilly, Beijing (2002)

13. Tidwell, D.: XSLT. O'Reilly, Beijing (2001)

14. Azzuri Ltd.: (JET tutorial part 1 (introduction to JET)) http://eclipse.org/articles/Article-JET/jet_tutorial1.html.

15. Azzuri Ltd.: (JET tutorial part 2 (write code that writes code)) http://eclipse.org/articles/Article-JET/jet_tutorial2.html.

Apoptosis and Self-Destruct:
A Contribution to Autonomic Agents?

Roy Sterritt[1] and Mike Hinchey[2]

[1] University of Ulster, School of Computing and Mathematics,
Jordanstown Campus, BT37 0QB, Northern Ireland
`r.sterritt@ulster.ac.uk`
[2] NASA Goddard Space Flight Center, Software Engineering Laboratory,
Greenbelt, MD 20771 USA
`michael.g.hinchey@nasa.gov`

Abstract. Autonomic Computing (AC), a self-managing systems initiative based on the biological metaphor of the autonomic nervous system, is increasingly gaining momentum as the way forward in designing reliable systems. Agent technologies have been identified as a key enabler for engineering autonomicity in systems, both in terms of retrofitting autonomicity into legacy systems and designing new systems. The AC initiative provides an opportunity to consider other biological systems and principles in seeking new design strategies. This paper reports on one such investigation; utilizing the *apoptosis* metaphor of biological systems to provide a dynamic health indicator signal between autonomic agents.

1 Introduction

One of the great things about being involved in the early days of development of a new paradigm is having the opportunity to look again at how things are done, and contemplate approaches not normally considered before the paradigm beds down into its evolutionary path.

Autonomic Computing is based on the biological metaphor of the Autonomic Nervous System (ANS) [1], taking the ANS as inspiration to achieve self-managing systems without 'conscious effort' from the user. IBM's initial set of self-properties (self-CHOP, configuration, healing, optimisation and protection) have been expanded to include many self-* properties leading to the adoption of the term *selfware*.

Biological systems inspire systems design in many other ways – reflex reaction and health signs [2, 3], nature-inspired systems (NIS) [4] – hive and swarm behaviour, fire flies, etc., for example.

At this stage in the emerging field of Autonomic Computing we are seeking inspiration for new approaches from (obviously, pre-existing) biological mechanisms. An obscure mechanism which is discussed in this paper is *Apoptosis* – the approach for cell self-destruction, which at first sight may seem a metaphor too far.

M.G. Hinchey et al. (Eds.): FAABS 2004, LNAI 3228, pp. 262–270, 2005.

2 Biological Apoptosis

The biological analogy of autonomic systems has been well discussed in the literature. While reading this the reader is not consciously concerned with his[1] breathing rate or how fast his heart is beating. Achieving the development of a computer system that can self-manage without the conscious effort of the user is the vision and ultimate goal [5]. Another typical biological example is that the touching of a sharp knife results in a reflex reaction to reconfigure the area in danger to a state that is out of danger (self-protection, self-configuration, and, if damage is caused, self-healing) [6].

If one cuts oneself and starts bleeding, good training results in washing the finger, applying a bandage and carrying on with one's tasks without any further conscious thought. Yet, often, the cut will have caused skin cells to be displaced down into muscle tissue [7]. If they survive and divide, they have the potential to grow into a tumour. The body's solution to dealing with this situation is cell self-destruction (with mounting evidence that cancer is the result of cells not dying fast enough, rather than multiplying out of control, as previously thought).

It is believed that a cell knows when to commit suicide because cells are programmed to do so – self-destruct (sD) is an intrinsic property. This sD is delayed due to the continuous receipt of biochemical retrieves. This process is referred to as *apoptosis* [8], meaning 'drop out', used by the Greeks to refer to the Autumn dropping of leaves from trees; i.e., loss of cells that ought to die in the midst of the living structure. The process has also been nicknamed 'death by default' [9], where cells are prevented from putting an end to themselves due to constant receipt of biochemical 'stay alive' signals (Figure 1).

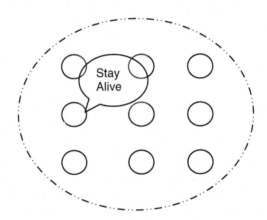

Fig. 1. Turning off the self-destruct sequence - cell receives 'stay alive' signal

Further investigations into the apoptosis process [10] have discovered more details about the self-destruct programme. Whenever a cell divides, it simultaneously receives orders to kill itself. Without a reprieve signal, the cell does indeed self-destruct. It is believed that the reason for this is self-protection, as the most dangerous

---

[1] Throughout this paper, for "his", read "his/her".

time for the body is when a cell divides, since if just one of the billions of cells locks into division the result is a tumour, while simultaneously a cell must divide to build and maintain a body.

The suicide and reprieve controls have been compared to the dual-key on a nuclear missile [7]. The key (chemical signal) turns on cell growth but at the same time switches on a sequence that leads to self-destruction. The second key overrides the self-destruct [7].

3 Autonomic Computing and Agents

Autonomic Computing is dependent on many disciplines for its success; not least of these is research in agent technologies. At this stage, there are no assumptions that agents have to be used in an autonomic architecture, but as in complex systems there are arguments for designing the system with agents [11], as well as providing inbuilt redundancy and greater robustness [12], through to retrofitting legacy systems with autonomic capabilities that may benefit from an agent approach [13].

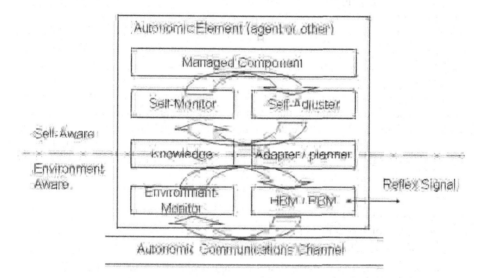

Fig. 2. Autonomic Element (agent or other) consists of a managed component and an autonomic manager. Control loops with sensors (self-monitor) and effectors (self-adjuster) together with system knowledge and planning/adapting policies allow the autonomic element to be self-aware and to self-manage. A similar scheme facilitates environment awareness (allowing self-managing if necessary, but without the immediate control to change the environment – this is effected through communication with other autonomic managers that have the relevant influence, through reflex or event messages)

Emerging research suggests that the autonomic manager may be an agent itself, for instance, an agent termed a *self-managing cell* (SMC) [14], containing functionality for measurement and event correlation and support for policy-based control.

Essentially, the aim of autonomic computing is to create robust dependable self-managing systems [15]. To facilitate this aim, fault-tolerant mechanisms such as a heart-beat monitor ('I am alive' signals) and pulse monitor (urgency/reflex signals) may be included within the autonomic element (Figure 2) [2, 16]. The notion behind the pulse monitor (PBM) is to provide an early warning of a condition so that preparations can be made to handle the processing load of diagnosis and planning a response, including diversion of load. Together with other forms of communications it creates dynamics of autonomic responses [17] – the introduction of multiple loops of control, some slow and precise, others fast and possibly imprecise, fitting with the biological metaphor of reflex and healing [2].

The major motivating factor for formal approaches to agent-based systems is to prevent race conditions and undesirable emergent behaviour. In this situation, Self-Destruction of the agent may be viewed as a last resort situation to prevent further damage; in other situations, such as security of the agent, Self-Destruction may be used as an intrinsic part of the process.

Agent destruction has been proposed for mobile agents to facilitate security measures [18]. Greenberg *et al.* highlighted the situation simply by recalling the situation where the server `omega.univ.edu` was decommissioned, its work moving to other machines. When a few years later a new computer was assigned the old name, to the surprise of everyone, email arrived, much of it 3 years old [19]. The mail had survived 'pending' on Internet relays waiting for `omega.univ.edu` to come back up.

Greenberg encourages consideration of the same situation for mobile agents; these would not be rogue mobile agents – they would be carrying proper authenticated credentials. This work would be done totally out-of-context due to neither abnormal procedure nor system failure. In this circumstance the mobile agent could cause substantial damage, e.g., deliver an archaic upgrade to part of the network operating system resulting in bringing down the entire network.

Misuse involving mobile agents comes in the form of:

- misuse of hosts by agents,
- misuse of agents by hosts, and
- misuse of agents by other agents.

From an agent perspective, the first is through accidental or unintentional situations caused by that agent (race conditions and unexpected emergent behaviour), the later two through deliberate or accidental situations caused by external bodies acting upon the agent. The range of these situations and attacks have been categorised as: damage, denial-of-service, breach-of-privacy, harassment, social engineering, event-triggered attacks, and compound attacks.

In the situation where portions of an agent's binary image (e.g., monetary certificates, keys, information, etc.) are vulnerable to being copied when visiting a host, this can be prevented by encryption. Yet there has to be decryption in order to execute, which provides a window of vulnerability [19]. This situation has similar overtones to our previous discussion on biological apoptosis, where the body is at its most vulnerable during cell division.

4 Autonomicity in NASA Missions

New paradigms in spacecraft design are leading to radical changes in the way NASA designs spacecraft operations [20]. Increasing constraints on resources, and greater focus on the cost of operations, has led NASA to utilize adaptive operations and move towards almost total onboard autonomy in certain classes of mission operations [21, 22].

NASA missions, particularly those to deep space, where manned craft will not at present be utilized, are considering the use of almost wholly autonomous decision-making to overcome the unacceptable time lag between a craft encountering new situations and the round-trip delay (of upwards of 40 (earth) minutes) in obtaining responses and guidance from mission control.

More and more NASA missions will, and *must*, incorporate autonomicity as well as autonomy [23, 27].

4.1 Previous Missions

Two of the first notable missions to use autonomy are DS1 (Deep Space 1) and the Mars Pathfinder [24].

The *Beacon Monitor* concept, first used in the DS1 mission work [25] automates the routine task of health monitoring and transfers the process of monitoring from ground to the spacecraft [16]. With beacon monitoring, the spacecraft sends a signal to the ground that indicates how urgent it is to track the spacecraft for telemetry.

This concept involved a paradigm shift for NASA from its traditional routine telemetry downlink and ground analysis, to onboard health determination and autonomous data summarization [25].

In terms of high-level concepts, the beacon monitor is analogous to the heartbeat monitor, but with the addition of a tone to indicate the degree of urgency involved: *nominal, interesting, important, urgent* and *no tone* [26].

Some long-term drawbacks of this approach have been discovered. Since one of the primary goals of beacon monitoring was to reduce the amount of data sent to the ground (achieved by eliminating the download of telemetry data), operators lost the ability to gain an intuitive feel for the performance and characteristics of the craft and its components, as well as losing the ability to run the data through simulations [20].

As such, to fully benefit from beacon monitoring, the *fast loop* of real-time health assessment must be supplemented by a *slow loop* to study the long-term behaviour of the spacecraft. This *engineering data summarization* is where the spacecraft creates a second set of abstractions regarding the sensor telemetry, which is then sent back to ground to provide the missing context for operators.

This dual approach has conceptually much in common with the reflex and healing approach [2, 16].

4.2 A Future Mission

The Autonomic Computing initiative has been identified by NASA as having potential to contribute to their goals of autonomy and cost reduction in future space exploration missions [22, 23, 27].

ANTS, Autonomous Nano-Technology Swarm, is a mission that will launch sometime between 2020 and 2030 ("any day now" in terms of NASA missions). The mission is viewed as a prototype for how many future unmanned missions will be developed and how future space exploration will exploit autonomous and autonomic behaviour.

The mission will involve the launch of 1000 pico-class spacecraft swarm from a stationary factory ship, on which the spacecraft will be assembled. The spacecraft will explore the asteroid belt from close-up, something that cannot be done with conventionally-sized spacecraft.

As much as 60% to 70% of the spacecraft will be lost on first launch as they enter the asteroid belt. The surviving craft will work as a swarm, forming smaller groupings of *worker* craft (each containing a unique instrument for data gathering), a coordinating *ruler*, that will use the data it receives from workers to determine which asteroids are of interest and to issue instructions to the workers and act as a coordinator, and *messenger* craft which will coordinate communications between the swarm and between the swarm and ground control. Communications with earth will be limited to the download of science data and status information, and requests for additional craft to be launched from earth as necessary.

A current project (FAST) is studying advanced technologies for the verification of this incredibly complex mission; the reader is directed to [22, 27] for a more detailed exposition of the ANTS mission and the FAST (Formal Approaches to Swarm Technologies) project.

5 The Role of Apoptosis

The discussions so far have established the concepts of:

- Heart-Beat Monitor (HBM) *I am alive*: a fault-tolerant mechanism which may be used to safeguard the autonomic manager to ensure that it is still functioning by periodically sending 'I am alive' signals.
- Pulse Monitor (PBM) *I am healthy*: extends the HBM to incorporate reflex/urgency/health indicators from the autonomic manager representing its view of the current self-management state. The analogy is with measuring the pulse rate instead of merely detecting its existence.
- Apoptosis *Stay alive*: a proposed additional construct used to safeguard the system and agent; a signal indicates that the agent is still operating within the correct context and behaviour, and should not self-destruct.

The title of this paper (purposely) raises the question of whether there is a role for the apoptosis metaphor within the development of autonomic agents. Additionally, in the introduction, we prompted the consideration of whether perhaps it is a metaphor too far.

Section 3 clearly highlights the general problem of agent security, whether from the agent's or host's perspective. In terms of generic contribution to autonomic agents development, with many security issues the lack of an agreed standard approach to agent-based systems prohibits further practical development for generic autonomic systems. As such, the proposal can only be 'put out there' as a concept.

Of course, within NASA missions, such as ANTS, we are not considering the generic situation. Mission control and operations is a trusted private environment. This eliminates many of the wide range of agent security issues discussed earlier, just leaving the particular concerns; is the agent operating in the correct context and showing emergent behaviour within acceptable parameters, where upon *apoptosis* can make a contribution.

For instance, in ANTS, suppose one of the *worker* agents was indicating incorrect operation, or when co-existing with other workers was the cause of undesirable emergent behaviour, and was failing to self-heal correctly. That emergent behaviour (depending on what it was) may put the scientific mission in danger. Ultimately the stay alive signal from the *ruler* agent would be withdrawn.

If a *worker*, or its instrument, were damaged, either by collision with another worker, or (more likely) with an asteroid, or during a solar storm, a *ruler* could withdraw the stay alive signal and request a replacement *worker* (from Earth, if necessary). If a *ruler* or *messenger* were similarly damaged, its stay alive signal would also be withdrawn, and a *worker* would be promoted to play its role.

All of the spacecraft are powered by batteries that are recharged by the sun using solar sails [22, 27]. Although battery technology has greatly advanced, there is still a "memory loss" situation, whereby batteries that are continuously recharged eventually lose some of their power and cannot be recharged to full power. After several months of continual operation, each of the ANTS will no longer be able to recharge sufficiently, at which point their 'stay alive' signals will be withdrawn, and new craft will need to be assembled or launched from Earth.

6 Conclusions

Autonomic Computing [1] has been gaining ground as a significant new paradigm to facilitate the creation of self-managing systems to deal with the ever increasing complexity and costs inherent in today's (and tomorrow's) systems.

In terms of the Autonomic Computing initiative, agent technologies have the potential to become an intrinsic approach within the initiative [28], not only as an enabler (e.g. ABLE agent toolkit [29]), but also in terms of creating autonomic agent environments.

Formal approaches to agent-based systems [30, 31] have a primary focus of identifying race conditions, highlighting undesirable emergent behaviour, and verifying the correctness of systems that are far too complex to ever test correctly. However, the practicality of mobile agents is predicated on the existence of realistic security techniques [19].

We have described the Heart-Beat Monitor (HBM) and Pulse Monitor (PBM) and proposed a logical addition which has an analogy from biological systems, *Apoptosis* and *Self-Destruct*, which we believe will be valuable in future autonomic systems.

Acknowledgements

The development of this paper was supported at University of Ulster by the Centre for Software Process Technologies (CSPT), funded by Invest NI through the Centres of

Excellence Programme, under the EU Peace II initiative. Acknowledgement is also due to the University of Ulster MSc Informatics student Gerry Clarke.

Part of this work has been supported by the NASA Office of Systems and Mission Assurance (OSMA) through its Software Assurance Research Program (SARP) project, Formal Approaches to Swarm Technologies (FAST), and by NASA Goddard Space Flight Center, Software Engineering Laboratory (Code 581).

References

1. P. Horn, "Autonomic computing: IBM perspective on the state of information technology," IBM T.J. Watson Labs, NY, 15th October 2001. Presented at AGENDA 2001, Scottsdale, AR (available at http://www.research.ibm.com/autonomic/), 2001.
2. R. Sterritt, "Pulse Monitoring: Extending the Health-check for the Autonomic GRID," Proceedings of IEEE Workshop on Autonomic Computing Principles and Architectures (AUCOPA 2003) at INDIN 2003, Banff, Alberta, Canada, 22-23 August 2003, pp 433-440.
3. T. Bapty, S. Neema, S. Nordstrom, S. Shetty, D. Vashishtha, J. Overdorf, P. Sheldon "Modeling and Generation Tools for Large-Scale, Real-Time Embedded Systems," Proceedings of IEEE International Conference on the Engineering of Computer Based Systems (ECBS'03), Huntsville, Alabama, USA, April 7-11 2003, IEEE CS Press, pp 11-16.
4. R. J. Anthony, "Natural Inspiration for Self-Adaptive Systems," Proceedings of IEEE DEXA 2004 Workshops - 2nd International Workshop on Self-Adaptive and Autonomic Computing Systems (SAACS 04), Zaragoza, Spain, August 30th - September 3rd 2004, IEEE, pp 732-736.
5. J. O. Kephart and D. M. Chess. "The vision of autonomic computing". Computer, **36**(1):41–52, 2003.
6. R. Sterritt, D.W. Bustard, "Towards an Autonomic Computing Environment," Proceedings of IEEE DEXA 2003 Workshops - 1st International Workshop on Autonomic Computing Systems, Prague, Czech Republic, September 1-5, 2003, pp 694-698.
7. J. Newell, "Dying to live: why our cells self-destruct," Focus, Dec. 1994.
8. R. Lockshin, Z. Zakeri, "Programmed cell death and apoptosis: origins of the theory," Nature Reviews Molecular Cell Biology. **2**:542-550, 2001.
9. Y. Ishizaki, L. Cheng, A.W. Mudge, M.C. Raff, "Programmed cell death by default in embryonic cells, fibroblasts, and cancer cells," Mol. Biol. Cell, 6(11):1443-1458, 1995.
10. J. Klefstrom, E.W. Verschuren, G.I. Evan, "c-Myc Augments the Apoptotic Activity of Cytosolic Death Receptor Signaling Proteins by Engaging the Mitochondrial Apoptotic Pathway," J Biol Chem,. 277:43224-43232, 2002.
11. N.R. Jennings, M. Wooldridge, "Agent-oriented Software Engineering," in J. Bradshaw (ed.), *Handbook of Agent Technology*, AAAI/MIT Press, 2000.
12. M.N. Huhns, V.T. Holderfield, R.L.Z. Gutierrez, "Robust software via agent-based redundancy," Second International Joint Conference on Autonomous Agents & Multiagent Systems, AAMAS 2003, July 14-18, 2003, Melbourne, Victoria, Australia, pp 1018-1019.
13. G. Kaiser, J. Parekh, P. Gross, G. Valetto, "Kinesthetics eXtreme: An External Infrastructure for Monitoring Distributed Legacy Systems," Autonomic Computing Workshop – IEEE Fifth Annual International Active Middleware Workshop, Seattle, USA, June 2003.
14. E. Lupu, et al., EPSRC AMUSE: Autonomic Management of Ubiquitous Systems for e-Health, 2003.

15. R. Sterritt, D.W. Bustard, "Autonomic Computing: a Means of Achieving Dependability?," Proceedings of IEEE International Conference on the Engineering of Computer Based Systems (ECBS'03), Huntsville, Alabama, USA, April 7-11 2003, pp 247-251.

16. R. Sterritt, "Towards Autonomic Computing: Effective Event Management," Proceedings of 27th Annual IEEE/NASA Software Engineering Workshop (SEW), Maryland, USA, December 3-5 2002, IEEE Computer Society Press, pp 40-47.

17. R. Sterritt, D.F. Bantz, "PAC-MEN: Personal Autonomic Computing Monitoring Environments," Proceedings of IEEE DEXA 2004 Workshops - 2nd International Workshop on Self-Adaptive and Autonomic Computing Systems (SAACS 04), Zaragoza, Spain, August 30 – 3 September, 2003.

18. J.D. Hartline, "Mobile Agents: A Survey of Fault Tolerance and Security," University of Washington, 1998.

19. M.S. Greenberg, J.C. Byington, T. Holding, D.G. Harper, "Mobile Agents and Security," IEEE Comms. Mag. July 1998.

20. M.A. Swartwout, "Engineering Data Summaries for Space Missions," SSDL, 1998.

21. J. Wyatt, R. Sherwood, M. Sue, J. Szijjarto, "Flight Validation of On-Demand Operations: The Deep Space One Beacon Monitor Operations Experiment," 5th International Symposium on Artificial Intelligence, Robotics and Automation in Space (i-SAIRAS '99), ESTEC, Noordwijk, The Netherlands, 1-3 June 1999.

22. W. Truszkowski, M. Hinchey, J. Rash and C. Rouff, "NASA's Swarm Missions: The Challenge of Building Autonomous Software," IEEE IT Professional mag., September/October 2004, pp 51-56.

23. W. Truszkowski, M. Hinchey, C. Rouff and J. Rash, "Autonomous and Autonomic Systems: A Paradigm for Future Space Exploration Missions," *submitted for publication*.

24. N. Muscettola, P. P. Nayak, B. Pell, and B. Williams, "Remote Agent: To Boldly Go Where No AI System Has Gone Before," Artificial Intelligence 103(1-2):5-48, 1998.

25. J.Wyatt, H. Hotz, R. Sherwood, J. Szijjaro, M. Sue, "Beacon Monitor Operations on the Deep Space One Mission," 5th Int. Sym. AI, Robotics and Automation in Space, Tokyo, Japan, 1998.

26. R. Sherwood, J. Wyatt, H. Hotz, A. Schlutsmeyer, M. Sue, "Lessons Learned During Implementation and Early Operations of the DS1 Beacon Monitor Experiment", Third International Symposium on Reducing the Cost of Ground Systems and Spacecraft Operations, Tainan, Taiwan, 1999.

27. W. Truszkowski, J. Rash, C. Rouff and M. Hinchey, "Asteroid Exploration with Autonomic Systems," Proceedings of IEEE Workshop on the Engineering of Autonomic Systems (EASe 2004) at the 11th Annual IEEE International Conference and Workshop on the Engineering of Computer Based Systems (ECBS 2004), Brno, Czech Republic, 24-27 May 2004, pp 484-490.

28. J. McCann, M. Huebscher, "Evaluation issues in Autonomic Computing," Imperial College, 2004.

29. J.P. Bigus, et.al., "ABLE: a toolkit for building multiagent autonomic systems," IBM Systems J., 41(3):350-371, 2002.

30. C.A. Rouff, M. G. Hinchey, W. Truszkowski, J.L. Rash and D. Spears, editors, *Agent Technology from a Formal Perspective*, NASA Monographs in Systems and Software Engineering, Springer Verlag, London, 2005.

31. W. Truszkowski, C.A. Rouff, H.L. Hallock, J. Karlin, J.L. Rash, M.G. Hinchey and R. Sterritt, *Autonomous and Autonomic Systems: With Applications to NASA Intelligent Spacecraft Operations and Exploration Systems*, NASA Monographs in Systems and Software Engineering, Springer Verlag, London, 2005.

Petri Nets as Modeling Tool for Emergent Agents

Margo Bergman

Penn State Worthington Scranton, 120 Ridge View Drive, Dunmore, PA 18410
mwb12@psu.edu
http://www.personal.psu.edu/mwb12

Abstract. Emergent agents, those agents whose local interactions can cause un-expected global results, require a method of modeling that is both dynamic and structured. Petri Nets, a modeling tool developed for dynamic discrete event system of mainly functional agents, provide this, and have the benefit of being an established tool. We present here the details of the modeling method here and discuss how to implement its use for modeling agent-based systems.

1 Introduction

Petri Nets have been used extensively in the modeling of functional agents, those agents who have defined purposes and whose actions should result in a known out-come. However, emergent agents, those agents who have a defined structure but whose interaction causes outcomes that are unpredictable, have not yet found a mod-eling style that suits them. A problem with formally modeling emergent agents that any formal modeling style usually expects to show the results of a problem and the results of problems studied using emergent agents are not apparent from the initial construction. However, the study of emergent agents still requires a method to ana-lyze the agents themselves, and have sensible conversation about the differences and similarities between types of emergent agents. We attempt to correct this problem by applying Petri Nets to the characterization of emergent agents. In doing so, the emer-gent properties of these agents can be highlighted, and conversation about the nature and compatibility of the differing methods of agent creation can begin.

1.1 Petri Nets

Petri Nets are a graphical modeling tool used mainly to analyze manufacturing proc-esses. The main strength of using Petri Nets lies in the fact that they can handle con-currency of events. For complex modeling the ability to allow several events to occur simultaneously and still analyze their effects on each other is a necessity. The classic Petri Net consists of four objects: *places, transitions, directed arcs* and *tokens*. A *place* is a state of existence for a model. Consider a traffic light which has three states, each of which indicates a different situation; red says stop, yellow indicates caution, and green allows forward motion. Each of these three states would be con-sidered a place in a Petri Net. Places are usually denoted by a circle. *Transitions* are the means by which the different places are reached. This would be the light chang-ing from red to yellow, yellow to green, etc. You must go through a transition in order

M.G. Hinchey et al. (Eds.): FAABS 2004, LNAI 3228, pp. 271–274, 2005.

to reach a place. Transitions take the form of a square, or a straight line. The *directed arc* links the places to the transition. If one must change from red to yellow, there would be an arc linking the place "red" to the transition "changing from red to yellow," and another linking this transition to the place "yellow." The *token* is the means by which the Petri Net is made active. It indicates at which *place* in the Petri Net the current process is, and allows for restrictions on the activity of the processes. If there were two traffic lights at an intersection the tokens would indicate which is green, which is red, and ensure that only one was green at any given time. Below is the traffic light example shown as a Petri Net (1).

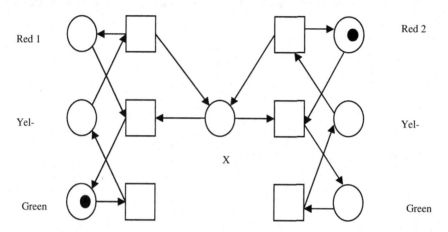

Fig. 1. This represents two traffic lights. The tokens in Green 1 and Red 2 show that the left traffic light is green currently, and the right one, red. Notice that both tokens have to be in spot X for the light to change. This is true of many actual traffic systems, where all lights at an intersection are briefly red before one turns green

This process of putting the Petri Net into action is referred to as *firing*. Each token allows for a single firing, which causes a token to move from place, through a transition, into another place. In order for a particular transition to be *enabled*, all of the places who have a directed arc leading to that transition must have a token. In the case above, the X place creates the situation where both lights must be red for one to turn green.

The model above is an example of a Petri Net that describes a well-defined system, with predictable results, and no emergent properties. Below we describe how this tool can be applied in a situation where emergence generates the interesting result.

2 Emergent Agent Modeling

We create a formal model of a classic agent-based model, the Schelling model of spatial segregation. In this model, there are two types of agents. Each agent has a threshold level for the number of similar agents they wish to have in their 'neighborhood', although no agent has a particular preference for segregation. When chosen by random, each agent takes an accounting of the percentage of each agent type in their neighborhood, and if the percentage of dissimilar agents is too high, they will move to another location. Although simple, the model's results stem from the emergent properties of the heterogeneous agents. Although this is not a detailed model, there are still many choices a researcher must make when programming the simulation, such as the type of neighborhood the agents live in, how the thresholds are determined, the method of location switching, etc. Each model has characteristics in common, however, and it is these characteristics that should be included in formal specification of the model. Below is a Petri Net model of the basic characteristics that should be included in every Schelling simulation, regardless of the individual choices made by the researchers.

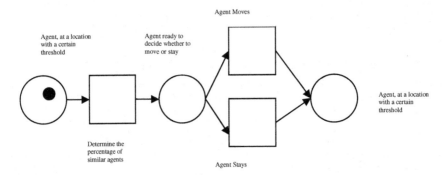

Fig. 2. A Petri Net of the Schelling model of spatial segregation [2]. Here, an agent has a certain threshold of similar agents they want in their "neighborhood". If that percentage falls beneath their threshold, they will choose to move. The Petri Net shows the basic model, without requiring knowledge of the specific parameters

Given this model of the process that the researcher is trying to analyze, and the specifics of the choices that she made in the design process, the original results should be replicable. In addition there is no need for every researcher to utilize the same programming language or software package in order to understand the workings of the model. Petri Nets are dynamic which makes them ideal for analyzing the structure of agent-based models, whose results usually rely on the dynamic interactions of their component parts.

3 Conclusion

These Petri Net models do not replace the agent-based model itself. The emergent nature of many agent-based results still requires a full computational simulation to be created. However, they do provide a method by which two modelers can discuss a single problem without being distracted by the particulars of their individual models. Since Petri Nets are mathematically based, issues of the efficiency of the model can also be analyzed. Finally, there is already an established body of work in the field of

Petri Nets, which prevents agent-based modelers from having to invent new systems of analysis. Just as economics and other fields adapted calculus for their own uses, agent-based modelers in all disciplines can use this technique.

References

1. http://tmitwww.tm.tue.nl/staff/wvdaalst/Petri_nets/pn_tutorial.html July 2004
2. Schelling, TC. Micromotives and Macrobehavior. W. W. Norton & Company, 1978

Massive Multi-agent Systems Control

Jean-Charles Campagne[1], Alain Cardon[1,2], Etienne Collomb[3],
and Toyoaki Nishida[3]

[1] LIP6 - UPMC - CNRS – 8, rue du Capitaine Scott - 75015 Paris - France
{jean-charles.campagne, alain.cardon}@lip6.fr
[2] IRD, Centre Ile-de-France – 32, rue Varagnat - 93143 Bondy Cedex - France
[3] The University of Tokyo – 7-3-1 Hongo, Bunkyo-ku, Tokyo 113-8656, Japan
{etienne, nishida}@kc.t.u-tokyo.ac.jp

Abstract. In order to build massive multi-agent systems, considered as complex and dynamic systems, one needs a method to analyze and control the system. We suggest an approach using morphology to represent and control the state of large organizations composed of a great number of light software agents. Morphology is understood as representing the state of the multi-agent system as shapes in an abstract geometrical space, this notion is close to the notion of phase space in physics.

1 Introduction

With the advent of new computer technologies new large-scale systems are now possible. However, methods for actually building such complex system are less frequently proposed. Existing common approaches include : "manual tuning", emergence-based theory approaches, genetic approaches.

Manual tuning is only feasible for a couple of agents. It is impracticable for bigger organizations.

Emergence-based theories seek the understanding of the requirements at the microscopic level (the agent) in the hope that the macroscopic (the system) level will eventually behave appropriately. Many of these theories suggest that the agents composing the system have to be cooperative : resolving local conflict is sufficient to yield a proper global behavior (eg [3]). This hypothesis seems too restrictive[2] ; natural self-adaptive systems composed of many entities are not all locally-cooperative.

Agent genetic approaches, which include non-necessarily cooperative agents (eg [5]), seem to be promising. However, they lack of the ability of analyzing and understanding the system. It is difficult to understand how the system works by only relying on the fitness function.

In order to build such a system one has to be able to analyze, maintain and control the behavior of the system. Deep understanding of the system workings is needed. And for such a system to be auto-adaptive, it needs to observe, analyze and control itself [4].

M.G. Hinchey et al. (Eds.): FAABS 2004, LNAI 3228, pp. 275–280, 2005.

Our proposal. For a system to be self-regulated, it has to have the ability to consider its internal state. We propose a way of describing the state of the agent organization in a problem-independent manner, by projecting the state of the agent organizations in an abstract geometrical space from various measurements made at the agent level (this is similar to the approach in physics as with phase-space), and letting the system access this representation in order to control itself. The underlying hypothesis is that the shapes representing the system's state are correlated to the system's behavior.

We describe the model, highlighting the important points, and then present an example of application of such an architecture applied to agent population control. We also discuss the advantages and current limitations based on the experiments with the implemented model.

2   Description of the Approach

2.1   Hypothesis

We seek to correlate the micro-level behavior (agent) with the macro-level behavior (organization) using a generic approach (morphology). The hypothesis is that the shapes should be correlated to the system's behavior, and that it is possible to attract the system toward another state using the morphological description if the system fails to behave appropriately.

2.2   General Description

The system is composed of three main organizations : the aspectual organization that represents a phenomena ; the morphological organization which describes the state of the aspectual organization in a geometrical way ; and the analysis organization controlling the aspectual organization relying on the description given by the morphological organization and following the guidelines provided by the system designer. A more detailed description can be found in [1].

2.3   Aspectual Organization

The aspectual organization, composed of many agents, represents a phenomena we want to study. This is the organization we seek to analyze and control. The term "aspectual" comes from the original agentification method proposed in [1].

In order to evaluate the system's state, the aspectual agents compute a value, called the "aspectual vector", as they run. This vector is a collection of values describing the agent's organizational state and its activity. The exact nature of these measures depend on the structure of the agent.

2.4   Morphological Organization

The whole collection of aspectual vectors make up the aspectual landscape of the aspectual organization which is then analyzed by the morphological agents.

Morphological agents attempt to describe what is happening in the aspectual organization in a geometrical way. The description does not take into account the ontology previously established : there is no semantics in the morphology space. Morphology space is only concerned with the activity and the organizational state of the agents. It points out structure, shapes, recurrent features, similarities, oppositions, dominant or recessive features... If we consider the aspectual measure as a mapping from a subset of the agent organizational state space to a numerical space (possibly multi-dimensional) ; the reciprocal is a function that modifies the agent behavior according to some target value so that the resulting aspectual vector of the agent would conform to that target value.

2.5   Analysis Organization

By using a proper way of computing the morphology, the shapes revealed by the morphology are correlated to the system's behavior. We intend to exploit this correlation.

The analysis agents use the morphological description to examine the aspectual organization and to orientate the system accordingly to some generic guidelines instructed by the designer (for example : "global variable X of the system should be around value Y"...). This is achieved by classifying and learning the morphology : as the system runs, typical shapes in the morphological spaces are revealed, these shapes are correlated to the system's behavior and categorized appropriately. Analysis agents can, following the designer's guidelines, influence the aspectual organization, either by direct injunctions on it, or by selecting appropriate shapes (learned from the system's past activity) and telling the morphological agents that this particular shape would be more appropriate than the current shape.

3   Example

We have developed an example using this approach in the context of agent population size control. The goal of this example is to illustrate how the global behavior of the system is correlated with its morphological description and how it is possible to exploit this correlation to control the system.

3.1   Aspectual Agents

The aspectual organization is subjected to population control. Aspectual agents reside in a common environment where they "see" each other and from which they can extract some "energy" in order to survive.

Agents have some limited social skills : an agent can ask another agent to give it some energy. The asked agent can either cooperate or refuse. In the event of refusal, the asker "fights" the non cooperative agent. A fight results in the loss of energy from both antagonists, however the initiator of the fight looses less energy than the other one (simulating the benefit of initiating the attack).

If the energy level of an agent drops below zero, the agent "dies" and is removed from the organization. If an agent collects enough energy it can clone, yielding another agent. Removal and cloning of agents enables the organization to change in size.

The behavior of each aspectual agent is parameterized a variable, called its "eagerness", it influences the agent's behavior in its choice on whether to attack or not other agents. This parameter can be updated by the agent itself when it receives a recommendation from the morphological agents.

3.2   The Morphology

To analyze the system, we chose to use only one characteristic of the agent's organizational state : its "supremacy". The idea of this measure is to relate the position of the agent within the organization : whether the agent is or not in a comfortable position. This is correlated to its energy level : the more energy the agent has, the more likely it is to survive. Hence, we chose to compute the agent's supremacy as equal to its energy level.

The shapes used to describe the organization's state are normalized and mean-centered histograms representing the agents' state distribution according to their supremacy. Histograms have the advantage of being easily comparable. It is possible to formulate a "reciprocal" of this mapping. An aspectual agent that is asked to change its vector value will try to do so by modified some of its variables (its eagerness) that alter its behavior accordingly.

3.3   The Analysis and Control

One analysis agent is used to control the system. This agent learns and classifies the histograms computed by the morphological agent. It can also directly know the actual number of agent in the aspectual organization, so it is able to determine, accordingly to rules defined by the system designer if the system is in a "good" or "bad" state and classify the corresponding shape properly (figure 1).

In this example, when the system behaves correctly there is no feedback. But when the population size is out of bounds, the analysis agent asks the

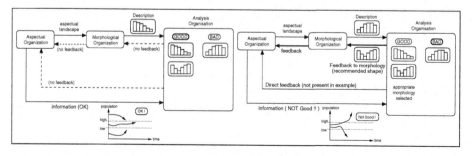

Fig. 1. The left figure shows the system when it is considered as fine. The right figure shows when the system is needs adjustment

morphological agent compute the appropriate feedback corresponding to the difference between the "good" and "bad" histograms.

3.4   Results and Discussion

The limitation of space does not permit us to discuss in details the results of all the simulations but we will mention the essential.

Figure 2 sums up the results of two tests : one without control and the other with control. The instantaneous error rate is a pseudo-distance of the population curve to the closest threshold (if the population curve is in-between both thresholds, the instantaneous error rate is zero). The error rate is the sum of all the instantaneous error rate in one 1000-cycles run. The average error rate for an analysis agent (if any) is computed over 10 such 1000 cycles run.

The reference test was done without control, it consisted of 500 tests. The other series of tests was done with control, over 200 tests (the difference of the number of tests is due to available time, the ones with control took longer to compute). In both cases the target population size was 50 agents with a margin of 0.2 (lower threshold and upper threshold are 40 and 60 respectively).

Figure 3 displays a couple of examples of the system's behavior, with and without control. These curves give a more palpable, qualitative, appreciation of control performance.

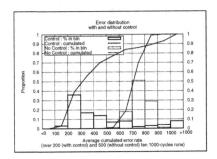

Fig. 2. Comparative plot with and without control. Cumulative curves show that this system allows control in most cases (85%)

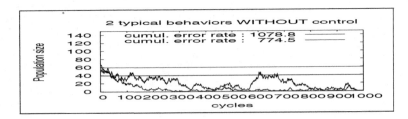

Fig. 3. Example of typical (more than 50% of the cases) system's behaviors (with and without control)

In most cases (85%) the control improves the system's behavior. However, the histograms reveal that in 15% of the cases it does worse. One possible explanation is that, in some cases, the initial configuration of the aspectual organization (ie when the analysis agent has learned nothing yet) does not permit the analysis agent to "discover" adequate shapes, and then learns inappropriately, thus badly controlling the population size.

We also have noted that in some cases, only less than half of the aspectual agents needed to comply with the morphological injunctions, so that the population level was maintained at an appropriate value.

Developing more elaborated morphological analyzes (augmenting the aspectual vector with other aspects of the agent's behavior and using trajectories by introducing the time dimension in the morphological space) and using more appropriate learning mechanisms would allow finer control of the system.

4   Conclusion

We seek to develop a general method to analyze and control multi-agent system, and to make them self-adaptive. We briefly described the model based on the morphology approach of representing the system's state. This representation is available to the system in order to make it self-adaptive.

We illustrated the workings of the system with a simple example consisting in population control. Shapes used in this example were histograms representing the relative distribution of one of the agent's properties. A simple learning mechanism permitted to outline and exploit a correlation between the micro-level behavior with macro-level behavior.

Other interests include developments of more elaborate morphology descriptions and understanding the needed properties of such description in order to be useful (toward formalization ?).

References

1. Alain CARDON. *Modéliser et concevoir une machine pensante. Approche constructible de la conscience artificielle.* Éditions Automates Intelligents, march 2003.
2. F. CHANTEMARGUE, T. DAGAEFF, M. SCHUMACHER, and B. HIRSBRUNNER. The emergence of cooperation in a multiagent system, 1996.
3. Institut de Recherche en Informatique de Toulouse (IRIT). Théorie AMAS. http://www.irit.fr/SMAC/.
4. Catriona Mairi KENNEDY. *Distributed reflective architectures for anomaly and autonomous recovery.* PhD thesis, School of Computer Science, Faculty of Science, University of Birmingham, 2003.
5. Samuel LANDAU. *Des AG vers les GA.* PhD thesis, Université de Paris VI, 2002.

Fuzzy Hybrid Deliberative/Reactive Paradigm (FHDRP)

Hengameh Sarmadi

Department of Electrical Engineering,
Vienna University of Technology (TU-Wien)
Karlsplatz 13, A-1040 Vienna / Austria

Abstract. This work aims to introduce a new concept for incorporating fuzzy sets in hybrid deliberative/reactive paradigm. After a brief review on basic issues of hybrid paradigm the definition of agent-based fuzzy hybrid paradigm, which enables the agents to proceed and extract their behavior through quantitative numerical and qualitative knowledge and to impose their decision making procedure via fuzzy rule bank, is discussed. Next an example performs a more applied platform for the developed approach and finally an overview of the corresponding agents architecture enhances agents logical framework.

1 Introduction

The definition of the agents world could be based on their social rules, communication cooperation and negotiation between them and their pursuance to achieve the defined (pre-given) goals. The central concerns on this area refer from one side to the individual design aspects for developing the design task and leading to an improved behavior and from another side to their social rules, cooperation of the individuals, and consideration based on the relationship between individual and overall social behaviors. The design aspect should be suitable for time constraint environment and interactions with environment in order to make agents capable to reconfigure and recover from changes due to environment and satisfy other flexible design criteria. Intelligent agents acquire information from the world interface and are able to perform tasks which are supposed to meet deadline on average [2].

In heterogonous approach the agents may differ from each other to ensure multi robot coordination. Self regulation agents concept is embedded within the environment and their autonomous action meets the design objectives. Part of the problem from individuality point of view refers to path planning and navigation, which implies the complexity of the problem and represents the physical limitations of robot platform.

In a human-based model of agents there is a close dependent between the deliberation, reaction and decision making and the agents relationship functional structure models these criteria [8]. If we face the unstructured or local environment the reactive planning is the most appropriate execution meanwhile in a knowledge rich environment (global/open world) a hierarchical paradigm works better based on deliberation process of global information and agent specific abstraction. Deliberation functions could not extend independently of reactive behavior and vice versa. Hybrid architectures benefit both concepts of reactive and deliberative paradigm [5].

M.G. Hinchey et al. (Eds.): FAABS 2004, LNAI 3228, pp. 281–286, 2005.

2 Incorporation of Fuzziness into Hybrid Paradigm

As stated above the agents should have some reactive design-base due to changes in environment, which in turn may effect and result in some limitation in the agents (local) goals and deliberation paradigm, which through reasoning procedure and intention tend to lead to action to achieve the goals (mean-ends). Therefore we need to balance between goal directed (deliberative) and reactive paradigm. Fuzzy approach can perform a suitable area for considering both these aspects, through which our decision function is a fuzzy one which proceeds the action design choice influenced by history and reconsideration and makes the agents enable to develop cognitive functions for evolution of intelligence.

Deliberation and reactivity face with problems of multiple conflicting criteria and multiple objectives. With incorporation of fuzziness into hybrid paradigm we can make decisions with vague, uncertain and inexact objects and extract the human knowledge in planning architecture without articulating an application-based world model and prepare a determinative interpretation from probability and randomness. Fuzzy approach profits knowledge representation about how the agents represent their world, plan and solve problems in close/open world and is appropriate to experiment on bold (never reconsider) to cautious (constantly reconsider) agents, since the decision procedure attempts to degree.

2.1 Development of the Concept of FHDRP

The decision making strategy will be based on a deliberation-reaction fuzzy rule bank with strategy acquisition in extracting the fuzzy rules incorporating with real time reasoning [3,4,6] (we can learn the fuzzy rules from experiences with numerical and/or linguistic sample data with enhancing the system profiting the ability of learning-base systems such as neural networks.).

The fuzzy deliberative/reactive rule bank could be defined as follow:

$- FRB:$ $\quad (X_i, C_i, A_i), \quad i = 1,2,...,n.$

Where:

-Deliberation state conditions: $X_i = \{X_{i1}, X_{i2},..., X_{im}\}.$

With deliberation fuzzy set: $D = \{(X, m_D(X)) | X \in \Omega\}$, Ω is the universe of the deliberations.

-Reaction state conditions: $C_i = \{C_{i1}, C_{i2},..., C_{is}\}.$

With reaction fuzzy set: $R = \{(C, m_R(C)) | C \in \Psi\}$, Ψ is the universe of the reactions.

-Action through deliberation and reaction: $A_i = \{A_{i1}, A_{i2},..., A_{ik}\}.$

With action fuzzy set: $A_C = \{(A, m_{A_C}(A)) | A \in \Gamma\}$, Γ is the universe of the actions.

The fuzzy hybrid paradigm can be defined as the projection:

$D \times R \rightarrow A_C$. Where D is deliberation, R is reaction and A_C is action power set. The i-th fuzzy rule seems like:

If (X_{i1} is f_{x1} and X_{i2} is f_{x2} ... and X_{in} is f_{xn}) and if (C_{i1} is f_{c1} and C_{i2} is f_{c2} ... and C_{is} is f_{cs}) then A_i is f_{ai}.

So the behavior could be written as: $Beh \subseteq R \times R \times R$, ordering on R and could be defined as:

$$- Beh = \left\{ (X_i, C_i, A_i) \middle| \ X \in D, C \in R, A \in A_C \right\}, \text{ or}$$

$$- Beh = R \bigcup_{i=1}^{P} X_{i1} \times X_{i2} \times ... \times X_{in} \times C_{i1} \times C_{i2} \times ... \times C_{is} \times A_{i1} \times A_{i2} \times ... \times A_{ik}.$$

The parallel associative inference will fire each fuzzy rule in parallel but to different degrees.

As a traditional defuzzifier approach the max-height method can be used:

$$- m_{A_C}(A_{max}) = \max_{i=1}^{p} \ m_{A_C}(A_i).$$

The defuzzification procedure can be completed with the priority rule or the subsumption theory developed by Brooks [1]. Therefore we get the inhibition relation in the hierarchy as follow:

$A_i \prec A_j$ if $(A_i, A_j) \in \prec$, and we read it:

A_i inhibits A_j or A_i is lower in hierarchie than A_j.

2.2   Example

As an example (modified from [7] and [8]):

Suppose the objective is to collect samples in an indoor environment of a particular type in a predefined place. The location of the samples is not known. A number of autonomous swarm agents are for this problem available which can go around and collect the samples. Furthermore the terrain is full of obstacles. This organized team of robots can in turn negotiate and cooperate together and divide up the task collaborating with a common coordinator and the individuals have autonomous decisions and navigations.

In this problem we face with a mix of path and deliberation/reaction planning and we need a path planning algorithm (for further detail see next section) as well, which sufficiently represents the terrain.

We could extract some fuzzy deliberation/reaction rules considering agents specific criteria and agents cooperation, for example:

– Deliberation (if there are more samples in one direction, Move-to-that-direction) and Reaction (if near a sample and obstacles or other agents are far, Speed-up-towards-the-sample).

– Deliberation (if the obstacle is far, Choose-the-up-gradient-direction-toward-the-sample, Move-to-sample) and Reaction (if the obstacle is very close, Change-the-direction).

– Deliberation (if your partner is closer than you to a sample and there are fewer obstacles in his way, Communicate-with-your-partner and Let-him-to-pick-it-up).

– Deliberation (if another agent near you has more frustration, Go-in-his-direction and Pick-up-the-sample).

– Deliberation (if many other agents in one direction, Choose-another-direction) and Social rule (if another agent near you Wait or Turn-to-the-left).

And so on.

We can interpret for all of the linguistic notions mentioned above, which can not be exactly described, such as: more, near, far, close, few,… the corresponding fuzzy sets and with defining the degree of indeterminacy articulate numerical data structure for partial occurrence of events or relations and have a quantitative interpretation from probability and randomness.

Finally pure reactive rules such as: Avoid-obstacle, finding a sample Take-it and carrying samples and at the base Drops-the-samples have the most priority and complete the decision procedure with inhibition characteristics.

3   FHDRP Control Architecture of the Agents

The design-base is appropriate for real time execution and the state hierarchy develops a layered intelligent structure, whereby each layer could be interpreted as software agent or function in order to develop logic based concepts of robots and assure a modular construction for replaning and adapting the configuration.

The control strategy suggested and the proactive behavior could be based on supervisory control on agent level. Coordinated decision must be suitable for application specific behavior on user level program embedded in the controller at run time. Application specific information, control algorithm and planning strategy specify the agent code. We should define the optimal in favor of our data base, rule base and changes that will occur in the environment.

The approach (fig. 1) is based on SENSE then PLAN ACT, whereby the sensed information goes through planning layer and by means of directives translates to actuator commands on a hierarchical paradigm. One of the inputs to the systems will be sequences of environment states or percepts through which the control rules will extract some deliberative/reactive behaviors. Action rules translate to the effectors via corresponding sensor system through pattern of motor schema action. The developed concept is applicable to both local and cooperative planning.

The fuzzy approach develops the social ability and satisfies the abilities due to uncertainty in the world model and provide a balance between goal directing and reactivity and interactions between the agents in order to coordinate and control them, using Quantitative numerical and qualitative knowledge. Fuzzy rules consider attention, reasoning, and information collection. We need real time processor to proceed with fuzzy reasoning about the global state to select the best behavior.

The behavioral manager plans which behavior to use in order to progress to the goal and there are assumptions mappings from global data structure (sensory inputs) to behavior generation.

Behaviors are inherently parallel and distributed and the goal directed approach is a sequence of generic behavior and updating the behavior.

Sequencer as in the general hybrid architecture generates the set of behaviors, adapts it with managerial style and subdivides the deliberation based on the control scope and enables the system to develop a behavior based control for coordinating planful activities with real time behavior for dynamic positioning, navigation (behavior based opportunity to change direction of navigation) and considering the goals, resources and timing constraints.

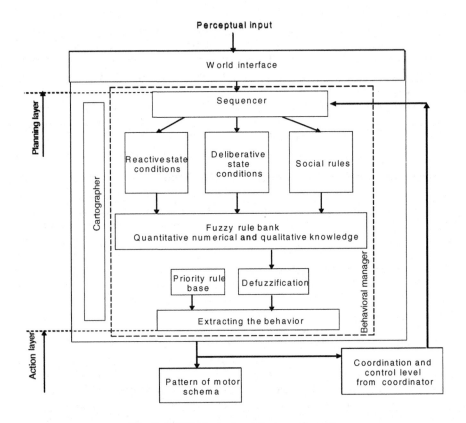

Fig. 1. FHDR Control architecture of agents

Cartographer is responsible for information collection (data structure) and path planning. Physical location of agents and the coordinated control program will be part of global knowledge and shared data structure. To find the optimal path in the configuration space, event noticeable by reactive system would trigger event-driven replaning.

Path generation algorithm will generate a pre completed path with a hill climbing algorithm to reach to the target position, where the target position is given by the strategy system and behavioral controlling of the moving direction of the agent is based on the direction of the target point and agent actual coordinates [6].

If an obstacle blocks the path, the path is replaned and optimized by computing the optimal route and decomposing it into waypoints [5], a goal to reach. After reaching each waypoint next goal is computed and cartographer gives the sequencer a set of waypoints to make a qualitative navigation possible.

Solutions to the problems such as interference, member unproductive or failed, agents interactions and communications, individuality and autonomy, emergent behavior and heterogeneity could be in collaborating with the coordinator. Coordinator defines new goals and sets the strategical plans, which lead to tactical instructions and coordinate the relationship between strategy and agents set of tactical behavior through social rules. Coordinator can modify the relationship between the behaviors of agents: one strategy and several tactics.

4 Future Directions

For a later work we could enhance the agents architecture with learning-base distributed AI systems such as neural networks to learn from experiences and new data and be able to improve the agents behavior. The learning system should develop the ability of on-line learning in time critical environment using qualitative abstraction, symbolic learning algorithm and rules generation and challenge the autonomous learning of sequential behavior and gain some skills to carry out the plan for more robustness and performance monitoring. We could also investigate on stability, redundancy and complementary of the system.

References

1. R. A. Brooks. A Robust Layered Control System for a Mobile Robot. *IEEE Journal on Robotics and Automation*, 2(1) (1986) 14-23
2. R. A. Brooks. Intelligence without Reason. *Proceeding of the Twelfth International Joint Conference on Artificial Intelligence* (IJCAI-91), Sydney, Australia (1991) 569-595
3. H. Hellendoorn. Reasoning with Fuzzy Logic. *University of Technology*, Delft (1992)
4. E.H. Mamdani. Application of Fuzzy Logic to Approximate Reasoning using Linguistic Synthesis. *IEEE Transaction on Computers C-26*, no.12 (1977) 1182-1191
5. R. R. Murphy. Introduction to AI Robotics. *Cambridge, Mass.: MIT Press* (2000)
6. H. Sarmadi. An Approach to a Supervised Neural Network-based Fuzzy Controller. *PhD Dissertation, Vienna University of Technology*, Vienna (1995)
7. L. Steels. Cooperation between Distributed Agents through Self Organization, Decentralized AI. *Proceeding of the First European Workshop on Modeling Autonomous Agents in a Multi-Agent World*, Amesterdam, the Netherlands (1990) 175-196
8. M. Wooldridge. Intelligent Agents: the Key Concepts. *Multi-Agent Systems and Application II, 9th ECCA-ACAI/EASSS 2001*, Prague, Czech Republic (2002) 3-43

Interaction and Communication of Agents in Networks and Language Complexity Estimates

Jan Smid[1], Marek Obitko[2], David Fisher[3], and Walt Truszkowski[4]

[1] Dept. of Computer Sci., Morgan State University, USA
jsmid@jewel.morgan.edu
[2] Dept. of Cybernetics, Czech Technical University
obitko@labe.felk.cvut.cz
[3] Software Eng. Inst., Carnegie Mellon University, USA
dfisher@sei.cmu.edu
[4] NASA/GSFC 588 , Greenbelt, MD 20771, USA
walt.truszkowski@gsfc.nasa.gov

1   Introduction

Knowledge acquisition and sharing are arguably the most critical activities of communicating agents. We report about our on-going project featuring knowledge acquisition and sharing among communicating agents embedded in a network [7, 8]. The applications we target range from hardware robots to virtual entities such as internet agents. Agent experiments can be simulated using a convenient simulation language. We analyzed the complexity of communicating agent simulations using Java and Easel [2]. Scenarios we have studied (see also our previous work [6]) are listed below. The communication among agents can range from declarative queries to sub-natural language queries.

- A set of agents monitoring an object are asked to build activity profiles based on exchanging elementary observations.
- A set of car drivers form a line, where every car is following its predecessor. An unsafe distance can create a strong wave in the line. Individual agents are asked to incorporate and apply directions how to avoid the wave.
- A set of micro-air vehicles form a grid and are asked to propagate information and concepts to a central server.

2   Knowledge Acquisition and Communication

For given knowledge representation language and agent communication language we follow several principles:

- The agent network is a graph that has short search paths [9], [1].
- The individual agent is a graph that has short search paths.
- All graphs can dynamically change, communities can be formed and communicate.
- The agent understanding substantially depends on the semantic information.

M.G. Hinchey et al. (Eds.): FAABS 2004, LNAI 3228, pp. 287–289, 2005.

For the knowledge acquisition of agents we use an algorithm that is based on the approach developed by J. Siskind [4]. In short, agents receive a sequence of utterances, each to be paired with a set of conceptual expressions. Conceptual expressions are assumed to be provided by e.g. the agent's cognitive system, and consist of conceptual symbols. The basic problem is to map words onto conceptual symbols. The thesis is that the natural language based knowledge representation is effective in representing the agent world.

3   Simulation Language Complexity

For knowledge processing as well as for other important agent-related tasks we have studied Easel property-based types (PBT) paradigm [2]. A type is a description of some class of objects, while a description is a set of properties. PBTs are intended to provide a foundation for automated systems that solve problems in ways analogous to those of humans. We further developed our initial comparison of Easel and Java presented in [6]. Java can be extended using special classes, such as Actor that is similar to an Easel actor type, which enables lower complexity of programming simulations, such as in a Jade [3] agent development environment. However, PBTs are not native structures in Java.

4   Conclusion

The presented knowledge acquisition method is promising for the next step of our project that deals with entities equipped with sensors. We have studied several examples of emergent agent systems and described knowledge acquisition and communication and the complexity of the implementation. The complexity of simulation using a specialized language such as Easel is lower compared with a general purpose language such as Java. The drawback of using a new language is the cost of mastering a special purpose language and its syntax rules.

References

1. A. Barabasi, E. Bonabeau: Scale-Free Networks. *Scientific American*. May 2003.
2. A. Christie, D. Durkee, D. A. Fisher and D. A. Mundie. Easel Language Reference Manual and Author's Guide. http://www.cert.org/Easel/lrm/index.html
3. F. Bellifemine, A. Poggi, G. Rimassa. JADE - A FIPA-compliant agent framework. *Proceedings of PAAM'99*. London. 1999
4. J. Siskind. Learning Word-to-meaning Mapping. *Models of Language Acquisition*. Edited by P. Broeder and J. Murre. Oxford University Press. 2000.
5. J. Smid, M. Obitko, W. Truszkowski. An Approach to Knowledge Exchange and Sharing Between Agents. *Innovative Concepts for Agent-Based Systems*. Springer Verlag. 2003. LNCS 2534.

6. J. Smid, M. Obitko, V. Snášel. Communicating Agents and Property-Based Types versus Objects. *SOFSEM — Current Trends in Computer Science*. 2004.
7. J. Smid. Knowledge Models for Network Environment Communications: An Overview. *IASTED AIA 2004, PSMP 2 Workshop*. Innsbruck. 2004.
8. J. Smid, M.Obitko, V. Snášel. Semantically Based Knowledge Representation. In B.d'Auriol (ed.): *The Proceedings of the CIC'04 Conference*. Las Vegas. 2004. http://jewel.morgan.edu/~jsmid/psmp3/index.php
9. D. Watts. *Small Worlds: The Dynamics of Networks between Order and Randomness*. Priceton University Pressq. 1999.

Author Index

Lecture Notes in Artificial Intelligence (LNAI)

Vol. 3070: L. Rutkowski, J. Siekmann, R. Tadeusiewicz, L.A. Zadeh (Eds.), Artificial Intelligence and Soft Computing - ICAISC 2004. XXV, 1208 pages. 2004.

Vol. 3068: E. André, L. Dybkjær, W. Minker, P. Heisterkamp (Eds.), Affective Dialogue Systems. XII, 324 pages. 2004.

Vol. 3067: M. Dastani, J. Dix, A. El Fallah-Seghrouchni (Eds.), Programming Multi-Agent Systems. X, 221 pages. 2004.

Vol. 3066: S. Tsumoto, R. Słowiński, J. Komorowski, J.W. Grzymała-Busse (Eds.), Rough Sets and Current Trends in Computing. XX, 853 pages. 2004.

Vol. 3065: A. Lomuscio, D. Nute (Eds.), Deontic Logic in Computer Science. X, 275 pages. 2004.

Vol. 3060: A.Y. Tawfik, S.D. Goodwin (Eds.), Advances in Artificial Intelligence. XIII, 582 pages. 2004.

Vol. 3056: H. Dai, R. Srikant, C. Zhang (Eds.), Advances in Knowledge Discovery and Data Mining. XIX, 713 pages. 2004.

Vol. 3055: H. Christiansen, M.-S. Hacid, T. Andreasen, H.L. Larsen (Eds.), Flexible Query Answering Systems. X, 500 pages. 2004.

Vol. 3048: P. Faratin, D.C. Parkes, J.A. Rodríguez-Aguilar, W.E. Walsh (Eds.), Agent-Mediated Electronic Commerce V. XI, 155 pages. 2004.

Vol. 3040: R. Conejo, M. Urretavizcaya, J.-L. Pérez-de-la-Cruz (Eds.), Current Topics in Artificial Intelligence. XIV, 689 pages. 2004.

Vol. 3035: M.A. Wimmer (Ed.), Knowledge Management in Electronic Government. XII, 326 pages. 2004.

Vol. 3034: J. Favela, E. Menasalvas, E. Chávez (Eds.), Advances in Web Intelligence. XIII, 227 pages. 2004.

Vol. 3030: P. Giorgini, B. Henderson-Sellers, M. Winikoff (Eds.), Agent-Oriented Information Systems. XIV, 207 pages. 2004.

Vol. 3029: B. Orchard, C. Yang, M. Ali (Eds.), Innovations in Applied Artificial Intelligence. XXI, 1272 pages. 2004.

Vol. 3025: G.A. Vouros, T. Panayiotopoulos (Eds.), Methods and Applications of Artificial Intelligence. XV, 546 pages. 2004.

Vol. 3020: D. Polani, B. Browning, A. Bonarini, K. Yoshida (Eds.), RoboCup 2003: Robot Soccer World Cup VII. XVI, 767 pages. 2004.

Vol. 3012: K. Kurumatani, S.-H. Chen, A. Ohuchi (Eds.), Multi-Agents for Mass User Support. X, 217 pages. 2004.

Vol. 3010: K.R. Apt, F. Fages, F. Rossi, P. Szeredi, J. Váncza (Eds.), Recent Advances in Constraints. VIII, 285 pages. 2004.

Vol. 2990: J. Leite, A. Omicini, L. Sterling, P. Torroni (Eds.), Declarative Agent Languages and Technologies. XII, 281 pages. 2004.

Vol. 2980: A. Blackwell, K. Marriott, A. Shimojima (Eds.), Diagrammatic Representation and Inference. XV, 448 pages. 2004.

Vol. 2977: G. Di Marzo Serugendo, A. Karageorgos, O.F. Rana, F. Zambonelli (Eds.), Engineering Self-Organising Systems. X, 299 pages. 2004.

Vol. 2972: R. Monroy, G. Arroyo-Figueroa, L.E. Sucar, H. Sossa (Eds.), MICAI 2004: Advances in Artificial Intelligence. XVII, 923 pages. 2004.

Vol. 2969: M. Nickles, M. Rovatsos, G. Weiss (Eds.), Agents and Computational Autonomy. X, 275 pages. 2004.

Vol. 2961: P. Eklund (Ed.), Concept Lattices. IX, 411 pages. 2004.

Vol. 2953: K. Konrad, Model Generation for Natural Language Interpretation and Analysis. XIII, 166 pages. 2004.

Vol. 2934: G. Lindemann, D. Moldt, M. Paolucci (Eds.), Regulated Agent-Based Social Systems. X, 301 pages. 2004.

Vol. 2930: F. Winkler (Ed.), Automated Deduction in Geometry. VII, 231 pages. 2004.

Vol. 2926: L. van Elst, V. Dignum, A. Abecker (Eds.), Agent-Mediated Knowledge Management. XI, 428 pages. 2004.

Vol. 2923: V. Lifschitz, I. Niemelä (Eds.), Logic Programming and Nonmonotonic Reasoning. IX, 365 pages. 2003.

Vol. 2915: A. Camurri, G. Volpe (Eds.), Gesture-Based Communication in Human-Computer Interaction. XIII, 558 pages. 2004.

Vol. 2913: T.M. Pinkston, V.K. Prasanna (Eds.), High Performance Computing - HiPC 2003. XX, 512 pages. 2003.

Vol. 2903: T.D. Gedeon, L.C.C. Fung (Eds.), AI 2003: Advances in Artificial Intelligence. XVI, 1075 pages. 2003.

Vol. 2902: F.M. Pires, S.P. Abreu (Eds.), Progress in Artificial Intelligence. XV, 504 pages. 2003.

Vol. 2892: F. Dau, The Logic System of Concept Graphs with Negation. XI, 213 pages. 2003.

Vol. 2891: J. Lee, M. Barley (Eds.), Intelligent Agents and Multi-Agent Systems. X, 215 pages. 2003.

Vol. 2882: D. Veit, Matchmaking in Electronic Markets. XV, 180 pages. 2003.

Vol. 2872: G. Moro, C. Sartori, M.P. Singh (Eds.), Agents and Peer-to-Peer Computing. XII, 205 pages. 2004.

Vol. 2871: N. Zhong, Z.W. Raś, S. Tsumoto, E. Suzuki (Eds.), Foundations of Intelligent Systems. XV, 697 pages. 2003.

Vol. 2854: J. Hoffmann, Utilizing Problem Structure in Planing. XIII, 251 pages. 2003.

Vol. 2843: G. Grieser, Y. Tanaka, A. Yamamoto (Eds.), Discovery Science. XII, 504 pages. 2003.

Vol. 2842: R. Gavaldá, K.P. Jantke, E. Takimoto (Eds.), Algorithmic Learning Theory. XI, 313 pages. 2003.

Vol. 2838: N. Lavrač, D. Gamberger, L. Todorovski, H. Blockeel (Eds.), Knowledge Discovery in Databases: PKDD 2003. XVI, 508 pages. 2003.

Vol. 2837: N. Lavrač, D. Gamberger, L. Todorovski, H. Blockeel (Eds.), Machine Learning: ECML 2003. XVI, 504 pages. 2003.

Vol. 2835: T. Horváth, A. Yamamoto (Eds.), Inductive Logic Programming. X, 401 pages. 2003.

Vol. 2821: A. Günter, R. Kruse, B. Neumann (Eds.), KI 2003: Advances in Artificial Intelligence. XII, 662 pages. 2003.